Acoustics

Acoustics

An introduction

Heinrich Kuttruff

Taylor & Francis
Taylor & Francis Group

LONDON AND NEW YORK

German edition first published 2004
by S. Hirzel Verlag

English edition published 2007
by Taylor & Francis
2 Park Square, Milton Park, Abingdon, Oxon OX14 4RN

Simultaneously published in the USA and Canada
by Taylor & Francis
711 Third Avenue, New York, NY 10017

Taylor & Francis is an imprint of the Taylor & Francis Group,
an informa business

© 2007 S. Hirzel Verlag

All rights reserved

Authorised translation from the German language edition published by
S. Hirzel Verlag, Birkenwaldstrasse 44, D-70191 Stuttgart, Germany

Typeset in Sabon by
Newgen Imaging Systems (P) Ltd, Chennai, India

The publisher makes no representation, express or implied, with regard to
the accuracy of the information contained in this book and cannot accept any
legal responsibility or liability for any efforts or omissions that may be made.

British Library Cataloguing in Publication Data
A catalogue record for this book is available
from the British Library

Library of Congress Cataloging in Publication Data
A catalog record for this book has been requested

ISBN10: 0–415–38679–9   ISBN13: 978–0–415–38679–1 (hbk)
ISBN10: 0–415–38680–2   ISBN13: 978–0–415–38680–7 (pbk)
ISBN10: 0–203–97089–6   ISBN13: 978–0–203–97089–8 (ebk)

Contents

List of symbols xi

1 Introduction 1

    1.1 What is sound? 2
    1.2 What is acoustics? 4

2 Some facts on mechanical vibrations 7

    2.1 A few examples 7
    2.2 Complex notation of harmonic vibrations 11
    2.3 Beats 12
    2.4 Forced vibrations, impedance 13
    2.5 Resonance 14
    2.6 Free vibrations of a simple resonator 18
    2.7 Electromechanical analogies 19
    2.8 Power 22
    2.9 Fourier analysis 23
    2.10 Transfer function and impulse response 29
    2.11 A note on non-linear systems 32

3 Acoustic variables and basic relations 34

    3.1 Acoustic variables 34
    3.2 Basic relations in acoustics 37
    3.3 Wave equations 42
    *3.4 Intensity and energy density of sound waves
in fluids 44*
    3.5 The sound pressure level 47

4   Plane waves, attenuation   48

    4.1   *Solution of the wave equation 48*
    4.2   *Harmonic waves 51*
    4.3   *A few notes on sound velocity 54*
    4.4   *Attenuation of sound 55*
    4.5   *Non-linear effects 65*

5   Spherical wave and sound radiation   69

    5.1   *Solution of the wave equation 69*
    5.2   *The point source 71*
    5.3   *The Doppler effect 74*
    5.4   *Directional factor and radiation resistance 76*
    5.5   *The dipole 79*
    5.6   *The linear array 81*
    5.7   *The spherical source 84*
    5.8   *Piston in a plane boundary 86*

6   Reflection and refraction   94

    6.1   *Angles of reflection and refraction 94*
    6.2   *Sound propagation in the atmosphere 96*
    6.3   *Reflection factor and wall impedance 98*
    6.4   *Absorption coefficient 103*
    6.5   *Standing waves 104*
    6.6   *Sound absorption by walls and linings 106*

7   Diffraction and scattering   118

    7.1   *Exact formulation of diffraction problems 120*
    7.2   *Diffraction by a rigid sphere 121*
    7.3   *Sound transmission through apertures 124*
    7.4   *Babinet's principle 132*
    7.5   *Multiple scattering, scattering from rough surfaces 134*

8   Sound transmission in pipes and horns   138

    8.1   *Sound attenuation in pipes 138*
    8.2   *Basic relations for transmission lines 141*
    8.3   *Pipes with discontinuities in cross section 143*
    8.4   *Pipes with continuously changing cross section (horns) 150*

8.5   *Higher order wave types 156*
8.6   *Dispersion 162*

9   Sound in closed spaces   166

9.1   *Normal modes in a one-dimensional space 166*
9.2   *Normal modes in a rectangular room with*
      rigid walls 169
9.3   *Normal modes in cylindrical and spherical*
      cavities 173
9.4   *Forced vibrations in a one-dimensional*
      enclosure 174
9.5   *Forced vibrations in enclosures of any shape 178*
9.6   *Free vibrations 182*
9.7   *Statistical properties of the transfer function 185*

10   Sound waves in isotropic solids   189

10.1   *Sound waves in unbounded solids 189*
10.2   *Reflection and refraction, Rayleigh wave 194*
10.3   *Waves in plates and bars 197*

11   Music and speech   209

11.1   *Simple and complex tones, noise 209*
11.2   *Pitch, intervals and scales 211*
11.3   *General remark on the function of musical*
       instruments 214
11.4   *String instruments 215*
11.5   *Wind instruments 223*
11.6   *The human voice 228*

12   Human hearing   233

12.1   *Anatomy and function of the ear 234*
12.2   *Psychoacoustic pitch 239*
12.3   *Hearing threshold and auditory sensation area 243*
12.4   *Loudness level and loudness, critical*
       frequency bands 244
12.5   *Auditory masking 248*
12.6   *Measurement of loudness 249*
12.7   *Spatial hearing 252*

13   Room acoustics   257

    13.1   *Geometric room acoustics 258*
    13.2   *Impulse response of a room 261*
    13.3   *Diffuse sound field 264*
    13.4   *Steady-state energy density and reverberation 268*
    13.5   *Sound absorption 271*
    13.6   *On the 'acoustics' of auditoria 277*
    13.7   *Special rooms for acoustic measurements 280*

14   Building acoustics   283

    14.1   *Characterisation and measurement of airborne sound insulation 284*
    14.2   *Airborne sound insulation of compound partitions 287*
    14.3   *Airborne sound insulation of single-leaf partitions 289*
    14.4   *Airborne sound insulation of double-leaf partitions 296*
    14.5   *Structure-borne sound insulation 301*

15   Fundamentals of noise control   309

    15.1   *Noise criteria 310*
    15.2   *Basic mechanisms of noise generation 311*
    15.3   *Primary noise control 316*
    15.4   *Secondary noise control 320*
    15.5   *Personal hearing protection 331*

16   Underwater sound and ultrasound   333

    16.1   *Acoustical detection and localisation of objects 333*
    16.2   *Sound propagation in sea water 335*
    16.3   *Strength of echoes 337*
    16.4   *Ambient noise, reverberation 338*
    16.5   *Transducer arrays 340*
    16.6   *General remarks on ultrasound 342*
    16.7   *Generation and detection of ultrasound 343*
    16.8   *Diagnostic applications of ultrasound 345*
    16.9   *Applications of high intensity ultrasound 349*
    16.10 *Generation of high and highest ultrasound frequencies 353*

17   Electroacoustic transducers   359

    17.1   Piezoelectric transducer 361
    17.2   Electrostatic transducer 365
    17.3   Dynamic transducer 368
    17.4   Magnetic transducer 371
    17.5   Magnetostrictive transducer 373
    17.6   The coupling factor 375
    17.7   Two-port equations and reciprocity relations 377

18   Microphones   379

    18.1   Principles of microphones for airborne sound 379
    18.2   Condensor microphone 382
    18.3   Piezoelectric microphones 387
    18.4   Dynamic microphones 389
    18.5   Carbon microphone 391
    18.6   Microphone directivity 392
    18.7   Hydrophones 396
    18.8   Vibration pickups 396
    18.9   Microphone calibration 399

19   Loudspeakers and other electroacoustic sound sources   403

    19.1   Dynamic loudspeaker 405
    19.2   Electrostatic or condensor loudspeaker 408
    19.3   Magnetic loudspeaker 410
    19.4   Improvement of loudspeaker efficiency 411
    19.5   Loudspeaker directivity 417
    19.6   Earphones 419
    19.7   Sound transmitters for water-borne sound and for
           ultrasound 421

20   Electroacoustic systems   426

    20.1   Stereophony 427
    20.2   Sound recording 433
    20.3   Sound reinforcement systems 443

    Literature   451
    Index   453

Symbols

Latin capital letters

A	constant, equivalent absorption area
B	constant, frequency bandwidth, irradiation density, bending stiffness, magnetic flux density
C	constant, clarity, electrical capacitance, specific heat
$C(\omega)$	spectral density
$C(z)$	Fresnel integral
C_n	Fourier coefficients, constants in eq. (9.30)
D	diameter, thickness, definition, bending moment, dielectric displacement, attenuation per metre
F	force
G	electrical admittance
$G(\omega)$	transfer function
H	transfer function, magnetic field strength
I	intensity, electrical current
$J_n(z)$	Bessel function
L	length, level, electrical inductance
M	transducer constant (M-transducers), microphone sensitivity
M_r	molecular mass
N	integer, number, loudness, transducer constant (E-transducers), in Subsection 15.4.3: frequency parameter
P	power
P_r	radiated power
Q	volume velocity, Q-factor, electrical charge
Q_r	backscattering cross section
Q_s	scattering cross section
R	reflection factor, molar gas constant, directional factor, electrical resistance
R_e	reciprocity parameter
R_A	sound reduction index or sound transmission loss
R_r	radiation resistance

S	area
$S(\omega)$	spectral density
$S(z)$	Fresnel integral
T	period of an oscillation, transmission factor, reverberation or decay time, absolute temperature
T_m	averaging time
U	circumference, electrical voltage
V	volume, speed
W	probability
$W(\omega)$	power spectrum
Y	admittance, Young's modulus
Z	impedance
Z_r	radiation impedance
Z_0	characteristic impedance

Latin lower case letters

a	radius, constant
b	width
c	sound velocity
d	thickness, distance, diameter
e	piezoelectric or piezomagnetic constant
f	frequency, arbitrary function
g	constant, arbitrary function
$g(t)$	impulse response
h	Planck's constant
j	imaginary unit
k	angular wavenumber, electroacoustic coupling factor
l	integer, length
m	integer, mass, attenuation constant
m'	specific mass, mass per unit area
m_r	radiation mass
n	integer, compliance, normal direction
p	sound pressure
q	ratio of diameters, amplifier gain
r	radius, distance, resistance
r_c	critical distance or diffuse-field distance
r_s	flow resistance
s	displacement or elongation, signal
t	time or duration
v	velocity, particle velocity
w	energy density, number of turns (of a coil)
x	coordinate
y	coordinate
z	coordinate

Greek capital letters

Δ difference
θ angle, sound temperature
Ξ specific flow resistance
ϕ angle
Ω solid angle

Greek lowercase letters

α angle, absorption coefficient
β angle
γ angle, gain
δ decay constant
$\delta(t)$ Dirac- or delta function
ε angle, relaxation strength, (relative) dielectric permittivity
ε permittivity of vacuum
ζ component of displacement, specific impedance
η component of displacement, imaginary part of the specific impedance, loss factor, viscosity constant
ϑ angle
κ adiabatic or isentropic exponent
λ wavelength, Lamé constant
μ mass per unit length, Lamé constant, permeability
μ_0 permeability of vacuum
ν heat conductivity, Poisson's ratio
ξ component of displacement, real part of the specific impedance
ρ density
σ elastic stress, porosity, standard deviation
τ transit time or delay time, relaxation time
φ phase angle
χ phase angle of reflection factor
ψ phase angle of impedance
ω angular frequency

Chapter 1

Introduction

Sound of any kind is an omnipresent companion during all our life. Early in the morning the alarm clock ends our sleep with a more or less enticing sound, and from thereon we perceive sounds of different kind throughout our day. In the densely populated areas where many of us are living most sound is produced by man, either intentionally or as an inevitable side effect of human activity. Each of us produces many sorts of sound: we talk with other people, we switch on the radio, the television or the stereo system, we drive a car or use noise producing tools or machines at our work.

Even in the country side we rarely find absolute quietness. In free air we hear the twittering of birds or the murmuring of the wind in the trees, or, if we are at the seaside, the sounds of the surf. Complete silence is very rare; it is so strange that we find it rather unpleasant or even unbearable. On the other hand, sound can be very annoying or may even damage our health. The former is by no means a matter just of the strength or the loudness of the sound. Although the faint noise of a dripping water tap is almost unmeasurable we may fly into a rage when we hear it at night. Very loud sounds, on the other hand, can be harmful to our hearing, that is, when exposed to intense sound our hearing organ can suffer temporary or even permanent damages leading to complete deafness. Even sound of medium intensity may lead to damage of the vegetative nerve system, manifesting itself in sleep irregularities, nervousness, elevated blood pressure, etc.

It is a remarkable fact that we cannot protect ourselves to any significant degree against sound in a natural way. We can close our eyes when we do not want to see anything; when falling asleep we do this involuntarily. In contrast, we do not stop receiving sounds, even during sleep we hear without becoming aware of this. Apparently, nature has given a particular warning function to sound. In the same direction points the fact that our visual field is quite limited whereas we perceive sound arriving from all directions, independently from the orientation of our head. So we cannot see a danger, for instance a motor vehicle, approaching from behind but we can hear it.

1.1   **What is sound?**

What is the physical nature of sound? At first we can state that the generation, propagation and perception of sound is connected with mechanical vibrations or oscillations. In some cases we can convince ourselves immediately of this fact, for instance, by touching our larynx when speaking or singing. Likewise, the vibrations of noise producing machines can often be felt with the hand, if the vibration stops no sound is heard. The vibration of the strings of a musical instrument can be seen with the naked eye, and in ancient times it was observed that the perceived pitch of a tone is related to the length of the string and hence to the number of oscillations per second or as we say nowadays: on the frequency of the vibration. However, in most cases these vibrations are so weak that it is impossible to see or feel them immediately. This is true, for instance, when sound penetrates a wall; in this case the vibrations can only be observed by means of special measuring devices.

Many sounds have a 'tonal' quality, that is, a certain pitch can be ascribed to them. Such sounds form the basic elements of music. Besides them, there are other sounds which although having a more general character such as 'bright' or 'muffled', do not have a distinct pitch. Imagine, as an example, a bang or the noise of an air stream. Such types of sounds can also be related to vibrations as we shall see later on.

Let us now consider the generation of sound by a vibrating body, for instance, by the corpus of a stringed musical instrument, the membrane of a loudspeaker or by some part of a machine in operation. In Figure 1.1 an element of its surface is sketched as a solid line. When it moves from the left side to the right as shown in the upper part of the figure, it cannot displace all the air in front of it but it will press some of it together. When moving in reverse direction the body will suck in some air, again not by moving the whole column of air but by expanding some of it (see middle figure). Now any density change of the air is associated with a change in air pressure. Hence the compressed air tends to transfer the pressure increase to the neighbouring air volume. Likewise, a decompressed air volume exerts underpressure to its vicinity. Generally, all pressure disturbances induced by the body's movement will travel into the resting air. Finally, we assume the surface of the body to move back and forth or, in other words, to oscillate. Then the alternating compressions and expansions of the air will detach from the body and travel into the medium (see below). The result is a sound wave. Gradually, it will reach larger and more remote areas, similar to a water wave issuing from a stone thrown into a pond. This is why we use the term 'sound waves' thus expressing the propagation of a state or process. The region filled with one or several sound waves is often referred to as 'sound field'.

The described changes of air state imply that the particles which air is thought to consist of are displaced from their resting position, and they

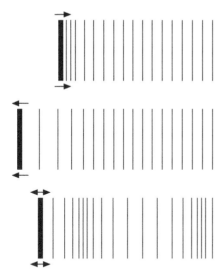

Figure 1.1 Radiation of sound waves from a moving body.

follow the vibrations of the body which emits the wave. Thus a sound wave may be conceived as a pressure disturbance or a sequence of disturbances on the one hand, or equally well as a large number of vibrating air particles. The same holds, of course, for waves in other gases or liquids.

In hearing sounds the reverse process takes place in a way: when a sound wave hits the head of a listener a tiny part of it enters the ear channel. At its end it impinges on the eardrum which is set into vibrations by the pressure fluctuation. These oscillations undergo further processing by the middle ear and the inner ear and are finally led to the brain.

Thus we can state that the propagation of sound is tied to the presence of a suitable medium, for instance air; in the empty space there is no sound. Furthermore, it is important to realise that the transfer of oscillations from one volume element to a neighbouring one cannot take place instantly but requires some time since masses must be accelerated which implies some delay. For this reason sound waves propagate with a finite velocity. Each of us knows from experience that in a thunderstorm the thunder arrives usually several seconds later than the flash of the lightning which caused it. The speed at which sound waves travel is called the sound speed or the sound velocity. It depends on the kind and state of the medium which carries the sound wave.

At this point it may be appropriate to compare sound waves with another kind of wave which dominates our everyday life, the electromagnetic waves.

Without them no broadcasting, no television, no telecommunication over large distances and no mobile telephone would exist. Light, too, consists of electromagnetic waves. Like sound waves they travel with a finite although much higher velocity. In contrast to sound waves, however, they are not tied to a material medium but can propagate in the empty space on account of their different nature. Likewise their formal description differs substantially from that of sound waves. While in the latter one the relevant physical quantity, namely, the pressure, is a scalar, the field quantities in electromagnetic waves, namely, the electric and the magnetic field strength, have vector character. From this point of view the formal description of sound waves is less complicated than that of electromagnetic waves, at least if we disregard sound in solids.

Despite all these differences there are many formal parallels and analogies between acoustical and electromagnetic waves. This is because the differential equation underlying both sorts of waves, the wave equation, has the same structure. Such parallels exist also between mechanical and electrical oscillations; many concepts such as the impedance or the energy are formally defined in the same way. In this book we shall often point out such analogies since many readers may be familiar with the basic concepts of electricity and thus can have recourse to known facts when it comes to mechanical vibrations and oscillatory systems.

1.2 What is acoustics?

Acoustics is the science of sound and deals with the origin of sound and its propagation, either in free space, or in pipes and channels, or in closed spaces. It is the basis of many fundamental phenomena and also of numerous practical applications. Some of them will be briefly touched upon here.

A first subdivision of the field can be based on the different media in which sound can propagate. In our everyday life sound waves are in air, or, somewhat more generally, in gases. From this we distinguish sound in liquids which has its most important application in underwater techniques, and, furthermore, sound in solid bodies. This subdivision intersects with another one based on the sound frequency. Again, sound waves with frequencies accessible to our hearing are in the foreground of interest. The frequency range of human hearing is roughly from 16 Hz to about 20 000 Hz. Here Hz is the unit of frequency, called Hertz (1 Hz means one period per second). These figures should not be taken too seriously; at low frequencies the limit between hearing and feeling is rather diffuse, and the upper limit shows wide individual differences and shifts with increasing age towards lower frequencies.

Below the range of audible sounds there is the infrasonic range. Sounds with very low frequencies can arise, for instance, from building vibrations or by industrial processes where large quantities of gas are moved. Very intense infrasound has quite unpleasant effects on human beings which may be associated with nausea; in extreme situations health damage can be caused by infrasound. A general lower-frequency limit of sound does not exist.

Sound waves with frequencies above the upper limit of hearing, that is, 20 000 Hz, roughly speaking, are known as ultrasound. Furthermore, sound with frequencies exceeding 1 Gigahertz ($=10^9$ Hz) is sometimes referred to as hypersound. Since ultrasound has many important and useful applications a particular chapter is devoted to its description – along with that of water-borne sound. In contrast to the situation at low frequencies there is indeed an upper frequency limit of all acoustic phenomena. This is due to the fact that all matter has a discrete structure since it is made up of atoms, molecules or ions. This upper limiting frequency depends on the kind of medium and is of the order 10 Terahertz $= 10^{13}$ Hz. In Chapter 16 this fact will be dealt with in some more detail.

The first task of acoustics is to formulate the physical laws governing sound when it propagates in free space. Equally interesting is the way in which its propagation is altered by obstacles of any kind, either by extended surfaces or by bodies of limited extension. Furthermore, sound can be conducted through channels of various sorts; it can travel in solid structures such as the walls and floors of a building and can be transmitted through windows and doors. In this context we have to deal with undesired sounds which are generally called noise although there is no clear-cut distinction between noise and other sounds. Since noise is an increasing problem in our society the techniques of noise control occupy broad space in practical acoustics. On the other hand, sound in the form of speech is the most important and the simplest way to communicate with each other since every healthy person can produce and understand speech.

Another equally important and mainly pleasant manifestation of sound is music which in all human cultures plays an outstanding role, probably of ritual origin. Today it serves mainly for enjoyment of a performance art or just entertainment. The acoustical aspects of music are dealt with in a particular discipline named musical acoustics, which on the one hand examines the production of tones with musical instruments, and on the other the perception of music by listeners. At this point musical acoustics blends with psychoacoustics the goal of which is the systematic investigation of the way in which sounds of any kind are processed and perceived by our hearing. It yields not only valuable insights into the performance of the human hearing organ, but also the yardstick for the subjective judgement of sound, for instance, for the assessment of telephone quality, or the tolerability of a certain noise situation.

A good deal of the sound which we perceive is produced by loud-speakers and other electroacoustic sound sources. By loudspeakers we are informed and entertained, and quite often, however, annoyed too. In any case the sound supply of large audiences in sports arenas, open-air performances, large convention halls, etc. would be impossible without electroacoustic reinforcement systems. Another important example of electroacoustic transmission is the telephone, and also ultrasound which has become an indispensable tool in medical diagnosis is produced by electroacoustic sources. Finally, we want to recall the possibility of storing sound events which are volatile by their very nature, and to revive them at any time and place. All of these problems form the subject of a particular field called electroacoustics.

As already mentioned the velocity of sound depends on the kind of the wave medium. This holds even more for the attenuation which sound waves undergo in the course of propagation. Reversely, valuable insights into the physical nature and internal structure of all kinds of matter can be derived from experimental data on sound propagation collected in different frequency ranges.

This brief review is far from exhaustive, as several other branches of acoustics have not even been mentioned. Nevertheless, it may give an idea of the great variety of acoustical phenomena and applications of sound. Moreover, it shows that acoustics is an interdisciplinary science being interconnected with many other fields – with physics, mechanical and electrical engineering, medicine, psychology, biology, architecture and building construction, music, etc. – a fact which makes the boundaries of acoustics somewhat unclear but contributes to the particular appeal of this science.

Chapter 2

Some facts on mechanical vibrations

As mentioned in Section 1.1, the basic process underlying any sound is vibration – mechanical vibration of the particles of which a material sound medium consists. For this reason the description of the various acoustical phenomena is preceded by a chapter in which the most important facts on vibrations will be briefly expounded. It will become clear that the term vibration is rather general in that it covers quite a wide variety of motions and variations.

At first we need a measure for the strength of vibrations. The quantity which suggests itself for this purpose is the displacement of a vibrating point or particle from its resting position. This is also called elongation. Of course, it is not sufficient to specify just the magnitude but also the direction of motion. This is most conveniently done by representing the displacement by a vector $\vec{s}$ depending on space and time. We denote the components with regard to a suitably defined Cartesian coordinate system with ξ, μ and ζ. A very common alternative is to specify vibrations by the velocity of the displacement rather than by the displacement itself. Like the displacement, the velocity of vibration is a vector and is related to displacement by differentiation with respect to time

$$\vec{v} = \frac{\partial \vec{s}}{\partial t} \tag{2.1}$$

2.1 A few examples

As an example of a simple mechanical vibrator we consider a mass m suspended from a spring, for instance from a spiral spring (see Fig. 2.1a). We assume that the spring alters its length proportionally to the force acting on it (Hooke's law). If one gives a short push to the mass which was at rest until this moment, it will start to move up and down, that is, it will perform vertical oscillations. Figure 2.1b shows the displacement of the mass as a function of time. Because of inevitable losses, for instance by air friction, the magnitude of the vibrations gradually reduces, and after some time the

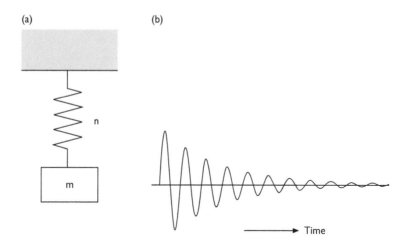

Figure 2.1 Simple oscillator consisting of a mass m and a spring with compliance n.
(a) schematically, (b) damped oscillation.

mass will come to rest again. This kind of motion is called a damped vibration or damped oscillation. It is typical of all simple vibrators which are set
in motion by some instantaneous energy supply and afterwards are left to
themselves. Another example is a struck tuning fork; likewise, plucked or
struck strings as well as bells perform damped vibrations although with a
more complex oscillation pattern than that shown in Figure 2.1b.

If the energy losses of the system are compensated by continuous energy
supply from outside, an undamped or stationary vibration can be maintained. The simplest case of a stationary vibration is a harmonic motion
with the instantaneous displacement given by a sine (or cosine) function of
time. Choosing the latter we have:

$$s(t) = \hat{s} \, \cos(\omega t + \varphi) \tag{2.2}$$

This is represented in Figure 2.2. The constant $\hat{s}$ is called the amplitude of
the vibration indicating the maximum and minimum values the function $s(t)$
can attain. Since the cosine function is periodic with the period 2π, a time
shift by $2\pi/\omega$ (or by an integral multiple of this) leads to the same value
of the function. We call this time shift the period T of the oscillation. Its
reciprocal indicates the number of oscillations per seconds and is called the
frequency of the vibration:

$$f = \frac{\omega}{2\pi} = \frac{1}{T} \tag{2.3}$$

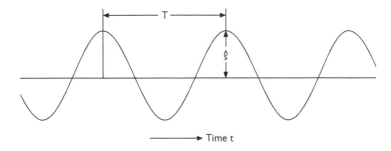

Figure 2.2 Harmonic oscillation.

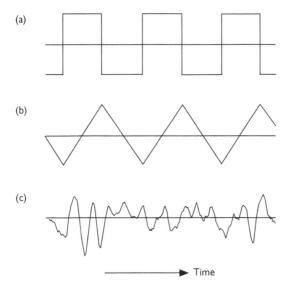

Figure 2.3 Various kinds of oscillations: (a) rectangular oscillation, (b) triangular oscillation and (c) random oscillation.

The quantity $\omega = 2\pi f$ is the angular frequency. Both the frequency and the angular frequency have the dimension s^{-1}; the unit of frequency is 1 Hertz, abbreviated as 1 Hz. The constant φ is called the phase angle – or simply the 'phase' of the oscillation. It takes regard of the fact that the vibration may be shifted with respect to the time $t = 0$, the choice of which is arbitrary.

The harmonic vibration is a special case of a more general class of vibrations, namely, periodic vibrations. Two further examples of periodic vibration are shown in Figure 2.3a and b. Again, the strength of these vibrations can be characterised by their maximum deviation $\hat{s}$ from the zero line. As a counter example Figure 2.3c shows a random oscillation which may represent, for instance, the swaying of a branch moved by the wind. Although such motions are unpredictable in detail they are nevertheless part of the concept of vibration.

The magnitude of vibrations including random oscillations may also be characterised by their quadratic average which is defined as follows:

$$\langle s^2 \rangle = \frac{1}{t_0} \int_0^{t_0} [s(t)]^2 dt = \tilde{s}^2 \tag{2.4a}$$

where t_0 is a sufficiently long time interval. Here it is assumed that $s(t)$ is free of a constant bias, that is, that the linear average of the oscillation

$$\langle s \rangle = \frac{1}{t_0} \int_0^{t_0} s(t) dt \tag{2.4b}$$

is zero. Such averages are only meaningful, however, if the oscillation is stationary, which means that it does not alter its general character at least during the averaging time t_0. This is certainly true for periodic vibrations for which t_0 is chosen equal to the period. The quantity $\tilde{s}$ is called the root-mean-square value of the displacement or elongation. Its value, divided by the maximum value $\hat{s}$, is shown in Table 2.1 for the type of periodic oscillations mentioned earlier. In the same way, we can define the effective value of any stationary quantity.

The harmonic vibration according to eq. (2.2) may be regarded as the prototype of all kinds of vibrations. One reason for its central position is the steadiness of the sine or cosine function describing it. An even more important reason will be discussed in Section 2.9.

The concepts introduced in this section are not restricted to the displacement of a vibrating body or particle which served just as an example. In fact, they can be applied to any physical quantity which varies in an oscillatory manner, for instance, to velocities, forces, gas pressures, temperatures, etc. The same holds for the contents of the subsequent sections.

Table 2.1 Root-mean-square of some periodic signals

Type of signal	$\tilde{s}/\hat{s}$
Sinus	$1/\sqrt{2}$
Symmetrical rectangle	1
Symmetrical triangle	$1/\sqrt{3}$

2.2   Complex notation of harmonic vibrations

A particularly useful and widespread representation of oscillatory processes is based upon the splitting of an exponential function with purely imaginary argument into its real and imaginary part (Euler's formula):

$$e^{jz} = \cos z + j \sin z \qquad (z \text{ real}) \tag{2.5}$$

From this relation it follows that

$$\cos z = \frac{e^{jz} + e^{-jz}}{2} \qquad \text{and} \qquad \sin z = \frac{e^{jz} - e^{-jz}}{2j} \tag{2.6a,b}$$

These formulae will be frequently used in the following. With eq. (2.5) we can write eq. (2.2) in the form

$$s(t) = \operatorname{Re} \left\{ \hat{s} e^{j(\omega t + \varphi)} \right\} \tag{2.7}$$

where Re means the real part of the subsequent complex number. We arrive at an even simpler representation by omitting the symbol Re:

$$s(t) = \hat{s} e^{j(\omega t + \varphi)} \tag{2.8}$$

As any complex quantity s(t) may be represented as a 'phasor' in the complex plane; the latter being formed by the real and the imaginary axis (see Fig. 2.4). The magnitude of the complex number, here the amplitude $\hat{s}$, represents itself as the length of this arrow whereas the angle subtended by the arrow and the real axis is the argument of the number, in the present case $\omega t + \varphi$.

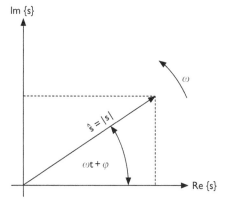

Figure 2.4 Phasor representation of a harmonic oscillation.

Accordingly, with increasing time the arrow rotates counterclockwise around the origin with angular velocity ω. Its projection onto the real axis generates the oscillation according to eq. (2.2).

The complex notation has the advantage that the time dependence is always given by a factor $e^{j\omega t}$ instead of $\cos \omega t$ and $\sin \omega t$. Very often this factor cancels; we shall even omit it provided this will not cause confusion. Another advantage becomes obvious if one wants to know the velocity with which the displacement varies. Differentiating eq. (2.2) with respect to time leads to

$$v(t) = -\omega \hat{s} \sin(\omega t + \varphi) \tag{2.9}$$

On the other hand we obtain by differentiating eq. (2.8):

$$v(t) = j\omega \hat{s} e^{j(\omega t + \varphi)} = j\omega s \tag{2.10}$$

It is evident that the real part of this expression agrees with eq. (2.9). Hence, in complex notation, differentiation with respect to time is equivalent to multiplication by $j\omega$. Conversely, indefinite integration of a quantity which varies sinusoidally with respect to time corresponds to dividing it by $j\omega$.

The complex representation fails, however, if we want to multiply quantities as in the calculation of power, energy density, etc. (see Section 2.8). In this case it is advisable to go back to the real notation by taking the real part of the complex quantities.

2.3   Beats

As a first example which immediately demonstrates usefulness of the complex notation we consider two harmonic vibrations of equal amplitudes but with slightly differing angular frequencies:

$$s_1(t) = \hat{s} e^{j(\omega - \Delta\omega)t} \quad \text{and} \quad s_2(t) = \hat{s} e^{j(\omega + \Delta\omega)t}$$

we assume that the difference $2\Delta\omega$ is significantly smaller than ω. If both vibrations are superimposed we obtain what is called beats, that is, a vibration or oscillation with periodically fluctuating amplitude. Indeed, the addition of s_1 and s_2 yields:

$$s_1(t) + s_2(t) = \hat{s}(e^{-j\Delta\omega t} + e^{j\Delta\omega t})\, e^{j\omega t} = 2\hat{s} \cos(\Delta\omega t)e^{j\omega t} \tag{2.11}$$

where we used eq. (2.6a). The vibration itself has the mean angular frequency ω, and its amplitude varies with the angular frequency $2\Delta\omega$. At certain instants both partial vibrations have equal phases and the amplitude of their sum shows a maximum. At instants halfway between such maxima both components will be superposed in opposite phases and will cancel each

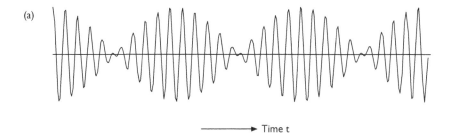

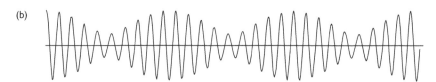

Figure 2.5 Beats between two harmonic oscillations for $\Delta\omega = \omega / 20$: (a) $\hat{s}_1 = \hat{s}_2$, (b) $\hat{s}_1 = 2\hat{s}_2$.

other. If the amplitudes of both partial vibrations are different the beats will be incomplete. In this case not only the amplitude but also the instantaneous frequency will fluctuate.

The upper part of Figure 2.5 shows the beats according to eq. (2.11). In the lower part incomplete beats are depicted, and the amplitude of the lower-frequency component is twice the amplitude of the other one. This diagram shows clearly that the frequency of the amplitude variation equals the difference frequency $2\Delta\omega$.

2.4   Forced vibrations, impedance

If the oscillation of a mechanical system is excited by an external force, we speak of forced vibration or forced oscillation in contrast to the free oscillation which we encountered in Figure 2.1b. Of course, the frequency of a forced oscillation is that of the exciting force. We represent the latter by

$$F(t) = \hat{F}e^{j\omega t} \tag{2.12}$$

where $\hat{F}$ denotes the force amplitude. The velocity v of the excited oscillation will not necessarily be in phase with the exciting force, therefore we represent it by

$$v(t) = \hat{v}e^{j(\omega t - \psi)} \tag{2.13}$$

The mechanical impedance which the force has to overcome is defined as the ratio of both quantities:

$$Z = \frac{F}{v} = \frac{\hat{F}}{\hat{v}} e^{j\psi} \tag{2.14}$$

Hence, the magnitude of the impedance is equal to the ratio of the force amplitude and the velocity amplitude; the phase angle ψ indicates how much the velocity is ahead of the exciting force (ψ negative) or lags behind it (ψ positive). This definition corresponds to that of electrical impedance, with the force replaced with the electrical voltage and the velocity with the electrical current. As in electricity, the reciprocal of the impedance is called the admittance:

$$Y = \frac{1}{Z} \tag{2.15}$$

In general, both the impedance and the admittance are functions of the frequency. When the impedance of a system is known, the response of the system expressed by its velocity can be determined by using eq. (2.14). To obtain an overall idea on how the complex impedance depends on the frequency, the impedance may be represented as a phasor in the complex plane – much in the same way as the displacement of a harmonic oscillation in Figure 2.4. The length and direction of this arrow are changed when the frequency is varied. The curve obtained by connecting the points of all phasors is called the locus of the impedance. In the next section we shall encounter a simple example.

Finally, it should be emphasised that a rational definition of the impedance requires the complex notation since only then the time dependence expressed by the time dependence of the involved quantities cancels in eq. (2.14).

2.5 Resonance

Now we return to the system depicted in Figure 2.1a. Figure 2.6 shows it in a slightly modified way, namely, upside down and with an additional element which is to represent the inevitable losses of the system. It suggests a piston which loosely fits a cylinder. When set in motion the piston will displace some air and some energy will be lost because of the viscosity of air. The frictional force F_r necessary to move the piston is proportional to the relative velocity between piston and cylinder:

$$F_r = r \cdot v = r \cdot \frac{ds}{dt} \tag{2.16}$$

r is often called the resistance constant or simply the resistance of the system. The compliance n of the spring – in the present case of the combination of

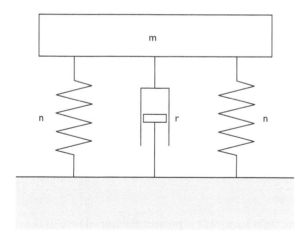

Figure 2.6 Simple resonance system (m: mass, n: spring, r: mechanical resistance).

two springs – is defined as the ratio of the elongation s of the spring and the force F_s acting on it, or

$$F_s = \frac{1}{n} \cdot s = \frac{1}{n} \int v dt \tag{2.17}$$

Alternatively, the spring can be characterised by its stiffness which is the reciprocal of the compliance. Finally, we have to account for the inertial force F_m by which the mass responds to any acceleration and which is proportional to that acceleration:

$$F_m = m\frac{dv}{dt} = m\frac{d^2 s}{dt^2} \tag{2.18}$$

An external force F(t) must balance these three forces: $F = F_m + F_r + F_s$, or, with the earlier equations:

$$F(t) = m\frac{d^2 s}{dt^2} + r\frac{ds}{dt} + \frac{1}{n}s \tag{2.19}$$

Any property of the system under consideration can be worked out by solving this second order differential equation. Suppose the external force F varies harmonically according to eq. (2.2). Since the differential equation (2.19) is linear in s with constant coefficients the system's reaction is also a harmonic oscillation with angular frequency ω. Thus we can apply the differentiation rule explained in Section 2.2 with the result:

$$F = -m\omega^2 s + j\omega r \cdot s + \frac{1}{n}s$$

or, after replacing s with $v/j\omega$:

$$F = \left(j\omega m + r + \frac{1}{j\omega n}\right)v \tag{2.20}$$

This expression yields immediately the impedance of this system after eq. (2.14):

$$Z = j\omega m + r + \frac{1}{j\omega n} \tag{2.21}$$

The three terms on the right hand side represent the impedance of the mass, of the resistance and of the spring, respectively. With the abbreviations

$$\omega_0 = \frac{1}{\sqrt{mn}} \tag{2.22}$$

and

$$Q = \frac{m\omega_0}{r} \tag{2.23}$$

this expression can be transformed into

$$Z = r\left[1 + jQ\left(\frac{\omega}{\omega_0} - \frac{\omega_0}{\omega}\right)\right] \tag{2.24}$$

The phase angle in eq. (2.14) is given by

$$\psi = \arctan\left[Q\left(\frac{\omega}{\omega_0} - \frac{\omega_0}{\omega}\right)\right] \tag{2.25}$$

Now the solution we were looking for reads

$$v(t) = \frac{\hat{F}e^{j(\omega t - \psi)}}{r\sqrt{1 + Q^2\left[(\omega/\omega_0) - (\omega_0/\omega)\right]^2}} \tag{2.26}$$

or

$$s(t) = \frac{\hat{F}e^{j(\omega t - \psi)}}{j\omega r\sqrt{1 + Q^2\left[(\omega/\omega_0) - (\omega_0/\omega)\right]^2}} \tag{2.26a}$$

The locus of the impedance of the oscillator is represented in Figure 2.7a. It is a vertical line; when the frequency varies from zero to infinity, the

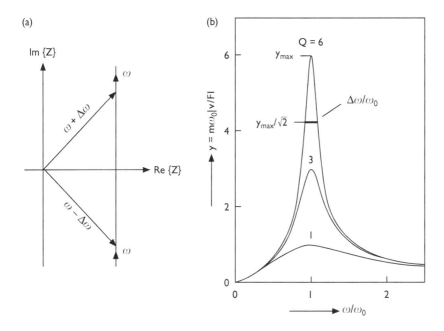

(a)

Im {Z}

Re {Z}

$\omega + \Delta\omega$

$\omega - \Delta\omega$

ω

ω

(b)

$Q = 6$

y_{max}

6

$y_{max}/\sqrt{2}$

4

$\Delta\omega/\omega_0$

3

2

1

0

0   1   2

$y = m\omega_0|v/F|$

ω/ω_0

Figure 2.7 Simple resonance system: (a) locus of the impedance, (b) resonance curves, parameter is the Q-factor.

locus runs along this line from $-j\infty$ to $+j\infty$. At the frequency $\omega = \omega_0$ the impedance is real and assumes its minimum. Accordingly, the velocity of the oscillatory motion for a given force is at its maximum at this frequency. This phenomenon is known as resonance and the considered system is called a resonator. The (angular) frequency ω_0 is its resonance frequency.

In Figure 2.7b the velocity amplitude divided by the force amplitude, that is, the magnitude of the admittance $|v/F|$ of the resonator, is plotted as a function of the frequency ratio ω/ω_0. Such curves are called 'resonance curves', and their parameter is the quantity Q introduced in eq. (2.23). Since the resonance becomes more pronounced with increasing Q, this quantity is called the 'Q-factor' of the resonator (Q stands for 'quality'). Alternatively, the width of the resonance peak can be characterised by the separation of the frequencies $\omega_0 \pm \Delta\omega$ at which the resonance curve is below its maximum by a factor of $1/\sqrt{2}$. If this is the case the vector representing the impedance in Figure 2.7a is inclined by $\pm 45°$ to the real axis. Therefore these frequencies are sometimes called the '45°-frequencies' of the resonator. Furthermore, the kinetic energy of the resonator, which is proportional to the square of the velocity v, is half its maximum value at $\omega_0 \pm \Delta\omega$. Therefore $2\Delta\omega$ is called the (double) 'half-width' of the resonator. It is related to the

Q-factor by the relation

$$\frac{2\Delta\omega}{\omega_0} = \frac{1}{Q} \approx \frac{r}{m\omega_0} \tag{2.27}$$

2.6   Free vibrations of a simple resonator

Equations (2.26) or (2.26a) represent the response of a simple oscillator to a stationary external force varying sinusoidally. The term 'stationary' means that the system is in a 'steady state', that is, it is assumed that all transient processes such as may be caused by some switching have died out. In contrast, it is these kinds of oscillations which we shall deal with in this section.

Our starting point is again the differential equation (2.19). But now we are looking for solutions without any external forces acting on the system; instead, the system is excited by its initial state. Accordingly, we set $F(t) = 0$. We try a solution of the form:

$$s(t) = s_0 e^{gt} \tag{2.28}$$

with g denoting some unknown constant. Inserting this expression into eq. (2.19) leads to a quadratic equation for g:

$$g^2 + \frac{r}{m}g + \frac{1}{nm} = 0$$

or, with ω_0 after eq. (2.22) and with the abbreviation $\delta = r/2m$:

$$g^2 + 2\delta g + \omega_0^2 = 0$$

Its solutions are:

$$\omega_{1,2} = -\delta \pm \sqrt{\delta^2 - \omega_0^2} = -\delta \pm j\sqrt{\omega_0^2 - \delta^2}$$

First we assume $\delta < \omega_0$. Introducing both roots into eq. (2.28) yields two partial solutions from which any other solution of the homogeneous differential equation may be obtained by linear combination. With the abbreviation

$$\omega' = \sqrt{\omega_0^2 - \delta^2}$$

the general solution reads:

$$s(t) = s_0 e^{-\delta t} \left(A e^{j\omega' t} + B e^{-j\omega' t} \right) \tag{2.29}$$

The constants A and B must be determined from the initial conditions, that is, from the state of the system at a given time, for instance, at $t = 0$. Let us suppose that the system is initially at its resting position but has a velocity $v_0 = v(0)$ caused, for instance, by a short stroke with a little hammer just at $t = 0$. At first, we conclude from $s(0) = 0$ that $A + B = 0$ and therefore

$$s(t) = s_0 A e^{-\delta t}(e^{j\omega' t} - e^{-j\omega' t}) = j2s_0 A e^{-\delta t} \sin \omega' t,$$

the latter expression is obtained with eq. (2.6b). To determine the constant A it is differentiated with respect to time

$$v(t) = j2s_0 A e^{-\delta t} \left(\omega' \cos \omega' t - \delta \sin \omega' t \right)$$

and for $t = 0$:

$$v_0 = j2s_0\omega' A$$

Hence the final solution of the differential equation for the given case reads:

$$s(t) = \frac{v_0}{\omega'} e^{-\delta t} \sin \omega' t \qquad (2.30)$$

It represents an oscillation with exponentially decreasing amplitude (see Fig. 2.1b) and with an angular frequency ω' which is smaller than the resonance frequency. The constant δ originally introduced as an abbreviation turns out to be the decay constant. After eq. (2.23) it is related to the Q-factor by

$$Q = \frac{\omega_0}{2\delta} \qquad (2.31)$$

For $\delta \geq \omega_0$ the angular frequency ω' will be imaginary or zero. Equation (2.29) tells us that there is no oscillation in this case; instead, the displacement of the mass will decrease exponentially for $t > 0$ without changing its sign. In the case of $\delta = \omega_0$ the oscillator is said to be critically damped; according to eq. (2.31) it corresponds to $Q = 0.5$.

2.7 Electromechanical analogies

In the preceding section we encountered three mechanical elements, namely, a loss element with the resistance r defined by eq. (2.16), a spring with the compliance n according to eq. (2.17), and the mass m as defined in eq. (2.18). They can be combined into more complicated mechanical systems than the simple resonator described earlier.

On the other hand, many linear and passive electrical systems are made up of three elementary components too: resistors, capacitors and inductances. Let U and I denote the electrical voltage and current, respectively. Then the quantities characterising these elements, namely, the resistance R, the capacitance C and the inductance L are defined by the following relations:

$$U_R = R \cdot I \qquad\qquad\qquad (2.16a)$$

$$U_C = \frac{1}{C} \cdot \int I dt \qquad\qquad\qquad (2.17a)$$

$$U_L = L \frac{dI}{dt} \qquad\qquad\qquad (2.18a)$$

Formally, these formulae agree with eqs. (2.16) to (2.18), provided voltage is considered the analog of force, and current the analog of velocity. In this way one arrives at the analogy I of electrical and mechanical quantities as shown in Table (2.2).

However, this is not the only way to translate mechanical elements and systems circuits into electrical ones. Equations (2.16a) to (2.18a) can be expressed as well in the form

$$I_G = G \cdot U \qquad\qquad\qquad (2.16b)$$

where G = 1/R is the conductance of a resistor. The capacitance C can be defined by

$$I_C = C \frac{dU}{dt} \qquad\qquad\qquad (2.17b)$$

and the inductance L by

$$I_L = \frac{1}{L} \cdot \int U dt \qquad\qquad\qquad (2.18b)$$

Comparing these equations with eqs. (2.16) to (2.18) leads to the analogy II as listed in Table 2.2, in which current is regarded as the analog of the force, and voltage the analog of velocity.

Table 2.2 Electromechanical analogies

Mechanical quantity	Electrical quantity (I)	Electrical quantity (II)
Force	Voltage	Current
Displacement	Charge	—
Velocity	Current	Voltage
Resistance	Resistance	Conductance
Compliance	Capacitance	Inductance
Mass	Inductance	Capacitance
Impedance	Impedance	Admittance
Admittance	Admittance	Impedance
Connection in series	Connection in parallel	Connection in series
Connection in parallel	Connection in series	Connection in parallel

Both analogies can be used to 'translate' mechanical systems into equivalent electrical circuits which may be particularly helpful for those readers who are more familiar with electrical networks than with mechanical systems. Since these analogies are purely formal, there is no point in arguing about which one is right and which is not; it is rather a matter of personal taste which one is preferred. The force–voltage analogy translates mechanical impedances into electrical impedances and vice versa, while the equivalent circuits derived with the force–current analogy have more similarity with the mechanical systems they represent. In this book the former one will be preferred.

As an example we consider the resonance system described in the Sections 2.5 and 2.6 which is depicted once more in Figure 2.8a. Its equivalent electrical circuit is shown in Figure 2.8b; the external force F is represented by a voltage U which is distributed among the three elements. In contrast, the velocity – equivalent to the electrical current – is the same for all elements. Therefore they all must be connected in series. When the system is not excited by an external force but instead by vibrations of the foundation on which it is mounted (see Fig. 2.9a) then the voltage source must be replaced with a 'current' source supplying the velocity v. Since the spring and the loss element oscillate with the same velocity they must be connected in series. This holds, however, not for the mass, which vibrates with a different velocity. On the other hand, the same force is acting on the mass and on the combination of the spring and the resistance. In this way we are led to the equivalent circuit of Figure 2.9b. It is easy to see that the velocity of the mass is

$$v_m = \frac{r + (1/j\omega n)}{j\omega m + r + (1/j\omega n)} \cdot v = \frac{1 + (j/Q)(\omega/\omega_0)}{1 + (j/Q)(\omega/\omega_0) - (\omega/\omega_0)^2} \cdot v \quad (2.32)$$

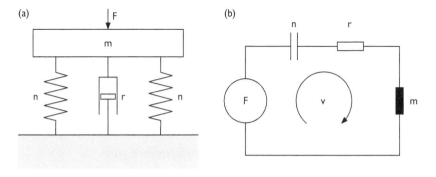

Figure 2.8 (a) Simple resonance system, (b) equivalent electrical circuit.

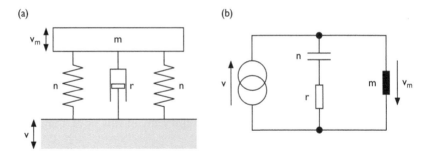

Figure 2.9 (a) Resonance system, excited by vibrations of its fundament, (b) equivalent electrical circuit.

where ω_0 as earlier is the resonance frequency according to eq. (2.22) and Q is given by eq. (2.23). Well above the resonance frequency of the system v_m will be much smaller than v. This is an efficient way of insulating sensitive instruments such as balances against vibrations underground. We shall come back to this equation in Chapter 15.

2.8 Power

To maintain the stationary oscillation of a system it is necessary to compensate for the inevitable losses. The energy required for this must be supplied by the exciting force.

If some force F displaces a point or a part of a system by the distance ds, then it supplies the mechanical work $dA = Fds$ to the system provided both the deflection and the force have the same direction. The work per second is the power; to obtain it we have to replace the deflection ds with the velocity v:

$$P = F \cdot v \tag{2.33}$$

Here it is advisable to leave the complex notation; accordingly, we insert for F and v not the expressions (2.12) and (2.13) but their real parts:

$$P = \hat{F}\hat{v}\cos(\omega t)\cos(\omega t - \psi) = \tfrac{1}{2}\hat{F}\hat{v}[\cos\psi + \cos(2\omega t - \psi)] \tag{2.34}$$

Hence the total power consists of a time-independent part

$$P_a = \tfrac{1}{2}\hat{F}\hat{v}\cos\psi \tag{2.35}$$

which is the active power, and a time-dependent part

$$P_{re} = \tfrac{1}{2}\hat{F}\hat{v}\cos(2\omega t - \psi) \tag{2.36}$$

called the reactive power. Only the first one is required for making up for the losses whereas the energy represented by P_{re} is being periodically exchanged between the source and the system. By invoking the impedance or the admittance from eqs. (2.14) and (2.15), we can express either $\hat{F}$ by $\hat{v}$ or vice versa:

$$P_a = \tfrac{1}{2}\hat{v}^2\mathrm{Re}\{Z\} = \tfrac{1}{2}\hat{F}^2\mathrm{Re}\{Y\} \tag{2.37}$$

or, by introducing root-mean-square values $\tilde{v} = \hat{v}/\sqrt{2}$ and $\tilde{F} = \hat{F}/\sqrt{2}$ as defined in eq. (2.4a):

$$P_a = \tilde{v}^2\mathrm{Re}\{Z\} = \tilde{F}^2\mathrm{Re}\{Y\} \tag{2.37a}$$

Finally, the active power can also be represented using complex notation after eqs. (2.12) and (2.13):

$$P_a = \tfrac{1}{4}\left(Fv^* + F^*v\right) \tag{2.38}$$

where the asterisk $*$ indicates the conjugate complex quantity. It is easily verified that this equation agrees with eq. (2.35).

2.9   Fourier analysis

The reason for the central significance of the harmonic oscillation is that virtually any kind of oscillation or vibration can be broken down into harmonic, that is, sinusoidal vibrations. This is true not only for periodic oscillations but for almost any kind of signal, for instance, for the damped oscillation considered in Section 2.6, or for single impulses, which we would not recognise as oscillations at first glance. The tool for doing so is the Fourier analysis which plays a fundamental role in all vibration and acoustics but also in many different fields as, for instance, signal or system theory.

2.9.1   Periodic signals

Here we consider a time function s(t) denoting not necessarily the displacement of some particle but may be a force or pressure, an electrical voltage, etc. At first it is assumed that s(t) is a periodic function with the period T:

$$s(t + T) = s(t)$$

It can be represented by a series, a so-called Fourier series, which in general contains an infinite number of terms. Each of them represents a harmonic oscillation (or simply 'harmonics'):

$$s(t) = \sum_{n=-\infty}^{\infty} C_n e^{j2\pi nt/T} \tag{2.39}$$

n is a positive or negative integer. Hence angular frequencies are integral multiples of a fundamental frequency $\omega_0 = 2\pi/T$. If s(t) is a given function the Fourier coefficients C_n which are complex in general can be calculated from it by:

$$C_n = \frac{1}{T} \int_0^T s(t) e^{-j2\pi nt/T} dt \tag{2.40}$$

If s(t) is real, this formula shows immediately that

$$C_{-n} = C_n^* \tag{2.41}$$

Sometimes a real representation of the Fourier series is preferable. From eq. (2.40) we obtain without difficulty:

$$s(t) = C_0 + \sum_{n=1}^{\infty} \left(C_n e^{j2\pi nt/T} + C_n^* e^{-j2\pi nt/T} \right)$$

$$= C_0 + 2 \cdot \sum_{n=1}^{\infty} \mathrm{Re}\left\{ C_n e^{j2\pi nt/T} \right\} \tag{2.42}$$

or, by putting

$$C_n = \hat{C}_n e^{j\varphi_n} \tag{2.43}$$

$$s(t) = C_0 + 2 \cdot \sum_{n=1}^{\infty} \hat{C}_n \cos\left(\frac{2\pi nt}{T} + \varphi_n \right) \tag{2.44}$$

Now each partial vibration is represented by a cosine function showing the particular phase angle in its argument, and its amplitude is $\hat{C}_n$. The partial with angular frequency $2\pi/T$ is called the fundamental vibration or briefly the fundamental. C_0 represents the mean value of s(t). The Fourier coefficients C_n form what is called the 'spectrum' of the time function s(t); it is an alternative description of the process which is completely equivalent to the description by the time function s(t).

As an example we regard the sawtooth vibration shown in Figure 2.10a as a thick line. Mathematically, it is given by:

$$s(t) = 2\hat{s} \cdot \frac{t}{T} \quad \text{for} - \frac{T}{2} < t < \frac{T}{2}, \tag{2.45}$$

outside this interval the function is periodically continued. Obviously, the zero of the time axis coincides with one of the zero passages of s(t).

Evaluation of the integral in eq. (2.40) yields

$$C_n = j\frac{\hat{s}}{n\pi}(-1)^n = \frac{\hat{s}}{n\pi}e^{j(n+1/2)\pi} \tag{2.46}$$

for $n \neq 0$; the coefficient C_0 is zero. Since

$$\cos(n\omega_0 t + n\pi + \pi/2) = -\sin(n\omega_0 t)\cos(n\pi) = \sin(n\omega_0 t) \cdot (-1)^{n+1}$$

the Fourier series of eq. (2.42) reads:

$$s(t) = \frac{2\hat{s}}{\pi}\left(\frac{\sin(2\pi t/T)}{1} - \frac{\sin(4\pi t/T)}{2} + \frac{\sin(6\pi t/T)}{3} + \cdots\right) \tag{2.47}$$

In Figure 2.10b the absolute values of the Fourier coefficients from eq. (2.46) are shown as vertical lines. Furthermore, the approximation given by the three first terms of the Fourier series is depicted in Figure 2.10a as a thin curve. It is obvious that this curve approximates the function s(t) more or less, and each additional term would improve the agreement. Particularly

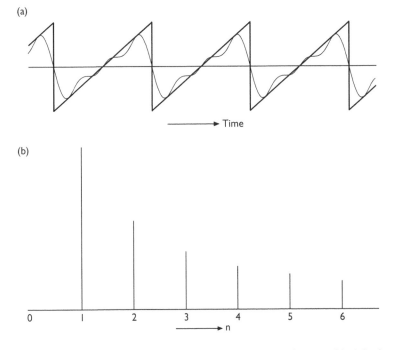

(a)

→ Time

(b)

0   1   2   3   4   5   6

→ n

Figure 2.10 Fourier analysis of a sawtooth oscillation: (a) original function (thick line) with an approximation including the first three Fourier components (thin line), (b) spectrum (magnitude).

large deviations occur in the vicinity of the discontinuities; an approximation which is satisfactory also in these areas would require many more terms in the series.

2.9.2 Non-periodic signals

If the period length T is increased over and over, the frequency increment $\omega_0 = 2\pi / T$ will tend towards zero and the sum in eq. (2.39) can be approximated by an integral. After replacing $n\omega_0$ in the exponent with ω this Fourier integral reads with $\Delta\omega = \omega_0$:

$$s(t) = \sum_{n=-\infty}^{\infty} \frac{C_n}{\omega_0} e^{j\omega t} \Delta\omega \rightarrow \int_{-\infty}^{\infty} C(\omega)e^{j\omega t}d\omega \quad \text{for T} \rightarrow \infty \tag{2.48}$$

This limiting process converts the sequence of discrete Fourier coefficients C_n into a continuous complex function of the frequency, the so-called spectral function or spectral density, often the simpler expression 'spectrum' is used:

$$C(\omega) = \lim_{T \rightarrow \infty} \frac{C_n}{\omega_0}$$

The same limiting process transforms eq. (2.40) into:

$$C(\omega) = \frac{1}{2\pi} \int_{-\infty}^{\infty} s(t)e^{-j\omega t}dt \tag{2.49}$$

For real functions s(t) we have, analogous to eq. (2.41):

$$C(-\omega) = C^*(\omega) \tag{2.50}$$

Again, the time function s(t) and its complex spectrum $C(\omega)$ are two completely equivalent representations of the same process.

To illustrate the Fourier integral by an example, we consider the exponentially decaying impulse shown in Figure 2.11a with decay constant δ. It is mathematically described by

$$s(t) = \delta e^{-\delta t} \quad \text{for t} \geq 0 \tag{2.51}$$

for negative time s(t) vanishes. The integral in eq. (2.49) yields:

$$C(\omega) = \frac{\delta}{2\pi} \cdot \frac{1}{\delta + j\omega} = \frac{1}{2\pi\sqrt{1 + (\omega/\delta)^2}} e^{j\phi(\omega)} \quad \text{with } \phi(\omega) = -\arctan\left(\frac{\omega}{\delta}\right)$$

$$\tag{2.52}$$

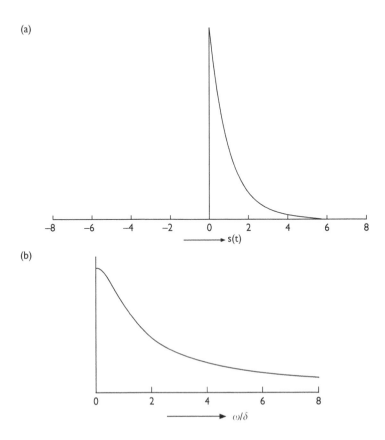

Figure 2.11 Fourier analysis of an exponential impulse: (a) time function, (b) spectrum (magnitude).

Figure 2.11b shows the absolute value of the spectral density $C(\omega)$ as a function of ω/δ.

Now suppose the decay constant δ grows beyond all limits, then the value of the function s(t) grows ad infinitum at $t = 0$ while it vanishes at all other times. The function with these properties is called the Dirac function or delta function and is abbreviated as $\delta(t)$:

$$\delta(t) = \lim_{\delta \to \infty} \delta e^{-\delta t} \tag{2.53}$$

According to eq. (2.52) with $\delta \to \infty$, its spectrum is given by

$$C(\omega) = \frac{1}{2\pi} \tag{2.54}$$

Inserting it into eq. (2.48) leads us to an alternative representation of the Dirac function:

$$\delta(t) = \lim_{\Omega \to \infty} \frac{1}{2\pi} \int_{-\Omega}^{\Omega} e^{j\omega t} d\omega \qquad (2.55)$$

2.9.3 Stationary signals

The application of eq. (2.49) to a time function s(t) fails, if the latter does not vanish with sufficient rapidity for $t \to \pm\infty$ since then the integral does not have a finite value. This holds for all stationary, aperiodic vibrational processes. Often such processes are random (see for instance Fig. 2.3c), that is, they cannot be described by a mathematical function. They are called random noise or simply noise. Examples of such 'signals' are all sorts of flow noise as, for instance, produced by a brook or by a water pipe, or the electrical noise generated in electronic circuits, in amplifiers, etc.

To analyse such vibrations as determined, for instance, by measurement, we consider a finite section of the time function s(t) of the duration t_0 and calculate the integral:

$$C_{t_0}(\omega) = \frac{1}{2\pi} \int_{-t_0/2}^{t_0/2} s(t) e^{-j\omega t} dt \qquad (2.56)$$

Obviously, the spectral function depends on the time interval t_0. If the latter is sufficiently long, $C_{t_0}(\omega)$ approximates the 'power spectrum' or the 'spectral power density' of the considered process defined by

$$W(\omega) = \lim_{t_0 \to \infty} \frac{1}{t_0} C_{t_0}(\omega) C_{t_0}^*(\omega) \qquad (2.57)$$

A stationary process the power spectrum of which is constant within a certain frequency range, for instance the range of human hearing, is called white noise. Of course, there must be some limiting frequency above which the power spectrum drops zero, otherwise the total power of the noise

$$P = \int_0^{\infty} W(\omega) d\omega$$

would be infinite.

2.9.4 Implementation of Fourier analysis

If the time function to be analysed is given in the form of a mathematical expression, then Fourier analysis is readily carried out by calculating the integrals in eqs. (2.40) and (2.49). If this is not the case, the spectral content of a sound signal can be experimentally determined. Probably the oldest, purely

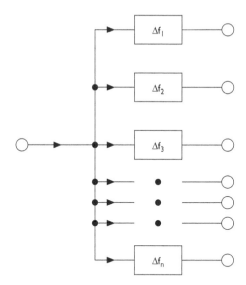

Figure 2.12 Fourier analysis with filter bank.

acoustical method is due to *Helmholtz*. It employed a set of cavity resonators (see Subsection 7.3.3) tuned to different frequencies. By holding one of their two openings close to the experimenter's ear the relative strength of the respective spectral component was subjectively assessed. Modern frequency analysers employ electrical means. They require, of course, that the sound signal to be analysed is previously converted with a microphone into an electrical signal.

For a coarse spectral analysis often a filter bank is used, that is, a set of bandpass filters (see Fig. 2.12) to which the signal under test is applied. In acoustics, the bandwidths are usually chosen to correspond to one octave or one-third octave (see Section 11.2). For a more precise examination electrical Fourier analysers may be employed. Their essential component is an automatically tuned narrowband filter which scans the signal spectrum.

A very elegant and efficient version of Fourier analysis is the Fast Fourier Transform (FFT), a smart algorithm which is based upon the symmetry of the trigonometric functions and which is performed by means of a digital computer.

2.10 Transfer function and impulse response

Although the subject of this section does not exactly belong to vibrations but rather to system theory, a few ideas shall be explained here which are as

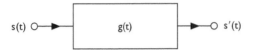

Figure 2.13 Transmission system.

important in acoustics as in other technical disciplines. More details may be found in relevant literature.[1]

The basis of the following discussion is the linear transmission system, a somewhat abstract concept of high generality and versatility. It may be imagined as a box with two ports as shown in Figure 2.13, an input and an output port. Any signal s(t) applied to its input port produces an output signal s'(t), both signals are unambiguously related to each other. Linearity means that increasing the input signal by an arbitrary factor leads to the output signal increased by the same factor. For such a system the principle of linear superposition applies: Suppose s_1' and s_2' are the output signals produced by the input signals s_1 and s_2, then the signal $s_1 + s_2$ fed to the input of the system will give rise to the output signal $s_1' + s_2'$. Examples of such systems are acoustical and electrical transmission lines, electrical filters, amplifiers, loudspeakers and microphones. Linearity is an ideal condition to which real systems often come close but which is never completely fulfilled.

The properties of a linear system are completely characterised by its frequency transfer function which is defined as follows: suppose the input signal is a harmonic vibration $s(t) = \hat{s}\exp(j\omega t)$, then the output signal will also be a harmonic vibration with the same angular frequency ω. Its amplitude, however, will differ from that of the input signal by a certain factor which depends on the frequency, and also the phase of the signal will be changed by passing through the system. Both these changes can be combined in a complex, frequency-dependent factor, called the transfer factor or the transfer function $G(\omega)$:

$$G(\omega) = |G(\omega)| \cdot e^{j\varphi(\omega)} \tag{2.58}$$

Using it, the output signal can be represented by

$$s'(t) = \hat{s}G(\omega)e^{j\omega t} = \hat{s}|G(\omega)|\,e^{j[\omega t + \varphi(\omega)]}$$

1 See for instance, A. V. Oppenheim and A. S. Willsky, *Signals and Systems*, Prentice Hall, Englewood Cliffs 1983.

Generally, the transfer function has a certain dimension. Thus the input signal may be an electrical quantity as with a loudspeaker whereas the output signal is mechanical or acoustical.

To determine the transfer function by experiment numerous amplitude and phase measurements at different frequencies are needed covering the whole frequency range of interest. This procedure can be carried out automatically today. A time-saving alternative would be to apply, so-to-speak, all frequencies simultaneously to the input of the system. According to eq. (2.55) this procedure is tantamount to excite the system at some time $t = 0$ with a vanishingly short signal, formally represented by a delta function $\delta(t)$. The corresponding output signal is called the impulse response $g(t)$ of the system. It is intuitively clear that

$$g(t) = 0 \quad \text{for } t < 0$$

since a system responding to a signal which has not yet been applied to it would contradict the principle of causality.

A mathematical expression for the impulse response of a system is obtained by multiplying each partial vibration in eq. (2.55) with a factor $2\pi G(\omega)$:

$$g(t) = \int_{-\infty}^{\infty} G(\omega)e^{j\omega t}d\omega \tag{2.59}$$

(Here the limits $\pm\Omega$ have been replaced with $\pm\infty$ since for all real systems the transfer $G(\omega)$ vanishes sufficiently fast when ω approaches infinity.) Obviously, the impulse response is the Fourier transform of the transfer function. Conversely, eq. (2.49) yields:

$$G(\omega) = \frac{1}{2\pi}\int_{-\infty}^{\infty} g(t)e^{-j\omega t}dt \tag{2.60}$$

Since the impulse response is a real function eq. (2.50) applies to G:

$$G(-\omega) = G^*(\omega) \tag{2.61}$$

This means that $|G(\omega)|$ is an even function of frequency whereas the phase function $\varphi(\omega)$ in eq. (2.58) is odd.

To calculate the reaction $s'(t)$ of the system to a given input signal $s(t)$ we note that every signal may be conceived as a dense succession of very short impulses with proper heights. Expressed with the help of the Dirac function:

$$s(t) = \int_{-\infty}^{\infty} s(\tau)\delta(t - \tau)d\tau \tag{2.62}$$

The corresponding expression of the output signal is created by replacing the Delta function in this integral with the system's response to it:

$$s'(t) = \int_{-\infty}^{\infty} s(\tau)g(t - \tau)d\tau = \int_{-\infty}^{\infty} g(\tau)s(t - \tau)d\tau \tag{2.63}$$

This operation is called the convolution of two functions s and g and is often abbreviated by

$$s'(t) = g(t) * s(t) \tag{2.64}$$

Transformed into the frequency domain, this equation reads:

$$S'(\omega) = G(\omega) \cdot S(\omega) \tag{2.65}$$

where $S(\omega)$ and $S'(\omega)$ are the spectral functions of s(t) and s'(t), respectively.

2.11 A note on non-linear systems

Many systems or elements of them show linear behaviour as long as the amplitudes of the vibrations do not exceed a certain limit. Therefore one speaks sometimes of a system's range of linearity.

To demonstrate the effect of a non-linearity we consider as an example a system which links the input and the output signal by the relation

$$s'(t) = e^{s(t)} - 1 = \frac{s}{1!} + \frac{s^2}{2!} + \frac{s^3}{3!} + \cdots \tag{2.66}$$

It is graphically represented in Figure 2.14. At the same time, this figure shows as input signals two harmonic oscillations with the amplitudes 0.25 and 1 (below) and also, on the right, the corresponding output signals. For the smaller amplitude the latter seems to be a replica of the input signal; obviously the signal is transmitted without any distortion. For the large amplitude input, however, the output oscillation has become asymmetric; the valleys are flattened but the peaks have become 'sharper'. This distortion of the waveform can also be described as a generation of spectral components which are not contained in the input signal. If we set $s(t) = \hat{s}\cos\omega t$ in eq. (2.66) the second term of the power series yields

$$\frac{\hat{s}^2}{2} \cos^2\omega t = \frac{\hat{s}^2}{4}(1 + \cos 2\omega t)$$

that is, a constant component and another component with twice the original frequency. From the third term we obtain

$$\frac{\hat{s}^3}{6} \cos^3\omega t = \frac{\hat{s}^3}{24} (3 \cos\omega t + \cos 3\omega t)$$

that is, the system produces among others a component with three times the input frequency. If the input signal consists of the superposition of two

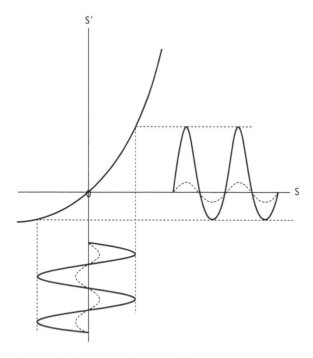

Figure 2.14 Distortion of a sinusoidal vibration by a non-linear transmission system (dotted line: $\hat{s} = 0.25$; solid line: $\hat{s} = 1$).

vibrations with frequencies ω_1 and ω_2, then the output signal will contain not only components with multiples of these frequencies but also partial components with the sum and difference frequencies

$$\omega_1 + \omega_2, \qquad \omega_1 - \omega_2$$
$$2\omega_1 + \omega_2, \quad 2\omega_1 - \omega_2, \quad 2\omega_2 + \omega_1, \quad 2\omega_2 - \omega_1, \text{ etc.}$$

If the signal is an audible sound the difference tone with the angular frequency $\omega_1 - \omega_2$ is often perceptible, for instance, if two strings of a stringed instrument are bowed simultaneously, or, still more clearly, if two flutes play tones with the interval of one-third. This is clear evidence that our hearing performs some non-linear signal processing.

Chapter 3

Acoustic variables and basic relations

Now we shall discuss in greater detail the changes of physical state mentioned in Section 1.1 which are brought about by a sound wave in a medium, and which in fact make up the very nature of sound. The instantaneous state of a medium can be described by specifying certain quantities like pressure, density, temperature, etc., and they are all altered when a sound wave is propagating in the medium. Conversely, the properties of a sound wave can be described in terms of these quantities. We call them the variables of a sound field. Some of them, as, for instance, the pressure, are particularly important because they are illustrative and, at the same time, lend themselves to direct measurement. A further subject of this chapter are the relations by which these quantities are connected to each other and from which important conclusions on the nature of sound can be drawn.

3.1 Acoustic variables

As mentioned in Chapter 2, the most obvious way to characterise the vibrational strength of an oscillating body or a part of it is by specifying the displacement vector $\vec{s}$ with its components ξ, μ and ζ in Cartesian coordinates. This simple concept can be applied as well to a sound field. Accordingly, $\vec{s}$ stands for the displacement of a medium particle from its normal position. When speaking of particles it is not the molecules of any gas or liquid we have in mind which perform rapid, irregular thermal motions even in absence of sound, instead, we think of particles imagined as small volumes each of them containing so many molecules that the effects of thermal agitation can be expected to be averaged out. As with oscillations of rigid bodies it is more common to use the velocity $\vec{v}$ with which the particle moves instead of the displacement itself. This 'particle velocity' has to be distinguished from the sound velocity which is the speed with which a disturbance travels through the medium. It is obtained from the displacement by differentiation with respect to time according to eq. (2.1). However, since $\vec{s}$ is a function of both space and time partial differentiation must now be applied indicated by round differentiation symbols:

$$\vec{v} = \frac{\partial \vec{s}}{\partial t} \tag{3.1}$$

If the medium is a fluid a sound wave can be conceived as a non-stationary flow with the space- and time-dependent flow velocity $\vec{v}$. If this flow is free of vortices which can usually be assumed in acoustics the particle velocity can be derived from a scalar potential Φ, the so-called velocity potential, by partial differentiation with respect to the coordinates:

$$v_x = -\frac{\partial \Phi}{\partial x}, \quad v_y = -\frac{\partial \Phi}{\partial y}, \quad v_z = -\frac{\partial \Phi}{\partial z} \tag{3.2}$$

In this book, however, no use will be made of this quantity.

The space dependence of $\vec{s}$ reflects the fact that the transfer of the oscillatory motion from one volume element to an adjacent one is associated with some delay. Therefore, different points of a sound field are generally at a different state at a given instant: while some particles are deflected to one side, the displacement at a different location may be zero or in the opposite direction (see for instance Fig. 1.1). Hence, a medium carrying a sound wave will undergo deformations.

Let us consider first the sound propagation in a gas or a liquid, or briefly: in a fluid. Here the deformations caused by non-synchronous motions of neighbouring particles consist of local compressions and rarefactions. Thus we arrive at another acoustical variable, namely, the variation $\rho_{\approx}$ of density. Furthermore, any change of fluid density is associated with a corresponding change $p_{\approx}$ of the pressure within the fluid. Hence the total density and pressure are

$$\rho_t = \rho_0 + \rho_{\approx} \quad \text{and}$$

$$p_t = p_0 + p_{\approx}$$

with ρ_0 and p_0 denoting the values of these quantities for the medium being at rest. The pressure variation is a particularly important variable because it is accessible to direct measurement. It is called the 'sound pressure' and has the dimension of Newton per square metre, abbreviated N/m^2. This unit is also called Pascal (Pa).

Finally, it should be mentioned that the variations of density and pressure are connected with variations of the local temperature since a fluid is heated when it is suddenly compressed and cooled down when rarefied. This temperature variation is sometimes referred to as 'sound temperature'.

As is well-known, the pressure within a fluid is a non-directional quantity acting perpendicularly on any boundary. The distribution of forces within a gas is completely described when the pressure is specified. This holds also

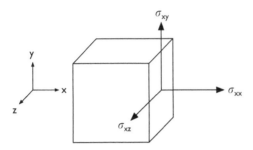

Figure 3.1 Tensile and shear stresses in a square volume element of a solid body.

for liquids as long as we disregard surface tension and viscosity. It is not true, however, for solids since a solid body tends to preserve not only its volume but also its shape. This involves more force components than just the pressure. Figure 3.1 gives an idea of the forces which may occur within a solid. It shows a material element, a small cube embedded in a solid. On its right face it may be exposed to a tensile force pointing into the x-direction. Apart from it, tangential forces may act on this face, so-called shear forces which would be inconceivable for a fluid since this would not produce elastic counter-forces to balance such shear forces.

As in the definition of the pressure in a fluid, it is useful to relate the forces in a solid to the area on which they are acting. They are called 'elastic stress' and are measured in N/m^2. One has to distinguish the normal or tensile stress from the tangential or shear stress. The latter can be split into two components according to the coordinate axes. Stresses are denoted by the symbol σ_{ik}, where the first subscript indicates the orientation of the face on which it is acting while the second one indicates the direction of the force. Accordingly, in the general case the stresses acting on the right face in Figure 3.1 are the normal stress σ_{xx} and the shear stresses σ_{xy} and σ_{xz}. Similar stresses act on the other faces of the volume element. Hence we obtain nine elastic stresses which form what is known as the stress tensor:

$$\begin{pmatrix} \sigma_{xx} & \sigma_{xy} & \sigma_{xz} \\ \sigma_{yx} & \sigma_{yy} & \sigma_{yz} \\ \sigma_{zx} & \sigma_{zy} & \sigma_{zz} \end{pmatrix} \tag{3.3}$$

The relations

$$\sigma_{xy} = \sigma_{yx}, \quad \sigma_{yz} = \sigma_{zy}, \quad \sigma_{zx} = \sigma_{xz} \tag{3.4}$$

reduce the number of independent stress components to six, which characterise the dynamic state at one point of a solid body.

3.2   Basic relations in acoustics

In this section we shall have a closer look at the way in which the acoustical variables mentioned earlier are related to each other. Our goal is to derive formulae which connect the motion of material particles with forces within the medium. This is achieved by applying the principles of conservation of momentum and matter to small volume elements within the medium under consideration.

3.2.1   Sound waves in fluids

We start by considering non-viscous gases and liquids because here just one non-directional quantity, namely, the pressure, suffices to characterise the dynamic state of the medium. Therefore we expect that the relations we are looking for are simpler than those for solids which will be dealt with in the next subsection. Since there is no risk of confusion we shall omit here and in the following the subscript $\approx$ of the symbols p and ρ as used in the preceding section.

At first we come back to the propagation of oscillatory states in a one-dimensional sound wave (see Fig. 1.1). Figure 3.2a shows a volume element which is embedded in the medium under consideration. The upper and the lower line may be imagined as the walls of a rigid tube or just as an imaginary boundary. The essential point is that the state of the medium depends only on one space coordinate called x in the following. The infinitesimal length of the volume element is denoted by dx while the cross-sectional area of the real or virtual tube is dS. The sound pressure at x is p(x); accordingly, the force acting from the left side on the shaded volume element is p(x)dS. Similarly, the force exerted to the right boundary is p(x + dx)dS.

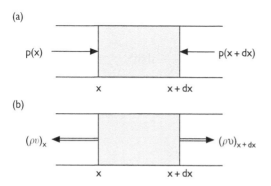

Figure 3.2 Derivation of the basic acoustical relations: (a) force balance, (b) mass balance.

The difference force

$$- \left[p(x + dx) - p(x) \right] dS = -\frac{\partial p}{\partial x} dx dS \tag{3.5}$$

accelerates the mass $\rho_t dx dS$ of the medium within the element and must overcome the inertial force which is $\rho_t dx dS$ multiplied by the acceleration dv_x/dt with v_x denoting the x-component of the particle velocity. Equating both forces leads to, after cancelling the volume $dxdS$ on both sides:

$$-\frac{\partial p}{\partial x} = \rho_t \frac{dv_x}{dt} \tag{3.6}$$

The differentiation symbol d/dt on the right side indicates the 'total acceleration' which an observer moving together with the volume element would experience. It can be written in the form:

$$\frac{dv_x}{dt} = \frac{\partial v_x}{\partial t} + v_x \frac{\partial v_x}{\partial x} \tag{3.7}$$

The first term on the right is the 'local acceleration' of the mass contained in the volume element, that is, the acceleration which a resting observer would see. After substituting this expression eq. (3.6) reads:

$$-\frac{\partial p}{\partial x} = \rho_t \left(\frac{\partial v_x}{\partial t} + v_x \frac{\partial v_x}{\partial x} \right) \tag{3.8}$$

Next we establish a mass balance for the considered volume element looking at Figure 3.2b. In contrast to the preceding derivation the volume element is now supposed to be fixed in space, that is, it does not move with the medium. At the left side the mass flowing per second into the volume element is $(\rho_t v_x)_x dS$ where the flow velocity in the positive x-direction is counted positive. Similarly, the rate with which mass flows across the right boundary is $-(\rho_t v_x)_{x+dx} dS$. The net influx of mass is

$$- \left[(\rho_t v_x)_{x+dx} - (\rho_t v_x)_x \right] dS = -\frac{\partial (\rho_t v_x)}{\partial x} dx dS \tag{3.9}$$

Because of the conservation of matter, it increases the density in the volume element $dxdS$:

$$-\frac{\partial (\rho_t v_x)}{\partial x} dx dS = \frac{\partial \rho}{\partial t} dx dS$$

or

$$-\rho_t \frac{\partial v_x}{\partial t} = \frac{\partial \rho}{\partial t} + v_x \frac{\partial \rho}{\partial x} \tag{3.10}$$

To close the circle we need a relation between the pressure and the density. This relation is called the equation of state. In general, it relates three physical quantities, namely, the pressure p_t, the density ρ_t of the fluid and its absolute temperature T. Now in acoustics we can assume – apart from rare exceptions – that compressions and rarefactions of the medium follow each other too rapidly to permit any heat exchange between adjacent volume elements. Changes of state of this kind are called 'adiabatic'. In this case the temperature is determined by the other two quantities and the equation of state simplifies to

$$p_t = p_t(\rho_t) \qquad (3.11)$$

Now the three quantities density, pressure and particle velocity are related to each other by eqs. (3.8), (3.10) and (3.11). In principle, we could use them to eliminate two of them to obtain one equation which the third variable has to satisfy. Unfortunately, these equations are non-linear; eqs. (3.8) and (3.10), for instance, contain products of these quantities. However, in a sound wave the particle velocities are so small that we can safely neglect the last terms on the right hand side of these equations. Furthermore, except in very intense acoustical situations, for instance, next to explosions, the variations of pressure and density are very small compared with their constant values ρ_0 and p_0:

$$\rho \ll \rho_0, \quad p \ll p_0 \qquad (3.12)$$

Thus, the variation of the air pressure caused by a tone of medium loudness (≈ 60 decibels) is smaller by at least seven powers of ten than the normal atmospheric pressure of 10^5 Pa. Therefore it is permissible to replace ρ_t with ρ_0 in eqs. (3.8) and (3.10). Finally, to linearise eq. (3.11) too we approximate the generally non-linear relationship (3.11) by a linear one. Accordingly, we apply $p_t - p_0 = p$ and $\rho_t - \rho_0 = \rho$:

$$p = \left(\frac{dp_t}{d\rho_t}\right)_{\rho_0} \cdot \rho = c^2 \rho \qquad (3.13)$$

Here the abbreviation

$$c^2 = \left(\frac{dp_t}{d\rho_t}\right)_{\rho_0} \qquad (3.14)$$

has been introduced which can be regarded as constant as long as the sound intensity is not excessively high. (An exception will be described in Section 4.5.) Hence, the linearised fundamental equations of a one-dimensional sound field read:

$$\frac{\partial p}{\partial x} = -\rho_0 \frac{\partial v_x}{\partial t} \tag{3.15}$$

$$\rho_0 \frac{\partial v_x}{\partial x} = -\frac{1}{c^2} \frac{\partial p}{\partial t} \tag{3.16}$$

In the latter equation the density variation ρ has been expressed as the sound pressure p using eq. (3.13). In Section 4.1 the constant c will turn out to be the sound velocity.

3.2.2 Isotropic solids

Now we turn to the somewhat more involved situation in solids. To avoid unnecessary complications we restrict our discussion to isotropic solids. A solid material is said to be isotropic when its physical properties are independent of the direction of an external influence, for instance, of an electrical field or a mechanical force. So glasses as well as most technically used metals can be assumed as nearly isotropic, whereas materials such as wood and all crystals are anisotropic.

As in the derivation of eq. (3.5) we calculate the net force acting on a material element $dV = dxdydz$ which is thought of as fixed within the solid. Figure 3.3 represents a section through such an element with the z-axis running perpendicular to the plane of the page. At first we consider the difference between the normal stresses σ_{xx} acting on the left and right face of the element, both faces being perpendicular to the x-axis. This difference forms the first term of the expression below which corresponds to the left side of eq. (3.5). Now we must remember that the normal stress is not the only force component directed in x-direction, others are the shear stresses

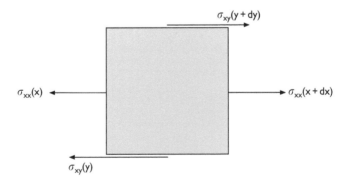

Figure 3.3 Forces in the x-direction, acting on a volume element of a solid body.

σ_{xz} acting on the faces perpendicular to the y-axis. They give rise to the net force shown in the second term. In an analogous way the shear stresses σ_{xz} (not shown in Fig. 3.3) on the faces perpendicular to the z-axis are taken into account yielding the third term. Hence the force component directed into x-direction is:

$$\left[\sigma_{xx}(x+dx) - \sigma_{xx}(x)\right] dydz + \lfloor \sigma_{xy}(y+dy) - \sigma_{xy}(y)\rfloor dxdz$$

$$+ \left[\sigma_{xz}(z+dz) - \sigma_{xz}(z)\right] dxdy = \left(\frac{\partial \sigma_{xx}}{\partial x} + \frac{\partial \sigma_{xy}}{\partial y} + \frac{\partial \sigma_{xz}}{\partial zx}\right) dxdydz$$

This expression is the analog of eq. (3.5). Again, this force must be balanced by the x-component of the inertial force which the volume element exerts when it is accelerated. This component is the mass $\rho_t dxdydz$ of the considered volume element multiplied by its acceleration. Assuming that the particle velocities within the solid are very small and that the variable part of the density is small compared to its equilibrium value we can replace the total acceleration with the local acceleration (see the discussion in preceding section); we arrive at a linearised equation corresponding to eq. (3.15). After cancelling dxdydz it reads:

$$\frac{\partial \sigma_{xx}}{\partial x} + \frac{\partial \sigma_{xy}}{\partial y} + \frac{\partial \sigma_{xz}}{\partial z} = \rho_0 \frac{\partial^2 \xi}{\partial t^2} \tag{3.17}$$

Again we have replaced ρ_t with ρ_0. Two similar equations for the y and the z components of the acceleration, $\partial^2 \eta / \partial t^2$ and $\partial^2 \zeta / \partial t^2$, are obtained by replacing the first subscript x of the stresses with y and z, respectively.

After having discussed the inertial forces in a solid we must now turn to the other component of any oscillatory motion, namely, the restoring forces. Whenever a solid body is deformed it will react with elastic forces which tend to re-establish its original state. These deformations can be described in terms of strains which are either fractional changes of the bodies' dimensions or changes of its shape. Formally, they can be expressed by partial derivatives of the displacement components ξ, η and ζ with respect to the coordinates x, y and z.

The stresses and relevant derivatives of the displacements are connected to each other by the relations:

$$\sigma_{xx} = 2\mu \frac{\partial \xi}{\partial x} + \lambda \operatorname{div} \vec{s}, \quad \text{similarly for } \sigma_{yy} \text{ and } \sigma_{zz} \tag{3.18}$$

$$\sigma_{xy} = \mu \left(\frac{\partial \xi}{\partial y} + \frac{\partial \eta}{\partial x}\right), \quad \text{similarly for } \sigma_{yz} \text{ and } \sigma_{xz} \tag{3.19}$$

Here the expression

$$\operatorname{div} \vec{s} \equiv \frac{\partial \xi}{\partial x} + \frac{\partial \eta}{\partial y} + \frac{\partial \zeta}{\partial z} \tag{3.20}$$

was introduced as a shorthand notation; it is called the divergence of the displacement vector $\vec{s}$. Physically, it describes a deformation known as the dilatation or – more precisely – the cubic dilatation which is the relative increase of volume or – except for the sign – the relative change of density. The symbols μ und λ (the latter must not be confused with an acoustical wavelength) are elastic constants of the material, the so-called Lamé constants. Equations (3.18) and (3.19) represent in a way the equation of state of the solid. Their linearity is ensured up to the highest mechanical stresses encountered in sound waves.

3.3   Wave equations

The relations developed earlier can be condensed into an equation which has to be satisfied by one acoustical variable, say, by the sound pressure if we consider first sound in a fluid. For this purpose the particle velocity v_x is eliminated from the linearised eqs. (3.15) and (3.16). This is achieved by partially differentiating eq. (3.15) with respect to x and eq. (3.16) with respect to t. This leads us to

$$\frac{\partial^2 p}{\partial x^2} = \frac{1}{c^2} \frac{\partial^2 p}{\partial t^2} \tag{3.21}$$

This differential equation is also known as the acoustical 'wave equation'; not only the sound pressure but also all other acoustical variables must satisfy it. It is of fundamental importance in acoustics as we shall see in the following chapters. Moreover, it describes the propagation of other kinds of waves, for instance, of transverse waves travelling in strings which are the basis of many musical instruments. In this case, the sound pressure p must be replaced with the transverse displacement of the string and the constant c is different from that in eq. (3.14), of course.

Equation (3.21) holds for one-dimensional sound waves only. However, real sound fields are more complex, they are three-dimensional. This means at first that we must consider also the components v_y and v_z of the particle velocity vector $\vec{v}$. Accordingly, two more equations of the type of eq. (3.15) must be established. To do this one just has to replace x in eq. (3.15) with y and z. The resulting three relations can be combined into one by applying vector notation:

$$-\operatorname{grad} p = \rho_0 \frac{\partial \vec{v}}{\partial t} \tag{3.22}$$

whereby grad p is a vector, called the gradient of p, with the components

$$\frac{\partial p}{\partial x}, \quad \frac{\partial p}{\partial y}, \quad \frac{\partial p}{\partial z}$$

Similarly, eq. (3.16) must be extended by replacing the differential quotient $\partial v_x / \partial x$ with

$$\frac{\partial v_x}{\partial x} + \frac{\partial v_y}{\partial y} + \frac{\partial v_z}{\partial z} \equiv \text{div } \vec{v} \tag{3.23}$$

(This expression is the time derivative of the dilatation introduced by eq. (3.20).) Hence, the three-dimensional form of eq. (3.16) reads:

$$\rho_0 \text{div } \vec{v} = -\frac{1}{c^2} \frac{\partial p}{\partial t} \tag{3.24}$$

From this the dilatation is obtained by integration with respect to time t:

$$\text{div } \vec{s} = -\frac{p}{\rho_0 c^2} \tag{3.24a}$$

Equation (3.22) contains an important statement on the nature of sound waves in fluids: A sound wave travels in that direction in which the interaction between adjacent volume elements, effected by the pressure differences, is strongest, that is, in the direction of the pressure gradient. This is also the direction in which the vector $\vec{v}$ representing the particle velocity points and hence the direction of the displacement vector $\vec{s}$. Waves of this kind are called longitudinal waves. Thus, eq. (3.22) tells us that sound waves in gases and liquids are longitudinal waves. Their counterparts are transverse waves which are encountered in solids. Schematic representations of plane longitudinal and transverse waves are shown in Figure 10.1.

As in the one-dimensional case one arrives at the wave equation in a more general form by combining eqs. (3.22) and (3.24). For this purpose the divergence operation is applied to eq. (3.22) with the result:

$$-\text{div grad } p = \frac{\partial}{\partial t} (\rho_0 \text{div } \vec{v})$$

Since the expression in the bracket is given by eq. (3.24), we obtain immediately

$$\Delta p = \frac{1}{c^2} \frac{\partial^2 p}{\partial t^2} \tag{3.25}$$

where the Laplace operator $\Delta \equiv \text{div grad}$ has been introduced as a kind of a shorthand notation. In rectangular coordinates the left hand side of eq. (3.25) reads

$$\Delta p \equiv \frac{\partial^2 p}{\partial x^2} + \frac{\partial^2 p}{\partial y^2} + \frac{\partial^2 p}{\partial z^2} \tag{3.26}$$

Although eqs. (3.22), (3.24) and (3.25) have been derived on the basis of cartesian coordinates, their validity is not confined to a particular coordinate system. If another coordinate system is chosen for some reason one has just to apply the respective expressions for the vector operations div, grad and Δ.

The wave equation for an isotropic solid is somewhat more complicated, due to the greater variety of quantities describing the dynamics of a solid. It is obtained, for instance, for the elongation component ξ by expressing in eq. (3.17) the elastic stresses by the derivatives of elongation components according to eqs. (3.18) und (3.19):

$$\mu \Delta \xi + (\mu + \lambda) \frac{\partial (\mathrm{div}\, \vec{s})}{\partial x} = \rho_0 \frac{\partial^2 \xi}{\partial t^2} \tag{3.27a}$$

Corresponding equations hold for the components η and ζ:

$$\mu \Delta \eta + (\mu + \lambda) \frac{\partial (\mathrm{div}\, \vec{s})}{\partial y} = \rho_0 \frac{\partial^2 \eta}{\partial t^2} \tag{3.27b}$$

$$\mu \Delta \zeta + (\mu + \lambda) \frac{\partial (\mathrm{div}\, \vec{s})}{\partial z} = \rho_0 \frac{\partial^2 \zeta}{\partial t^2} \tag{3.27c}$$

We will postpone the discussion of these wave equations to Chapter 10.

3.4 Intensity and energy density of sound waves in fluids

Any oscillatory motion contains energy, namely, kinetic energy stored in the mass element, and potential energy stored in the spring representing the restoring force. The same holds for the motion of particles in a sound wave.

The energy content of a sound wave per unit volume is called the energy density. Since the wave travels through the medium, it transports the energy contained in it. This energy flow is characterised by the 'sound intensity', sometimes also called 'energy flux density', which is perhaps even more illustrative. We understand by this quantity the energy passing per second through an imaginary window of unit area perpendicular to the direction in which the wave travels. Figure 3.4 shows a window with area dS; the energy travelling through the window in time t is contained in the volume element dV = cdtdS (dashed) behind the window. Expressed in terms of the energy

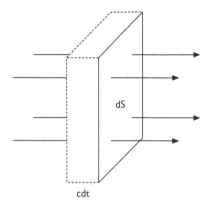

dS

cdt

Figure 3.4 Definition of sound intensity.

density w it is wdV. Dividing this energy by dS and dt yields immediately a simple relation between the intensity and the energy density:

$$I = c \cdot w \tag{3.28}$$

More precisely, we must conceive the intensity as a vector pointing in the direction of sound propagation. It can be expressed by an equation similar to eq. (2.33) by replacing the force F with the sound pressure p. Since we are only interested in the 'active intensity' (see Section 2.8) the product on the right side is averaged with respect to time, marked by a horizontal dash:

$$\vec{I} = \overline{p\vec{v}} \tag{3.29}$$

Another important relation is found from the fact that the energy leaving a fixed volume V per second must be equal to that of the rate of energy reduction in this volume. The former is the normal component I_n of the intensity with respect to the boundary of this volume, integrated over the boundary area. Hence we obtain

$$\iint_S I_n dS = - \iiint_V \frac{\partial w}{\partial t} dV$$

Here we supposed that there are no sound sources within V. On the other hand, Gauß's theorem of vector analysis tells us that

$$\iint_S I_n dS = \iiint_V \operatorname{div} \vec{I} dV$$

Since this equations must be valid for arbitrarily small volume elements, equating both volume integrals yields

$$\text{div}\,\vec{I} = -\frac{\partial w}{\partial t} \qquad (3.30)$$

To express the energy density by the sound pressure and the particle velocity we insert $\vec{I}$ from eq. (3.29) into the left side of the above relation. According to the rules of vector analysis we set:

$$\text{div}\,(p\vec{v}) = \vec{v}\,\text{grad}\,p + p\,\text{div}\,\vec{v}$$

The gradient and the divergence on the right can be expressed by the time derivatives of the particle velocity and the pressure, using eqs. (3.22) and (3.24). The result is:

$$\text{div}\,(p\vec{v}) = -\rho_0\vec{v}\frac{\partial\vec{v}}{\partial t} - \frac{1}{\rho_0 c^2}p\frac{\partial p}{\partial t} = -\frac{\rho_0}{2}\frac{\partial v^2}{\partial t} - \frac{1}{2\rho_0 c^2}\frac{\partial p^2}{\partial t}$$

with v denoting the magnitude of the vector $\vec{v}$. By time averaging we obtain from this expression:

$$\text{div}\,\left(\overline{p\vec{v}}\right) = \text{div}\,\vec{I} = -\left(\frac{\rho_0}{2}\right)\frac{\overline{\partial v^2}}{\partial t} - \frac{1}{2\rho_0 c^2}\frac{\overline{\partial p^2}}{\partial t}$$

According to eq. (3.30) the right side of this equation is the negative time derivative of the energy density w . Hence the energy density itself is:

$$w = \frac{\rho_0}{2}\overline{v^2} + \frac{\overline{p^2}}{2\rho_0 c^2} \qquad (3.31)$$

or, using the root-mean-square values of v and p (see Section 2.1):

$$w = \rho_0\frac{\tilde{v}^2}{2} + \frac{\tilde{p}^2}{2\rho_0 c^2} = w_{\text{kin}} + w_{\text{pot}} \qquad (3.32)$$

For harmonic signals both the kinetic and the potential energy density are equal. If the sound pressure and the particle velocity are given as complex quantities, the intensity can be represented as:

$$\vec{I} = \tfrac{1}{4}\left(p\vec{v}^* + p^*\vec{v}\right) = \tfrac{1}{2}\text{Re}\,\{p\vec{v}^*\} \qquad (3.33)$$

in analogy to eq. (2.38).

3.5   The sound pressure level

In Section 3.3 it was mentioned that the pressure fluctuations occurring in sound waves are very small compared to the normal atmospheric air pressure. Thus, the faintest pressure variation which is just perceivable as sound is about $2 \cdot 10^{-5}$ Pa $= 2 \cdot 10^{-5}$ N/m^2. The upper limit of useful hearing sensation is the so-called threshold of pain. The sound pressure corresponding to this value is of the order of 20 Pa which is still very small in comparison to the atmospheric pressure ($\approx 10^5$ Pa). It is quite remarkable that our hearing has such a high sensitivity and that it is capable of processing sound signals the sound pressures of which cover about six orders of magnitude. Therefore it is more useful to employ the logarithm of the sound pressure as a measure of the strength of a sound:

$$L = 20 \cdot \log_{10} \left(\frac{\tilde{p}}{p_b} \right) \text{dB} \tag{3.34}$$

As before, $\tilde{p}$ denotes the root-mean-square pressure, and $p_b = 2 \cdot 10^{-5}$ Pa is an internationally standardised reference pressure. The quantity defined in this way is called the sound pressure level, and its 'unit' is given the name decibel, abbreviated as dB (the quotation marks are to indicate that the decibel is not a unit in the physical sense). The sound pressure level can be formed for any stationary sound signals, that is, for all signals for which the effective value is meaningful at all.

The main advantage of the sound pressure level is that it leads to manageable numerical values. However, it is not directly related to the loudness with which we perceive sounds. Nevertheless, it has become the most common measure of sound strength, not only in acoustical measuring techniques but also in daily life, for instance, when noise annoyance by a highway, a factory or a discotheque is being debated.

If two sounds with root-mean-square pressures $\tilde{p}_1$ und $\tilde{p}_2$ are to be compared to each other with respect to their strengths, this is usually done in terms of level difference

$$\Delta L = 20 \cdot \log_{10} \left(\frac{\tilde{p}_1}{\tilde{p}_2} \right) \text{dB} \tag{3.35}$$

Because of the widespread use of the sound pressure level one should keep in mind that doubling the sound pressure corresponds to a level increase of 6 dB, while a ten-fold sound pressure means a level increase of 20 dB. Since, on the other hand, the intensity is proportional to the square of the sound pressure, doubling the sound intensity means a level increase of 3 dB; when the intensity is increased by a factor of 10 the level will be raised by 10 dB.

Chapter 4

Plane waves, attenuation

In the simplest types of sound waves the acoustical variables such as the sound pressure or the particle velocity depend on one spatial coordinate only. If at any given instant all variables are constant on any plane perpendicular to this coordinate we arrive at the plane wave. While it propagates perpendicular to these planes its shape remains unaltered provided all loss effects are disregarded. Therefore, the formal treatment of this kind of wave is quite simple. For this reason the plane wave is well-suited to explain particular propagation phenomena such as the attenuation of sound which will be discussed later in this chapter, or the reflection, refraction and diffraction of sound which will be dealt with in Chapters 6 and 7. Furthermore, the plane wave is elementary in the sense that more complicated wave forms can be decomposed into plane waves in analogous to Fourier analysis which represents almost any vibrations in terms of harmonic vibrations (see Section 2.9).

However, the plane wave represents a very idealised form of sound propagation. In our everyday life it is almost never encountered, at least not in its pure form. At best more realistic sound fields as that of a spherical wave can be approximated by plane waves in restricted regions. Moreover, with certain precautions one can generate and propagate plane sound waves in tubes with rigid walls.

4.1 Solution of the wave equation

For the sake of simplicity we assume that the plane wave travels along the x-axis of a rectangular coordinate system. Then we can take advantage of the one-dimensional wave equation (3.21) for fluids as derived in Section 3.4:

$$\frac{\partial^2 p}{\partial x^2} = \frac{1}{c^2} \frac{\partial^2 p}{\partial t^2}$$

It is easy to see that any function $p = f(x, t)$ with existing second derivatives is a solution of this partial differential equation provided it contains the variables x and t in the combination $x - ct$. To demonstrate this, we introduce the variable u = x − ct to form the first derivatives:

$$\frac{\partial p}{\partial x} = \frac{dp}{du} \cdot \frac{\partial u}{\partial x} = \frac{dp}{du} \quad \text{and} \quad \frac{\partial p}{\partial t} = \frac{dp}{du} \cdot \frac{\partial u}{\partial t} = -c\frac{dp}{du} \tag{4.1}$$

The second derivatives are calculated in the same way:

$$\frac{\partial^2 p}{\partial x^2} = \frac{d^2 p}{du^2} \cdot \left(\frac{\partial u}{\partial x}\right)^2 + \frac{dp}{du} \cdot \frac{\partial^2 u}{\partial x^2} = \frac{d^2 p}{du^2}$$

and

$$\frac{\partial^2 p}{\partial t^2} = \frac{d^2 p}{du^2} \cdot \left(\frac{\partial u}{\partial t}\right)^2 + \frac{dp}{du} \cdot \frac{\partial^2 u}{\partial t^2} = c^2 \frac{d^2 p}{du^2}$$

Inserting these expressions into the wave equation proves immediately that p = f(x − ct) is one possible solution of it.

Another solution is p = g(x + ct) with g denoting another arbitrary function. This is proved in the same way. Hence the general solution of the one-dimensional wave equation reads:

$$p(x, t) = f(x − ct) + g(x + ct) \tag{4.2}$$

To illustrate the meaning of this expression we consider the first part of eq. (4.2), that is, we set p(x, t) = f(x − ct). This expression tells us that an initial disturbance given by a function f(x) at time t = 0 will be shifted by the distance ct towards increasing x after t seconds without changing its shape. This fact is represented in Figure 4.1. It is obvious that the disturbance travels with the speed c from left to right. Since the pressure disturbance is a sound wave, the quantity introduced in eq. (3.13) originally as an abbreviation has turned out to be the sound velocity. Furthermore, this discussion shows that the notion of a sound wave is by no means restricted to a more or less regular succession of wave hills and valleys.

The same result applies to the second part of eq. (4.2) with the difference that the combination x + ct indicates that the disturbance g propagates in the opposite direction, that is, in direction of decreasing x-values.

The waves considered here are called plane waves since all volume elements which have the same x-coordinate and hence are situated within a plane perpendicular to the x-axis are at the same vibrational state at any given instant, that is, they are compressed or dilated to the same degree. Generally, such surfaces of equal vibrational state or equal vibrational phase are called wave surfaces or surfaces of constant phase, and the lines perpendicular to them are the wave normals. Thus we can say, the wave surfaces of a plane wave are parallel planes (see Fig. 4.2).

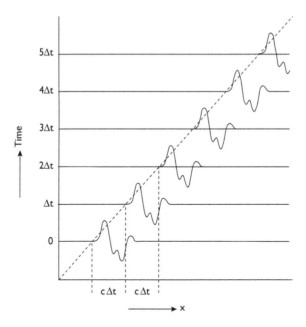

Figure 4.1 Propagation of a pressure disturbance.

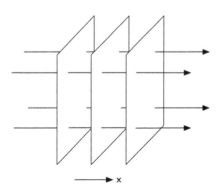

Figure 4.2 Plane wave.

Now we return to the positive-going plane wave represented by $p = f(x - ct)$. The particle velocity v_x associated with this pressure distribution varies also with $u = x - ct$. It is obtained by using eq. (3.15); the calculation of the derivatives is similar as in eq. (4.1). Then,

$$-c\frac{dv_x}{du} = -\frac{1}{\rho_0}\frac{dp}{du}$$

Indefinite integration yields, apart from a constant which we set equal to zero:

$$v_x = \frac{1}{\rho_0 c} p \qquad (4.3)$$

Hence the particle velocity in a progressive plane wave is proportional to the sound pressure. The ratio of the sound pressure and the particle velocity is called the 'characteristic impedance' and is denoted with the symbol Z_0:

$$\frac{p}{v_x} = \rho_0 c = Z_0 \qquad (4.4a)$$

It is the most significant acoustical property of a medium, and we shall encounter it in many of the subsequent relations. Its numerical value in air at 20°C and at normal pressure is 416 Ns/m^3. For a wave travelling in the negative x-direction we find in the same way:

$$\frac{p}{v_x} = -\rho_0 c = -Z_0 \qquad (4.4b)$$

By using eq. (4.3) alternative expressions for the energy density and for the intensity in a sound wave are obtained. Thus eq. (3. 32) can be simplified as

$$w = \frac{\tilde{p}^2}{\rho_0 c^2} \qquad (4.5)$$

whereas from eq. (3.29) the magnitude

$$I = \frac{\tilde{p}^2}{Z_0} = Z_0 \tilde{v}_x^2 \qquad (4.6)$$

of the intensity vector is obtained which, of course, points in the direction of wave propagation.

Since the particles of the medium oscillate parallel to the direction of wave propagation, the wave field will not be disturbed by a rigid and non-porous surface running parallel to that direction. We conclude from this that plane sound waves can propagate within rigidly walled tubes. For generating such a wave a reciprocating piston can be employed as depicted in Figure 4.3.

4.2 Harmonic waves

Now we consider a plane wave proceeding in the x-direction with the sound pressure varying according to a harmonic time law. Consequently, we set

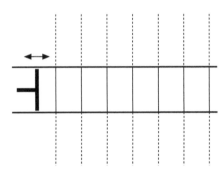

Figure 4.3 Plane wave in a rigid tube, generated by an oscillating piston.

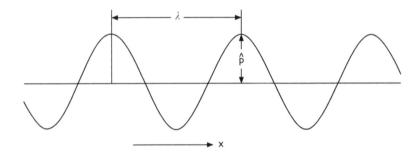

Figure 4.4 Spatial pressure distribution in a plane harmonic wave.

$g \equiv 0$ in eq. (4.2) while the function f is specified as a cosine function. Therefore

$$p(x, t) = \hat{p} \cos \left[k(x - ct)\right] = \hat{p} \cos(\omega t - kx) \qquad (4.7)$$

with k as an arbitrary constant with the dimension metre^{-1}. In complex notation this reads:

$$p(x, t) = \hat{p} e^{j(\omega t - kx)} \qquad (4.8)$$

For a given distance x this expression represents a harmonic vibration after eq. (2.8) with the amplitude $\hat{p}$, the angular frequency $\omega = kc$ and the phase angle kx. Considered as a function of space coordinate x eq. (4.7) represents a spatial vibration. It is sketched in Figure 4.4 which agrees with Figure 2.2 with the difference that now x plays the role of the independent variable. Accordingly, the temporal period T is replaced with the spatial period λ: proceeding in the wave by a distance $\lambda = 2\pi/k$ (or by an integral multiple of it) leads us to the same state of vibration as before. And the constant k turns out to be just the spatial analog of the angular frequency ω. It is called the 'angular wave number'. Combining the formulae in this paragraph yields the relations

$$k = \frac{2\pi}{\lambda} = \frac{\omega}{c} \tag{4.9}$$

and

$$c = f \cdot \lambda \tag{4.10}$$

Of course, plane waves can travel in directions other than that of the x-axis. To find a mathematical expression for this more general case we mark the orientation of the wave planes by their common normal vector $\bar{n}$ with the Cartesian components $\cos\alpha$, $\cos\beta$ and $\cos\gamma$. The latter satisfy the condition $\cos^2\alpha + \cos^2\beta + \cos^2\gamma = 1$; α, β and γ are the angles which this vector subtends with the three axes (see Fig. 4.5). It points in the direction of wave propagation. The distance of one of these planes from the origin of the coordinate system is

$$\bar{n}\,\vec{r} = x \cdot \cos\alpha + y \cdot \cos\beta + z \cdot \cos\gamma$$

$\vec{r}$ denotes the vector with components x, y and z, and $\bar{n}\,\vec{r}$ indicates the scalar product of both vectors. Hence the plane wave travelling in the direction of

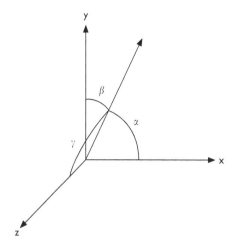

Figure 4.5 Definition of angles α, β and γ.

ñ is represented by

$$p(x, y, z, t) = \hat{p}e^{j[\omega t - k(x \cos \alpha + y \cos \beta + z \cos \gamma)]} \tag{4.11}$$

For a somewhat more formal representation we can define the wave number vector $\vec{k} = k\vec{n}$ with the components $k \cdot \cos \alpha, k \cdot \cos \beta$ and $k \cdot \cos \gamma$ and arrive at

$$p(x, y, z, t) = \hat{p}e^{j(\omega t - \vec{k}\vec{r})} \tag{4.12}$$

4.3 A few notes on sound velocity

Now we return once more to eq. (3.14) and the discussion in Section 4.1, according to which the sound velocity is given by

$$c = \sqrt{\left(\frac{dp_t}{d\rho_t}\right)^{(ad)}_{\rho_0}} \tag{4.13}$$

The notation (ad) is to remind us of the assumption that all changes of state take place adiabatically. For a perfect gas the adiabatic equation of state (3.11) reads

$$\frac{p_t}{p_0} = \left(\frac{\rho_t}{\rho_0}\right)^{\kappa} \tag{4.14}$$

From this we obtain by differentiation:

$$c^2 = \left(\frac{dp_t}{d\rho_t}\right)^{(ad)}_{\rho_0} = \frac{\kappa p_t}{\rho_t} \approx \frac{\kappa p_0}{\rho_0} \tag{4.15}$$

The constant κ is the adiabatic or isentropic exponent; its value for air is 1.4. Furthermore, the general equation of state of a perfect gas tells us that

$$\frac{p_t}{\rho_t} = \frac{RT}{M_r} \tag{4.16}$$

T denotes the absolute temperature in Kelvin (K), $R = 8.31$ Nm/mol·K is the molar gas constant and M_r the molecular mass of the considered gas. Hence the sound velocity of a perfect gas can be written as well as

$$c = \sqrt{\frac{\kappa RT}{M_r}} \tag{4.17}$$

Table 4.1 Sound velocity and characteristic impedance of fluids

Material	Temperature (°C)	Density (kg/m³)	Sound velocity (m/s)	Characteristic impedance (Ns/m³)
Gases				
Argon	0	1.783	319	569
Helium	0	0.178	965	172
Oxygen	0	1.429	316	452
Nitrogen	0	1.251	334	418
Hydrogen	0	0.090	1284	116
Ammonia	0	0.771	415	320
Carbon dioxide	0	1.977	259	512
Air	0	1.293	331	429
Liquids				(in 10^6)
Water	20	998	1483	1.48
Mercury	20	13 500	1451	19.6
Ethyl alcohol	20	790	1159	0.92
Glycerine	20	1228	1895	2.33
Carbon tetrachloride	20	1594	938	1.50
Benzine	20	878	1324	1.16
Acetone	20	794	1189	0.94
Diesel oil	20	800	1250	1.0
Helium	−272.15	145	239	0.035
Hydrogen	−252.7	355	1127	0.40
Oxygen	−183	1143	909	1.04

It may be noted that the differential quotient in eq. (3.13) is the adiabatic compression modulus $K^{(ad)}$ divided by the density ρ_0. So we arrive at

$$c = \sqrt{\frac{K^{(ad)}}{\rho_0}}, \tag{4.18}$$

an expression which is more commonly used for liquids.

In Table 4.1 the sound velocity and the characteristic impedance of some gases and liquids are listed.

4.4   Attenuation of sound

Up to now we have neglected all losses occurring in a sound wave. In reality, the energy of every sound wave is continuously reduced by certain dissipative processes in the course of its propagation. The lost energy is ultimately converted into heat. This process is called attenuation or absorption. (In this text we prefer the expression attenuation to avoid confusion with wall

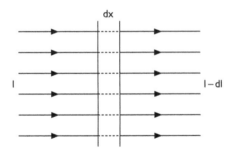

Figure 4.6 Attenuation of a plane wave.

absorption as discussed in Chapter 6.) Prior to discussing the physical facts underlying attenuation we shall make a few notes on the formal description of attenuation.

Figure 4.6 shows a plane wave indicated by parallel lines. We consider a layer with thickness dx through which a wave passes. The decrease of sound intensity in the wave due to dissipation is proportional to the thickness of the layer, of course, and furthermore to the intensity of the wave incident onto the layer:

$$-dI = mIdx$$

m is called the attenuation constant of the medium. This simple differential equation has the solution:

$$I(x) = I_0 e^{-mx} \tag{4.19}$$

with the arbitrary constant I_0. Of course, the decrease of sound intensity is tantamount to a reduction of sound pressure. Since the root-mean-square sound pressure and hence the sound pressure amplitude, according to eq. (4.6), is proportional to the square root of the intensity, the sound pressure in a harmonic plane is

$$p(x, t) = p_0 e^{-mx/2} e^{j(\omega t - kx)} \tag{4.20}$$

if attenuation is taken into account. This formula corresponds to eq. (4.8). The formal similarity with this equation becomes even clearer by combining m/2 and k into a complex angular wave number:

$$\underline{k} = k - j\frac{m}{2} = \frac{\omega}{c} - j\frac{m}{2} \tag{4.21}$$

Then, eq. (4.8) can be retained with the extended meaning of the angular wave number.

Concerning the physical causes of sound attenuation, several different mechanisms can be held responsible for it, the number of which is generally the greater the more complex the interior structure of the medium.

4.4.1 Attenuation in gases

At first the reader may be reminded that the compressions and rarefactions of a gas carrying a sound wave are accompanied with variations of the local temperature. To calculate the latter we combine eqs. (4.14) and (4.16) taking into regard that eq. (4.16) applies to the constant values p_0, ρ_0 and T_0. This leads to

$$\frac{T}{T_0} = \left(\frac{p_t}{p_0}\right)^{(\kappa-1)/\kappa} \tag{4.22}$$

or

$$\frac{T_0 + \theta}{T_0} = \left(\frac{p_0 + p}{p_0}\right)^{(\kappa-1)/\kappa} \approx 1 + \frac{\kappa-1}{\kappa}\frac{p}{p_0}$$

with θ denoting 'sound temperature', that is, the temperature variation. The final result is:

$$\frac{\theta}{T_0} = \frac{\kappa-1}{\kappa}\frac{p}{p_0} \tag{4.23}$$

Hence the sound temperature varies in phase with the sound pressure p.

As mentioned in Section 3.2, these formulae are based on the assumption that all changes of state are adiabatic. In reality, there will occur at least a small heat exchange between adjacent volume elements since there is no material without heat conductivity. As a consequence, a spatially and temporally varying heat flux will develop which is proportional to the negative gradient of the sound temperature θ (see Fig. 4.7) with the tendency to smooth the temperature distribution. This process is not reversible, hence heat conduction has the tendency to level out all temperature differences

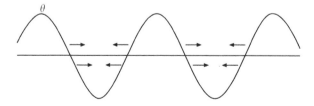

Figure 4.7 Heat flux in a sound wave.

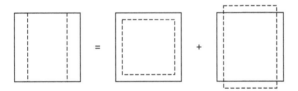

Figure 4.8 Decomposition of a unilateral deformation into a uniform compression and a shear deformation.

and, linked to them, all pressure differences. The corresponding component of the attenuation constant is

$$m_{th} = \frac{\kappa - 1}{\kappa} \frac{\nu\omega^2}{\rho_0 c C_v c^3} \tag{4.24}$$

ν denotes the heat conductivity of the medium and C_v its specific heat at constant volume.

Another component is due to the kind of deformation which a gaseous medium element undergoes in a longitudinal sound wave (see Fig. 10.1a). It consists of unidirectional compressions and expansions. As illustrated in Figure 4.8 this deformation can be imagined as consisting of a uniform compression and a shear deformation of an initially cubical volume element. Now any shear flow within a fluid is associated with frictional losses because of the viscosity of the medium. As m_{th} the viscous component of the attenuation constant increases with the square of the frequency:

$$m_{vis} = \frac{4\eta\omega^2}{3\rho_0 c^3} \tag{4.25}$$

η is the viscosity constant of the gas.

Both these absorption mechanisms together are usually referred to as 'classical attenuation' or 'classical absorption' since they can be explained on the basis of continuous matter. They describe the actual attenuation correctly for monatomic gases, that is, for rare gases.

For gases with the molecules consisting of more than one atom an additional effect is observed, namely, the so-called 'molecular attenuation' which often considerably exceeds the classical component. To understand this one should realise that the heat content of a gas or liquid is identical with the energy of the random motion of its molecules. Now a gas molecule composed of at least two atoms has several possibilities of storing motional and hence thermal energy. First, it can be displaced as a whole, second, it can rotate like a rigid body around its centre of gravity, and finally, its constituents may

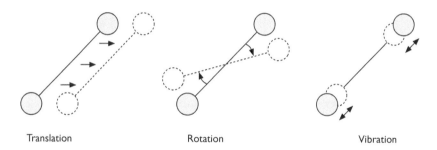

Translation   Rotation   Vibration

Figure 4.9 Possible motions of a polyatomic molecule. (By courtesy of S. Hirzel Verlag, Stuttgart.)

vibrate against each other. These three kinds of motion – translation, rotation and interior vibration – are sketched in Figure 4.9 for a two-atomic molecule. If the gas is in thermal equilibrium, its total heat content is distributed amongst the various energy stores according to a certain key.

If some thermal energy is suddenly supplied to the gas from outside, for instance, by a sudden compression, the whole additional energy will first be stored in translational motion, thereafter a gradual re-distribution among all energy stores will take place until a new equilibrium is arrived at. The delay with which the new equilibrium is established is due to the fact that the re-distribution of energy is achieved by collisions of the molecules, a process which, of course, takes some time.

If the energy supply varies sinusoidally then the energy content of the translational store follows instantaneously the changes imposed externally while the energy contents of the rotational and the vibrational stores are lagging behind. This process is known as relaxation or, more precisely, thermal relaxation.

An electrical model of a relaxation process is the simple circuit shown in Figure 4.10. One of the capacitors is directly connected with the terminals of the circuit while the other one is connected in series with a resistor and hence is not immediately accessible from outside. Accordingly, the charge on the left capacitor, which represents the translational store, reacts instantaneously to the changes of the voltage at the terminals whereas the charge on the right one follows the variations of the voltage with, however, some delay and to a reduced degree, just as the energy content of the molecular rotational or vibrational store does. At very high frequencies the right hand capacitor does not respond to any noticeable degree to the voltage fluctuations. The losses of the circuit are caused by the current flowing through the resistor which is proportional to the difference between the voltages of both capacitors. The energy lost in one oscillation period is negligible at very low

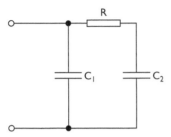

Figure 4.10 Electrical relaxation model.

and very high frequencies, in the former case because both capacitors are nearly at equal voltage while in the latter case the period is just too short. Hence the losses per period are largest at some medium frequency $\omega \approx 1/\tau$ with the time constant $\tau = RC_2$.

This consideration tells us that thermal relaxation can be thought of as some kind of internal, frequency-dependent heat conduction which is associated with energy losses much as in the same way as normal heat conduction. On account of these losses, the intensity of a plane wave will decrease within one wavelength by a factor $\exp(-m_{rel} \cdot \lambda)$ with:

$$m_{rel} \cdot \lambda = 2\pi\varepsilon \, \frac{\omega\tau}{1 + (\omega\tau)^2} \left(\frac{c}{c_0}\right)^2 \qquad (4.26)$$

Additionally, relaxation is accompanied with some 'stiffening' of the gas with increasing frequency which manifests itself as an increase of the sound velocity according to

$$\left(\frac{c_0}{c}\right)^2 = 1 - \varepsilon \, \frac{(\omega\tau)^2}{1 + (\omega\tau)^2} \qquad (4.27)$$

The time constant τ occurring in these formulae is called the relaxation time, the constant ε is the 'relaxation strength' which characterises the fraction of energy lagging behind and which is mostly in the range of a few per cent.

The content of both equations is represented in Figure 4.11. The frequency dependence of the sound velocity, the so-called dispersion, is most significant in the vicinity of the angular relaxation frequency $1/\tau$; the same holds for the wavelength-related attenuation $m_{rel}\lambda$. The wide frequency range which is covered by this attenuation curve is characteristic of relaxation.

Both constants ε and τ depend on the type of the relaxing energy store. More complicated molecules have three different possibilities of rotation – according to the three coordinates – and maybe much more possibility of

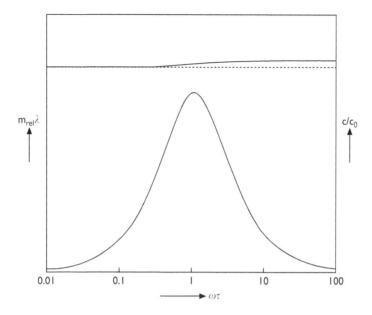

Figure 4.11 Attenuation per wavelength (lower curve, left axis) and dispersion (upper curve, right axis) by relaxation.

interior vibrations; several or even many relaxation processes may occur in one and the same gas with quite different relaxation times τ_n. Matters are even more complicated in mixtures of different gases.

The attenuation of sound waves in air is caused by the vibrational relaxation of the oxygen and the nitrogen molecules. For pure oxygen the relaxation frequency is about 50 Hz. The other constituents of air shift the frequency of this relaxation towards higher values. A particularly strong effect in this respect is due to the water dissolved in air. For this reason the attenuation of sound in air shows a pronounced dependence on humidity. In Figure 4.12 the decrease of the sound pressure level per kilometre is plotted logarithmically as a function of the frequency, the parameter of the curves is the relative humidity of the air. It should be noted that the quantity plotted in this diagram is the attenuation D per unit length and not, as in eq. (4.26) and Figure 4.11, the attenuation per wavelength. It is related to the attenuation constant m by

$$D = 10\,m \cdot \log_{10} e \approx 4.34 \cdot m \quad dB/m$$

where m is expressed in metre^{-1}.

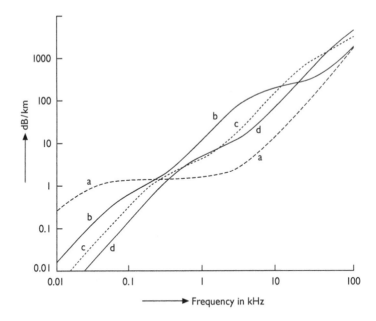

Figure 4.12 Attenuation of air (in dB/km) at 20°C and normal pressure. Parameter: relative humidity: (a) 0%, (b) 10%, (c) 40%, (d) 100%.

4.4.2 Attenuation in liquids

Regarding their absorptive properties, most liquids can be classified into three groups. For liquids of the first group all sound absorption is 'classical' according to eqs. (4.24) and (4.25). Members of this group are monatomic liquids like mercury and other molten metals, furthermore liquified rare gases but also some other liquified gases as, for instance oxygen, hydrogen or nitrogen. The latter do have rotational and vibrational heat in principle, but because of the low temperature these stores are 'frozen' so-to-speak and cannot participate in the storage of heat.

The liquids of the second group are characterised by high attenuation constants which exceed the classical one by orders of magnitude in some cases. A further characteristic of this group is a positive temperature coefficient of attenuation. The excess absorption of these materials is attributed – as with polyatomic gases – to thermal relaxation of its molecules. Examples of such liquids are benzene, toluene, hexane, carbon tetrachloride and carbon disulphide.

Finally, the third group comprises liquids whose molecules show some tendency to association, that is, to form larger complexes. Examples of such materials are liquids in which a monomer modification is in equilibrium with a dimer one, or the molecules of which have polar character, favouring a certain order similar to that in crystals. In highly associated liquids there may even be two different kinds of order or states of packing which are in equilibrium. The most important example of the latter kind is water, further examples are alcohols, phenols or aniline. The attenuation constant is typically twice to three times the classical value; furthermore, it has a negative temperature coefficient.

The excess attenuation of materials of this group is attributed to 'structure relaxation' by which term the following is meant: different molecular orders are accompanied by different packing densities of the molecules. Hence, an externally caused compression or expansion, as, for instance, by a sound wave, will shift the equilibrium between the ordered and the unordered domains, or between different states of order which is tantamount to an internal reorganisation of the liquid structure. This process takes place at high but nevertheless finite speed. Therefore the establishment of a new equilibrium occurs with some delay. This fact is responsible for a frequency dependence of the excess attenuation as given in eq. (4.26) or as sketched in Figure 4.11.

Apart from the absorption processes discussed so far there are numerous special effects from which we just mention the sound absorption in aqueous electrolytes because of its importance in underwater sound (see Chapter 16). It is caused by the electrolytic dissociation of the salts dissolved in sea water which takes place in several steps. The various dissociation products are in pressure- and temperature-dependent equilibria which are shifted periodically in one or other direction by a sound wave. Therefore electrolytes offer another example of structure relaxation; magnesium sulphate, for instance, shows maximum attenuation (per wavelength) at about 100 kHz and 100 MHz.

4.4.3 Attenuation in solids

Although this chapter is devoted to sound propagation in gases and liquids while sound in solids will be treated to some length in Chapter 10 we shall anticipate sound attenuation in solids here. Because of the great variety of forms in which solids occur we expect quite a number of different absorption processes in them. Therefore this discussion must be restricted to the treatment of a few cases of particular importance. For more detailed information the reader is referred to the relevant literature.

Starting with single crystals one should note that no real crystal is free of defects of the crystal lattice. Among them the so-called dislocations play a particular role. Dislocations are line defects in which either a plane of the lattice does not find its continuation (edge dislocation) or where lattice planes are not correctly linked to each other (screw dislocation). Both kinds of defects are shown in Figure 4.13, and the dotted line marks the so-called dislocation lines.

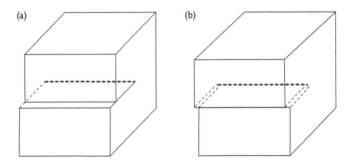

Figure 4.13 Dislocations in a crystal: (a) edge dislocation, (b) screw dislocation.

A free dislocation line can be moved perpendicular to its extension by relatively low elastic stresses until it gets stuck at an isolated lattice defect or at another dislocation line. A real crystal can be thought of as interspersed with a network of dislocation lines connected to each other. Any shear stress will deform each section of a line into a curved loop. Since molecular masses are moved in this process, we can attribute a certain mass to a dislocation line. Furthermore, a dislocation is a region of increased elastic energy striving to attain an energy minimum and hence to remain straight. Thus it behaves in a similar way as a stretched string. When oscillating around its resting position a dislocation line radiates secondary sound waves the energy of which is withdrawn from the energy of the primary wave.

Technically important metals are not homogeneous but consist of numerous small crystallites of different shape, size and orientation which may also differ in chemical composition. If a sound wave penetrates such a polycrystalline material as shown in Figure 4.14, at each grain boundary a certain portion of its energy is deflected from its original direction. If the grain dimensions are large compared with the acoustical wavelength this process consists mainly of reflection and refraction (see Chapter 6), however, if they are smaller or much smaller than the wavelength the prevailing process is diffraction or scattering (see Chapter 7). The secondary waves are reflected or scattered again and again from other grain boundaries, etc., eventually forming some background noise which has nothing to do with the original wave; their energy is just lost. (Similar effects occur with light penetrating fog.) The frequency dependence of the attenuation constant when the relevant effect is scattering is given by

$$m = B_1 f + B_2 f^4 \tag{4.28}$$

with B_1 and B_2 denoting material constants.

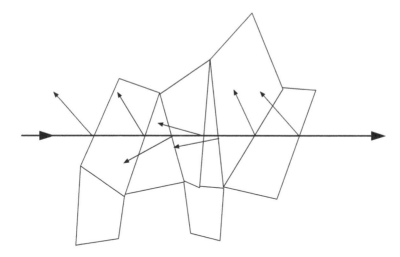

Figure 4.14 Sound scattering in a polycrystalline material.

4.5   Non-linear effects

As was shown in Section 3.5 the low sound pressures and particle velocities of everyday sounds justify the linearisations which we applied in the derivation of the wave equation. However, there may be situations where these simplifications are not permitted. So the non-linear terms neglected in Chapter 3 will certainly have an essential influence on the propagation of pressure waves from explosions, for instance, at least in the vicinity of their origin.

But even in less dramatic cases marked deviations from linear sound propagation are observed under certain conditions. In particular, it can be stated that a sufficiently intense wave does not keep its shape in the course of its propagation but will become steeper in regions of positive-going pressure and flatter in the other parts. In the limit this process may lead to a triangular wave with (nearly) vertical pressure jumps. After *Eisenmenger,* this phenomenon can be easily demonstrated with a garden hose of a few metres length, which is connected at one end to a compressed-air bottle via a hand valve, and with a funnel placed in its other end to improve the sound radiation. If the valve is suddenly opened for a brief instant, a sharp pop is heard while without the hose only a hissing noise is produced.

To describe this process in a more quantitative manner an approximation for the sound velocity is needed which is somewhat more precise than in eq. (4.17). At first, it is clear that a sound wave in a moving medium travels with a velocity which is the sum of the sound velocity c in the resting medium and the flow velocity. Even if the medium itself is at rest it will be set into motion by a sound wave, now the particle velocity v_x plays the role of the flow velocity. This consideration leads to a local sound velocity:

$$c_1 = c + v_x \tag{4.29}$$

Second, we know that the sound velocity in an ideal gas depends on the temperature including its variations, that is, also on the 'sound temperature' θ as calculated in Subsection 4.4.1 Hence according to eq. (4.17):

$$c = \sqrt{\frac{\kappa R(T_0 + \theta)}{M_m}} \approx c_0 \left(1 + \frac{\theta}{2T_0} \right) \tag{4.30}$$

Here the sound velocity for vanishingly small sound amplitudes is denoted by c_0. On the other hand we obtain from eq. (4.23) with $p = \rho_0 c_0 v_x$ and $\kappa p_0 = \rho_0 c^2$:

$$\frac{\theta}{T_0} = (\kappa - 1)\frac{v_x}{c_0} \tag{4.31}$$

or, after combining this expression with eq. (4.29):

$$c_1 = c_0 + \frac{\kappa + 1}{2}v_x \tag{4.32}$$

This velocity depends not only on time and space but also on the strength of the sound wave. The difference $\kappa - 1$ is often referred to as 'non-linearity parameter' in the literature. For air which is basically a diatomic gas ($\kappa = 1.4$) it is 0.4. For water it has been determined by experiment as about 6.

The effect of a sound velocity which depends on the instantaneous particle velocity is demonstrated in Figure 4.15. It shows as a dotted line a sine wave at the instant of its generation. The dashed line represents the same wave after having travelled a certain distance. Since the domains of positive particle velocity have moved a little faster, those with negative particle velocity a little slower, the positive-going flanks have become steeper, while the falling have been flattened. The change of the wave form is equivalent to the generation of higher harmonics (see Section 2.11). In the course of further propagation the steepening process leads to a discontinuity in the leading flank, ultimately the sine wave will have morphed into a sawtooth wave. In principle, this effect will take place with any sound wave however weak it may be, provided it has travelled a sufficiently long distance. In reality, however, the steepening process will be counteracted by the attenuation which has not been taken into account so far and which affects the higher harmonics more than the fundamental component. Practically, the steepening process is only observed with sufficiently strong waves.

Another non-linear effect which can easily be observed with ultrasound is the radiation pressure. This is a constant pressure which a sound wave exerts

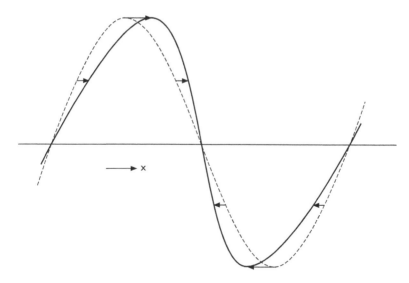

Figure 4.15 Steepening of a plane wave.

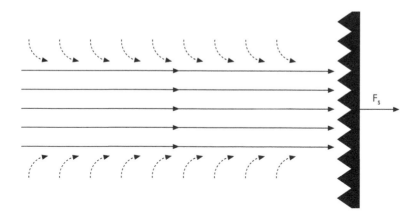

Figure 4.16 Radiation pressure ($F_s = p_s \cdot S =$ radiation force).

when it hits an obstacle (see Fig. 4.16). It is explained by the fact that a sound wave carries not only energy but as well momentum, which is the product of the velocity of a moving body with its mass. Hence the momentum per unit volume in a sound wave is $\rho_g v_x = (\rho_0 + \rho)v_x$.

Let us now consider a plane sound wave with particle velocity $v_x = p/\rho_0 c$ arriving perpendicularly at a plane by which it is totally absorbed.

The momentum arriving per second at the plane is

$$\rho_g v_x \cdot c = \rho_0 c v_x + c \rho v_x$$

this is at the same time the pressure acting onto the plane. If this expression is averaged over the time, the first term on the right will vanish, the second one yields $p_s = c\overline{\rho v_x}$ which is the expression for the radiation pressure we are looking for. Finally, one can express the particle velocity v_x and the density fluctuation ρ by the sound pressure using eq. (3.13) :

$$p_s = \frac{\overline{p^2}}{\rho_0 c^2} \tag{4.33}$$

According to eq. (3.32) this is equal to the energy density in the wave. If the plane does not absorb the arriving sound but reflects it instead, the radiation pressure equals twice the energy density.

Chapter 5

Spherical wave and sound radiation

Another elementary wave is arrived at if the distance of some point from a given point is regarded as the relevant coordinate on which the acoustical variables depend. It is called a spherical wave since the wave surfaces, that is, the surfaces of constant sound pressure are concentric spheres. Figure 5.1 shows a section through some of these surfaces, along with some wave normals. It is obvious that this wave originates from the common centre of all wave surfaces where we must imagine some sound source of vanishing extension. This somewhat abstract object is called a point source, sometimes also a simple source. Its properties will be discussed in Section 5.2. Although the point source does not exist in the real world, it is a very useful concept which helps us to understand sound radiation from more complicated sound sources as will also be shown in Sections 5.5 to 5.8.

5.1 Solution of the wave equation

The very structure of the spherical wave suggests the adoption of spherical polar coordinates (see Fig. 5.2) for its quantitative description. In this system one coordinate is distance r of some point from the origin. The other coordinates are two angles defining the direction of the line connecting that point with the origin, namely, the polar angle θ subtended by that line and a fixed axis, and the azimuth angle ϕ between the projection of that line into a plane perpendicular to the polar axis and a further fixed line situated in this plane.

Expressed in these coordinates the Laplace operator occurring in the wave equation (3.25) is

$$\Delta p \equiv \frac{\partial^2 p}{\partial r^2} + \frac{2}{r}\frac{\partial p}{\partial r} + \frac{1}{r^2 \sin^2\theta}\frac{\partial^2 p}{\partial\phi^2} + \frac{1}{r^2}\frac{\partial^2 p}{\partial\theta^2} + \frac{\cot\theta}{r^2}\frac{\partial p}{\partial\theta}$$

$$(5.1)$$

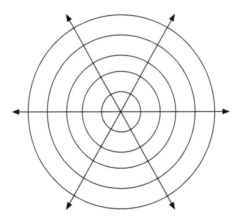

Figure 5.1 Spherical wave (Section).

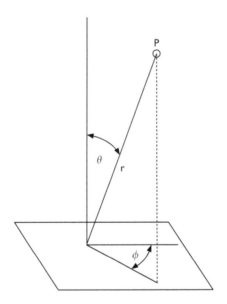

Figure 5.2 Spherical polar coordinates.

Since the sound pressure in each sphere is assumed as constant, that is, independent of the angular coordinates the three last terms of the expression above are zero. Then the wave equation reads:

$$\frac{\partial^2 p}{\partial r^2} + \frac{2}{r}\frac{\partial p}{\partial r} = \frac{1}{c^2}\frac{\partial^2 p}{\partial t^2} \tag{5.2}$$

By substituting $p(r,t) = f(r,t)/r$ it is transformed into the well-known one-dimensional form

$$\frac{\partial^2 f}{\partial r^2} = \frac{1}{c^2} \frac{\partial^2 f}{\partial t^2}$$

(see eq. (3.21)), which is solved in the same way as for a plane wave, that is, by any function with existing second derivatives and with the argument containing the variables in the combination $r - ct$. Therefore the sound pressure in a spherical wave is

$$p(r,t) = \frac{1}{r} f(r - ct) \qquad (5.3)$$

As in a plane wave, an initial disturbance of the pressure travels in the direction of increasing r-values, that is, away from the origin. However, with increasing distance its strength is gradually diminished. This behaviour is quite understandable since in the course of propagation the wave has to cover larger and larger regions and hence becomes more 'diluted'. In principle, a combination $r + ct$ would also lead to a solution of the wave equation (5.2). It would represent a spherical wave contracting to the point $r = 0$. Since this process is not very realistic we shall disregard it in the following.

5.2 The point source

Now we come back to the point source already mentioned in the introduction to this chapter. It is a logical consequence of a wave spreading out from some point, the origin of the spherical coordinate system. Its only function is to expel or to inhale a certain volume of the medium. It is characterised by its volume velocity $Q(t)$ which is the volume of the medium the source adds or removes per second, expressed in m^3/s. To establish a relation of the function $Q(t)$ and the function f we must calculate the particle velocity in a spherical wave by applying eq. (3.22). Its only non-vanishing component is the radial velocity v_r, directed outward. The radial component of the vector grad p is $\partial p/\partial r$. Then eq. (5.3) yields

$$\frac{\partial v_r}{\partial t} = -\frac{1}{\rho_0} \frac{\partial p}{\partial r} = \frac{1}{\rho_0} \left(\frac{f(r - ct)}{r^2} - \frac{f'(r - ct)}{r} \right) \qquad (5.4)$$

the dash ′ means differentiation with respect to the whole argument. Now we imagine a sphere with a very small radius r circumscribed around the origin. If the earlier equation is multiplied with the area $4\pi r^2$ of the sphere, the left side of this equation represents by definition the time derivative $\dot{Q}$ of the volume velocity where the point above the symbol Q denotes differentiation with respect to time. Since r is very small, the second term in the bracket on the right, can be neglected compared with the first one, leading to

$$\dot{Q}(t) = 4\pi r^2 \left(\frac{\partial v_r}{\partial t}\right)_{r\to 0} = \frac{4\pi}{\rho_0} f(-ct)$$

which is equivalent to

$$f(r - ct) = \frac{\rho_0 \dot{Q}(t - r/c)}{4\pi}$$

With this result the sound pressure in a spherical wave can be represented as

$$p(r,t) = \frac{\rho_0}{4\pi r}\, \dot{Q}\left(t - \frac{r}{c}\right) \tag{5.5}$$

Particularly important is the propagation law according to which the sound pressure is inversely proportional to the distance r from the origin. This means that the sound pressure level is diminished by 6 dB when the distance is doubled; it is reduced by 20 dB when the distance is increased by a factor of ten (see Section 3.6).

We arrive at a harmonic spherical wave by presuming that the volume velocity varies according to $Q(t) = \hat{Q}e^{j\omega t}$. Then

$$p(r,t) = \frac{j\omega\rho_0\hat{Q}}{4\pi r}\, e^{j(\omega t - kr)} \tag{5.6}$$

Again we have introduced here the angular wave number $k = \omega/c$ (see eq. (4.9)). The factor $j = \exp(j\pi/2)$ indicates an additional phase shift of 90° between the sound pressure and the volume velocity.

If the position of the sound source is interchanged with that of the observation point the ratio of the sound pressure p and the volume velocity remains unchanged. This almost trivial fact holds even when reflecting walls or sound scattering obstacles are within the sound field, or if we consider sound transmission in a pipe or in a closed room. In this more general form it is the content of the important principle of reciprocity.

For calculating the particle velocity in a spherical wave we use the first part of eq. (5.4), replacing the time differentiation with a multiplication by $j\omega$:

$$j\omega\rho_0 v_r = -\frac{\partial p}{\partial r}$$

The derivative of p obtained from eq. (5.6) reads:

$$\frac{\partial p}{\partial r} = \frac{j\omega\rho_0\hat{Q}}{4\pi r}\left(-\frac{1}{r} - jk\right) e^{j(\omega t - kr)} = -p\left(jk + \frac{1}{r}\right)$$

Hence the particle velocity is

$$v_r = \frac{p}{\rho_0 c} \cdot \left(1 + \frac{1}{jkr}\right) \tag{5.7}$$

In contrast to the plane wave, the sound pressure and the particle velocity in a spherical wave are not in phase. For very small distances r the second term in the bracket prevails and the phases of both variables differ by 90°. If, on the other hand, $kr \gg 1$, that is, the distance r is large compared with the acoustical wavelength $\lambda = 2\pi/k$, the ratio of sound pressure and particle velocity approaches the characteristic impedance $Z_0 = \rho_0 c$. This is understandable since with increasing distance the curvature of the wave surfaces will become smaller and smaller, the wave becomes more and more plane. It is remarkable that the limit between 'very small' and 'very large' distances depends on the wavelength and hence on the sound frequency. Thus, for $kr = 4$ the magnitude of the ratio of the sound pressure and the particle velocity differs only by about 3% of the characteristic impedance. For air and at a frequency of 100 Hz this holds true at a distance of about 2 metres from the sound source. If the frequency is raised to 5000 Hz, however, this critical distance is just slightly more than 4 cm.

The intensity in a spherical wave can be determined from eq. (3.33). Inserting v_r from eq. (5.7) yields the same expression as that for a plane wave:

$$I = \frac{|p|^2}{2Z_0} = \frac{\tilde{p}^2}{Z_0} \tag{5.8}$$

(compare to eq. (4.6)), and with $Z_0 = \rho_0 c$ and p after eq. (5.6):

$$I = \frac{\rho_0 \omega^2 \hat{Q}^2}{32\pi^2 c r^2} \tag{5.9}$$

Hence, the sound intensity in a spherical wave is inversely proportional to the square of the distance.

To determine the total acoustical output power of the point source this expression is integrated over the surface of a sphere with arbitrary radius r, circumscribed around the origin. In the present case this is equivalent to a simple multiplication of eq. (5.9) with the area $4\pi r^2$ of the sphere; therefore we obtain:

$$P_r = 4\pi r^2 I = \frac{\rho_0 \omega^2 \hat{Q}^2}{8\pi c} \tag{5.10}$$

Of course, a point source is a highly idealised sound generator which in its pure form is never encountered in the real world. Its importance stems from the possibility of imagining more realistic sound sources as being composed of point sources and to calculate the sound field produced by them just by adding the contributions of all of those point sources. In the subsequent sections we shall repeatedly apply this method. Moreover, many real sound sources come close to a point source as far as the sound field produced is concerned. For instance, any loudspeaker built in a closed box will produce a spherical wave provided all dimensions of the box are sufficiently small. Since in acoustics the sound wavelength is the natural yardstick, 'very small' means in our context 'very small compared with the wavelength'. A possibility to generate spherical waves with a more extended sound source will be described in Section 5.7.

5.3   The Doppler effect

Up to this point it was tacitly presumed that the sound source as well as the observer (or the sound receiver) are at rest, not only relative to each other but also relative to the medium surrounding them. In this section this assumption will be omitted in order to explain an everyday observation, namely, the change in frequency which occurs when a sound source and/or the observer are moving with respect to one another. This phenomenon is known as the Doppler effect.

A moving sound source runs behind the sound wave which it is emitting, so-to-speak. If the source is a point source the wave fronts are spherical as with a source at rest. But they are not concentric as in Figure 5.1, instead they are arranged asymmetrically as shown in Figure 5.3a. Suppose the circles

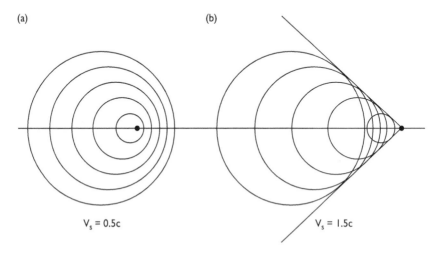

(a)   (b)

$V_s = 0.5c$ $V_s = 1.5c$

Figure 5.3 Moving point source. Left side: (a) $V_s < c$; right side: (b) $V_s > c$ ($V_s =$ speed of sound source).

in this figure mark pressure maxima of a harmonic wave, the time interval between the emission of two subsequent maxima is the period T. During this time the source moves a distance $V_s T$ with V_s denoting the speed of the sound source. Therefore the spatial distance between two wavefronts and hence the wavelength, both on the right side of the figure, is reduced to $\lambda' = (c - V_s)T$. The sound frequency $f' = c/\lambda'$ on the right is thus increased to

$$f' = \frac{f}{1 - V_s/c} \tag{5.11}$$

If the source moves away from the observer, that is, if the latter is to the left hand side of the source in the figure V_s is negative and eq. (5.11) indicates a diminished frequency. This kind of Doppler effect is frequently observed, for instance, if a motor vehicle is passing the observer: in the moment of passing the pitch of the sound falls noticeably, which is particularly impressive for the attendants of a motor race. Another common experience is the continuous decrease of pitch which is heard when a plane is flying over the observer. Figure 5.3b depicts, by the way, the case $V_s > c$, that is, when the sound source is moving with supersonic speed and therefore is overtaking the wavefronts it emits. Then all circles have an envelope consisting of a cone the aperture of which is given by an angle α with

$$\sin \frac{\alpha}{2} = \frac{c}{V_s} \tag{5.12}$$

The sound field is confined within the cone which moves with the velocity V_s from left to right.

Now we consider the opposite case, namely, that of an observer moving with speed V_r towards a resting sound source. This situation is by no means equivalent to the one described above since the observer moves not only with respect to the sound source but also to the medium. It is equivalent, however, to a medium with an embedded sound source flowing towards the stationary observer with speed V_r. Viewed from the observer the speed at which the wavefronts arrive is the regular sound velocity plus the flow velocity of the medium; accordingly, he will experience the frequency $f' = (c + V_r)/\lambda$. Inserting the wavelength $\lambda = c/f$ yields the frequency experienced by the stationary observer or, what is the same, by an observer moving towards a resting source while the medium is at rest:

$$f' = \left(1 + \frac{V_r}{c}\right) f \tag{5.13}$$

This kind of Doppler shift could be observed, for instance, by somebody standing at the open window of a train which is passing a playing music band (a rather rare event).

If the direction of the listener's motion makes an angle ϑ with the direction in which he sees (or rather hears) the source, then in eq. (5.13) a factor $\cos \vartheta$ must be inserted in the second term of the bracket.

5.4   Directional factor and radiation resistance

Now we consider an arrangement of several point sources producing sine signals with equal frequency but with different amplitudes and phases. Together they form a more complex sound source which radiates a spherical wave only when the volume velocities of elements have equal phase and when the whole arrangement is small compared with the acoustical wavelength. In general, however, the magnitude and the phase of the sound pressure will depend in a more or less involved way on the position of the observation point. This behaviour is caused by interference: the spherical waves originating from the various point sources (see Fig. 5.4) are superimposed at the observation point, and depending on the paths they travelled they reach this point with different phases. If these are predominantly equal the resulting sound pressure will be particularly high. On the other hand, the various contributions can mutually cancel each other, partially or totally; in this case the sound pressure in the observation point will be very small or will even vanish. It is evident that the result of this summation depends on the location of the observation point. The same consideration applies for extended sound sources, that is, for radiating surfaces such as vibrating plates or membranes since these can be imagined as being covered with an infinite number of densely packed point sources.

The situation becomes somewhat less complicated at large distances from the sound source, that is, in the so-called far field. It is characterised by the sound pressure amplitude varying with inverse proportion to the distance r from the source, as in a simple spherical wave. At fixed distance, however, the sound pressure depends generally on the direction of radiation indicated

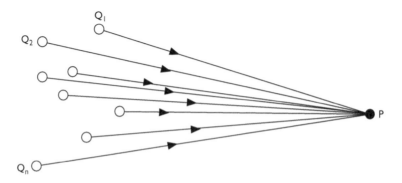

Figure 5.4 Several point sources (P: point of observation).

by angles $R(\theta, \phi)$:

$$p(r, \theta, \phi, t) = \frac{A}{r} R(\theta, \phi) e^{j(\omega t - kr)} \tag{5.14}$$

with some constant factor A. The function $R(\theta, \phi)$ is the so-called directional factor. Usually, it is normalised in such a way that its maximum value is 1. Its absolute value – plotted as a polar diagram over the relevant angle – is called the directional diagram of the sound source. According to eq. (5.8) the intensity of the radiated sound is

$$I(r, \theta, \phi) = \frac{\tilde{p}^2}{Z_0} = \frac{A^2}{2Z_0 r^2} |R(\theta, \phi)|^2 \tag{5.15}$$

Many sound sources concentrate the radiated sound energy into a limited range of solid angle, that is, their directional diagram as shown schematically in Figure 5.5 contains a pronounced main lobe. A quantitative measure of the directionality of such a sound source, that is, for the sharpness of this lobe is the 'gain' γ , borrowed from antenna techniques. It is defined as the ratio of maximum intensity I_{max} and the intensity averaged over all directions denoted by $\langle I \rangle$, both at the same distance r:

$$\gamma = \frac{I_{max}}{\langle I \rangle} = 4\pi r^2 \frac{I_{max}}{P_r} \tag{5.16}$$

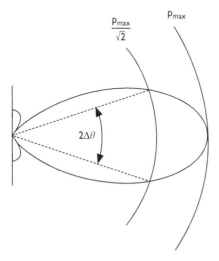

Figure 5.5 Definition of half-width in a polar plot.

In the latter expression P_r is the total power output of the source:

$$P_r = \iint_S I(\theta, \phi) dS = \frac{A^2}{2Z_0} \iint_{4\pi} |R(\theta, \phi)|^2 d\Omega \tag{5.17}$$

In the first integral integration is performed over the surface of a large sphere, in the latter the range of integration is the full solid angle 4π; the element of the solid angle is $d\Omega = \sin\theta d\theta d\phi$ when expressed in spherical polar coordinates. With this expression the gain can be represented in the form:

$$\gamma = \frac{4\pi}{\iint |R(\theta, \phi)|^2 d\Omega} \tag{5.16a}$$

An alternative measure for the directivity of a sound source is the half-width of the main lobe, that is, the angular distance $2\Delta\theta$ between both directions in which the sound pressure amplitude has fallen by a factor $1/\sqrt{2}$, the intensity by the factor $1/2$ from the maximum value (see Fig. 5.5).

Often a sound source consists of a rigid surface which oscillates as a whole with a given velocity v_0. To maintain this vibration a certain force F_r is needed. This force can also be understood as the reaction of the surrounding medium on the vibration. For harmonic vibrations, this reaction can be described by an impedance:

$$Z_r = \frac{F_r}{v_0} \tag{5.18}$$

called the radiation impedance of the source. It depends on the shape of the surface and on the medium. If this impedance is known the power output of the source can be expressed by

$$P_r = \tfrac{1}{2}\hat{v}_0^2 \text{Re}\{Z_r\} = \tfrac{1}{2}\hat{v}_0^2 R_r \tag{5.19}$$

according to eq. (2.37). The real part of the radiation impedance occurring in this formula is named the radiation resistance R_r.

In acoustical measuring techniques the radiated power of a sound source is often characterised by a logarithmic measure derived from the sound power P_r. This is the power level which must not be confused with the sound pressure level according to eq. (3.34):

$$L_P = \log_{10}\left(\frac{P_r}{P_0}\right) \tag{5.20}$$

The reference power P_0 is 10^{-12} W.

5.5   The dipole

As a first example of a directive sound source we regard the dipole source or briefly the dipole. It can be modelled by two point sources separated by distance d which produce equal sine signals apart from the sign:

$$Q_{1,2}(t) = \pm \hat{Q} e^{j\omega t}$$

In Figure 5.6a they are represented by small circles. Their distances from the observation point P are called r_1 and r_2. Then the sound pressure in P is:

$$p(r,t) = \frac{j\omega \rho_0 \hat{Q}}{4\pi} \left(\frac{e^{-jkr_1}}{r_1} - \frac{e^{-jkr_2}}{r_2} \right) e^{j\omega t} \tag{5.21}$$

Furthermore, we assume $r_{1,2} \gg d$. Accordingly, we use the approximation

$$r_{1,2} \approx r \mp \frac{d}{2} \cos \theta \tag{5.22}$$

where r is the distance of point P from the midpoint between both sources and θ is the angle subtended by the radius vector r and the line connecting both point sources. Before inserting this into eq. (5.21) we note that r_1 and r_2 can be replaced with r in the denominators since the differences in magnitude become negligibly small for somewhat larger distances r. However, these differences are not negligible when it comes to the exponentials in eq. (5.12) since the path difference $r_1 - r_2$ and hence the corresponding phase difference does not vanish at any distance r. Then we obtain from eq. (5.21):

$$p(r,t) = \frac{j\omega \rho_0 \hat{Q}}{4\pi r} \left(e^{j(kd/2)\cos\theta} - e^{-j(kd/2)\cos\theta} \right) e^{j(\omega t - kr)}$$

$$= -\frac{\omega \rho_0 \hat{Q}}{2\pi r} \sin \left(\frac{kd}{2} \cos \theta \right) e^{j(\omega t - kr)}$$

If d is small compared with the acoustical wavelength, that is, if $kd \ll 1$, the sine function can be replaced with its argument, and the expression on the right becomes:

$$p(r,t) = -\frac{\omega^2 \rho_0 \hat{Q} d}{4\pi r c} \cos \theta \cdot e^{j\omega t - kr} \tag{5.23}$$

Hence, the directivity function of the dipole as introduced in eq. (5.14) reads $R(\theta) = \cos \theta$. It is represented in Figure 5.6b as a polar diagram; its three-dimensional extension is obtained by rotating it around the dipole axis ($\theta = 0$). No sound is radiated into all directions perpendicular to this axis since the contributions of both point sources cancel each other completely

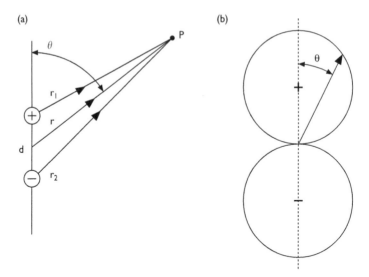

Figure 5.6 Dipole: (a) arrangement (schematically), (b) polar plot.

in the plane $\theta = 90°$. For other directions this cancellation is incomplete, and it is more effective at low frequencies than at high ones. This fact is responsible for the unfamiliar rise of the sound pressure amplitude with the square of frequency in eq. (5.23) which means that the dipole is a particularly ineffective sound source at low frequencies. This cancellation is often referred to as 'acoustic short-circuit'.

Any rigid body oscillating about its resting position can be regarded as a dipole source provided it is small compared to the acoustic wavelength. When being deflected from its resting position it tends to compress the medium at its one side, to rarefy it on the other. Examples of such dipole sources are a vibrating string, or a small loudspeaker membrane which can radiate sound from both back and front. For instance, a small rigid sphere with radius a which oscillates with the velocity amplitude $\hat{v}_0$ produces the sound pressure at large distances:

$$p(r, \theta, t) = -\frac{\omega^2 \rho_0 a^3 \hat{v}_0}{2cr} \cos\theta \; e^{j(\omega t - kr)} \tag{5.24}$$

In the same way as we can imagine a dipole as a combination of two point sources, more complex 'multipoles' can be constructed by combining dipoles. Thus, two dipoles of opposite polarity arranged close to each other form a quadrupole. This can be done in two ways: either the dipoles are combined lengthwise, or they are arranged in parallel. The directional characteristics for both cases are quite different. An example for a quadrupole of the former kind is the tuning fork: each of its prongs is a dipole, and both prongs vibrate in opposite phase. Therefore a tuning fork held up in the air produces only a very faint tone. Only when its foot is pressed upon a table or something similar, a clear tone is heard, since then the vibration of the foot is transferred to a surface with a relatively high radiation resistance.

5.6 The linear array

A further example of a combination of several point sources, which is of considerable practical interest, is the linear group or the linear array. It consists of a number of point sources arranged equidistantly along a straight line as depicted in Figure 5.7. In contrast to the preceding section all elements of the group are assumed to produce equal volume velocities Q including their phases. The total sound pressure in a field point is obtained by adding the contributions from each source, using eq. (5.6). For distant observation points the differences in their amplitude are negligible, hence the individual distances r_n in the denominators can be replaced with a typical distance r. Furthermore, the lines connecting the field point with the sources can be regarded as nearly parallel. Then the lengths of adjacent paths differ approximately by $d \cdot \sin \alpha$ with d denoting the distance of two elementary

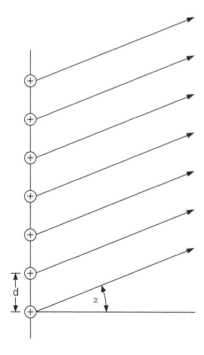

Figure 5.7 Linear array of point sources.

sources; α is the elevation angle characterising the direction of radiation. Hence the phase difference between the contributions of adjacent point sources is $kd \cdot \sin \alpha$ and one obtains:

$$p(r, \alpha, t) = \frac{j\omega\rho_0 \hat{Q}}{4\pi r} \sum_{n=0}^{N-1} e^{jkdn\sin\alpha} \cdot e^{j(\omega t - kr)}$$

N is the total number of point sources comprising the group. We note that each term of the sum is the nth power of $\exp(jkd \cdot \sin\alpha)$. Therefore the summation rule for geometric series can be applied leading to

$$p(r, \alpha, t) = \frac{j\omega\rho_0 N\hat{Q}}{4\pi r} \cdot \frac{e^{jNkd\sin\alpha} - 1}{N\left(e^{jkd\sin\alpha} - 1\right)} \cdot e^{j(\omega t - kr)}$$

Comparison with eq. (5.14) tells us that the second fraction in this formula is the directional factor $R(\alpha)$. Its absolute value is

$$|R(\alpha)| = \left| \frac{\sin(Nkd/2 \sin\alpha)}{N\sin(kd/2 \sin\alpha)} \right| \tag{5.25}$$

As may be seen from the limiting process $\alpha \to 0$ this function assumes the value 1 for $\alpha = 0$ which means that the highest intensity is emitted in all directions perpendicular to the axis of the array.

Figure 5.8 plots the magnitude $|R|$ of the directional factor of an array with eight elements; the abscissa is $kd \cdot \sin\alpha/2$. Whenever this quantity is an integral multiple of π, that is, whenever $\sin\alpha$ is an integral multiple of λ/d, the function $|R|$ attains a main maximum; between each two of these peaks there are N-2 satellite peaks. However, since $-1 \le \sin\alpha \le 1$, the range of meaningful abscissa values is limited by $\pm kd/2$ as indicated in Figure 5.8. By converting this section into a polar diagram with α as angular variable the directional diagram for this particular kd-value is obtained. In Figure 5.9 directional diagrams of this kind for $kd = 0.75$ and $kd = 2$ are depicted, the number N of elements is six. The diagrams are extended into the third dimension by rotating it around the array axis. Accordingly, the sound is not concentrated into one particular direction but into a plane perpendicular to the array.

Under the assumption that the array has significant directivity the following simple expression for the half-width of the main lobe can be derived:

$$2\Delta\alpha \approx \frac{\lambda}{Nd} \cdot 50° \tag{5.26}$$

The linear array is often used to direct the sound of a public address system, for instance, in a large hall, towards those areas where it is needed,

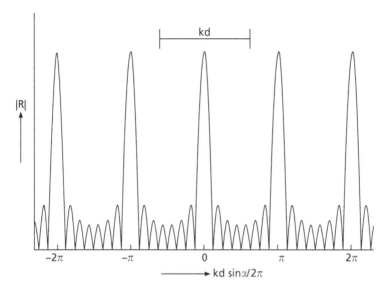

Figure 5.8 Directional factor (magnitude) of a linear array with eight elements.

Figure 5.9 Directional diagrams of a linear array with six elements.

namely, to those occupied by audience. Here the elementary sources are loudspeakers fed with the same electrical signal. Usually, the single loudspeaker will itself have a certain directivity on account of its construction and size. Then the effective directivity of the whole arrangement results from multiplying the directivity of the element with that of the array, the latter according to eq. (5.25). Other applications of linear arrays are used in ultrasonic diagnostics and in underwater sound where it serves for the insonification of limited angular ranges (see Chapter 16).

The directional characteristics of a linear array may be widely changed by varying the volume velocities of its elements according to a particular scheme which depends on the desired goal. So directional diagrams can be achieved which are free of secondary lobes, or the directivity of the array can be almost perfectly removed.

5.7 The spherical source ('breathing sphere')

The conceptually simplest sound source with finite extension is the spherical source, often referred to as 'pulsating' or 'breathing sphere'. We can use it to put some life into the somewhat abstract concept of the radiation impedance as introduced in Section 5.4.

The spherical source, shown in Figure 5.10a, consists of a solid sphere the radius of which varies sinusoidally with small amplitude. Because of its shape we expect it to produce a spherical wave, with its surface representing one of the wavefronts on which a particle velocity v_0 is imposed.

The radiation impedance and hence radiated power is easily determined from eq. (5.7) by replacing v_r with v_0 and r with the rest radius a of the source. Multiplying the sound pressure on the surface of the sphere as obtained from that equation with the area $S = 4\pi a^2$ yields the total force which the medium exerts on the surface. Divided by the velocity v_0 it leads immediately to the radiation impedance

$$Z_r = \frac{S\rho_0 c}{1 + 1/jka} \tag{5.27}$$

Its real part is the radiation resistance of the spherical source:

$$R_r = S\rho_0 c \frac{(ka)^2}{1 + (ka)^2} \rightarrow \frac{\rho_0 S^2}{4\pi c} \cdot \omega^2 \quad \text{for } ka \ll 1 \tag{5.28}$$

from which its power output can be calculated by using eq. (5.19).

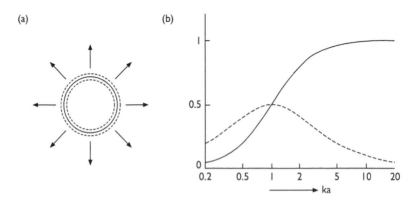

Figure 5.10 Breathing sphere: (a) schematically, (b) normalized radiation impedance (solid line: real part R_r/SZ_0, broken line: imaginary part X_r/SZ_0).

The real and the imaginary part of the radiation impedance of the pulsating sphere are plotted in Figure 5.10b as a function of ka. At low frequencies the real part, that is, the radiation resistance, increases with the square of the frequency, since in this range the sphere is small compared with the wavelength and virtually acts like a point source (compare eq. (5.10)). At more elevated frequencies the radiation resistance approaches a constant value $S\rho_0 c$ which would also be valid for a very extended plane and oscillating plane (see Section 5.8).

This frequency dependence of the radiation resistance can be explained by the fact that at low frequencies the pulsating sphere shifts some of the surrounding medium to and fro without compressing it and hence without producing sound. At higher frequencies the inertia of the medium resists its displacement, and the medium reacts to the oscillation of the surface by tolerating some compression.

For further illustration we regard the reciprocal of the radiation impedance, the 'radiation admittance' which is from eq. (5.27):

$$\frac{1}{Z_r} = \frac{1}{S\rho_0 c} + \frac{1}{j\omega m_r} \tag{5.29}$$

Here we introduced the 'radiation mass', $m_r = 4\pi a^3 \rho_0$, which represents the medium mass with which the vibrating surface is loaded. It is three times the mass of the fluid displaced by the resting sphere. If Z_r were an electrical impedance we would represent the content of eq. (5.29) by two circuit elements connected in parallel, namely, a resistance $S\rho_0 c$ and an inductor m_r (see Fig. 5.11). From this equivalent electrical circuit it is easily seen that at low frequencies the inductor draws almost the whole current flowing into the terminal and that almost no current is left for the resistance which represents the radiation. At high frequencies we have the reverse situation.

For some acoustical measurements, for instance, in room acoustics, it would be desirable to have a uniformly radiating sound generator to hand. However, its realisation in form of a pulsating sphere offers considerable

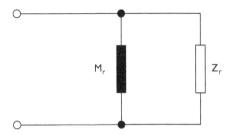

Figure 5.11 Equivalent electrical circuit for the radiation impedance of a breathing sphere.

difficulties. At least its radiation characteristics can be approximated within certain limits by a regular dodecahedron or icosahedron composed of 12 or 20 regular polygons, respectively, each of them fitted out with a loudspeaker in its centre.

5.8   Piston in a plane boundary

The types of sound sources discussed up to this point – point source, dipole and 'breathing sphere' – are highly idealised models, serving in the first place for explaining the basic process of sound radiation, which, however, describe the behaviour of real sound sources only partially and mostly for limited frequency ranges.

   In contrast, the piston source to be treated in this section comes much closer to real sound sources. A piston source is a plane, rigid plate vibrating with uniform amplitude. In order to keep the formal treatment as simple as possible we imagine this plate set flush in an infinite rigid baffle wall as shown in Figure 5.12.

   If the piston performs harmonic vibrations according to

$$v_0(t) = \hat{v}_0 e^{j\omega t} \tag{5.30}$$

each of its area elements dS may be considered as a point source with the volume velocity $v_0 dS$ producing the sound pressure $j\omega\rho_0\hat{v}_0 dS e^{j(\omega t - kr')}/2\pi r'$ in the point of observation, according to eq. (5.6). The symbol r' is the distance of the area element dS from the observation point P. The factor 2 instead of 4 in the denominator accounts for the fact that the sound is radiated into the right half space only which benefits from the whole volume velocity and which is separated from the rear of the piston by the baffle.

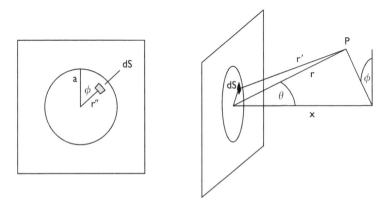

Figure 5.12   Circular piston; definition of coordinates.

The total sound pressure in the field point P is obtained by integrating this expression over the active area S of the radiator:

$$p(r,\theta,t) = \frac{j\omega\rho_0\hat{v}_0}{2\pi}e^{j\omega t}\iint\limits_{S}\frac{e^{-jkr'}}{r'}dS \tag{5.31}$$

For further discussion we assume that the piston is circular with radius a. Then the sound field generated by it is rotationally symmetric with respect to the perpendicular from the piston centre. It is useful to choose this axis as the polar of a spherical polar coordinate system. Hence, the position of a point P is determined by its distance r from the centre of the piston and by its polar angle θ. The coordinates of an area element dS on the piston are given by its distance r'' from the centre and the angle φ. Then $dS = r''dr''d\varphi$, and we obtain from eq. (5.31):

$$p(r,\theta,t) = \frac{j\omega\rho_0\hat{v}_0}{2\pi}e^{j\omega t}\int_0^a r''dr'' \int_{-\pi}^{\pi}\frac{e^{-jkr'}}{r'}d\varphi \tag{5.32}$$

To express the distance r' by r'' we may set $\phi = 0$

$$r' = \sqrt{r^2 + r''^2 - 2rr''\cos\varphi\sin\theta} \tag{5.33}$$

In general, the second integral in eq. (5.32) cannot be evaluated in closed form. However, there are two important special cases for which a closed solution can be found.

5.8.1   Sound pressure on the centre axis of the piston

If the field point P is located on the middle axis of the piston, that is, if $\theta = 0$, eq. (5.33) simplifies to $r' = \sqrt{r''^2 + r^2}$. The integration over φ is reduced to a multiplication with the factor 2π. The remaining integration with respect to r'' in eq. (5.32) can be carried out without difficulty by changing to r' as an integration variable, noting that $r''dr'' = r'dr'$. The result is:

$$p(r,t) = \rho_0 c\hat{v}_0\left(e^{j(\omega t-kr)} - e^{j(\omega t-k\sqrt{r^2+a^2})}\right) \tag{5.34}$$

This expression represents two plane waves of equal amplitude but with opposite phase, one originating from the centre of the piston, the other from its rim. They interfere with each other: if their paths differ by half a wavelength or an odd multiple of it the pressure will be at maximum. If, on the contrary, the path length difference equals a multiple of the wavelength both waves will completely cancel each other.

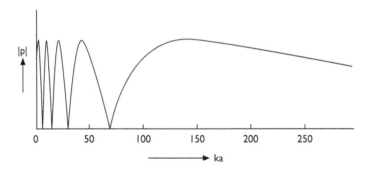

Figure 5.13 Magnitude of the sound pressure on the centre axis of a rigid piston with ka = 30.

Figure 5.13 plots, according to eq. (5.34), the absolute value of the sound pressure along the axis of the piston for ka = 30 which means that the circumference of the piston equals 30 wavelengths. Next to the piston the sound pressure amplitude shows rapid fluctuations; they characterise the near field of the source. With increasing distance the curve becomes smoother and finally passes into a monotonous decay. In fact, for r ≫ a we obtain:

$$\sqrt{r^2 + a^2} \approx r + \frac{a^2}{2r}$$

Consequently, the second exponential in eq. (5.34) reads $\exp(j(\omega t - kr)) \cdot \exp(-j\,ka^2/2r)$. If the additional condition $r \gg ka^2/2$ is fulfilled the last exponential can be approximated by $1 - j\,ka^2/2r$. With $a^2 = S/\pi$ this leads finally to:

$$p(r, t) \approx \frac{j\rho_0 \omega \hat{v}_0 S}{2\pi r} e^{j(\omega t - kr)} \tag{5.35}$$

Because $S\hat{v}_0 = \hat{Q}$ this expression agrees with eq. (5.6) – apart from the factor 2 in the denominator. Hence, the far field sound pressure along the piston axis depends in the same way on distance as that in a simple spherical wave. This was first mentioned in Section 5.4. Now we are in a position to indicate a little more precisely the distance which separates the far field from its counterpart, the near field: we can define it as the outermost point at which the pressure amplitude in eq. (5.34) assumes the value $\rho_0 c \hat{v}_0$. For sufficiently large ka-values this is at:

$$r_f \approx \frac{S}{\lambda} \tag{5.36}$$

which we call the far field distance. This simple rule of thumb can be applied also to pistons of different shape.

5.8.2 Directional characteristics

The following discussion applies to the far field of a circular piston characterised by $r \gg r_f$ (see eq. 5.36). Since we consider large distances one can neglect the variation of r' in the denominator of eq. (5.32) and replace this r' with r. Furthermore, $r \gg a$ implies $r \gg r''$, therefore we obtain from eq. (5.33):

$$r' \approx \sqrt{r^2 - 2rr'' \cos \varphi \sin \theta} \approx r - r'' \cos \varphi \sin \theta \tag{5.37}$$

to be substituted in the argument of the exponential. With these approximations eq. (5.32) reads:

$$p(r,\theta,t) = \frac{j\omega\rho_0 \hat{v}_0}{2\pi r} e^{j(\omega t - kr)} \int_0^a r'' dr'' \int_{-\pi}^{\pi} e^{jkr'' \sin \theta \cos \varphi} d\varphi \tag{5.38}$$

Now we take advantage of the integral representation of the Bessel function of order n:[1]

$$\frac{1}{2\pi} \int_{-\pi}^{\pi} e^{jx(\cos \varphi - n\varphi)} d\varphi = J_n(x) \tag{5.39}$$

(The Bessel functions J_0 und J_1 are shown in Fig. 8.15b.) Therefore the second integral in eq. (5.38) can be written in the form $2\pi J_0(kr'' \sin \theta)$. Furthermore, we use the relation:

$$\int x^{n+1} J_n(x) dx = x^{n+1} J_{n+1}(x)$$

or applied to the present case:

$$\int_0^a J_0(kr'' \sin \theta) r'' dr'' = \frac{a}{k \sin \theta} J_1(ka \sin \theta)$$

With these relations eq. (5.38) is transformed into:

$$p(r,\theta,t) = \frac{j\omega\rho_0 \hat{v}_0 S}{2\pi r} \cdot \frac{2J_1(ka \sin \theta)}{ka \sin \theta} e^{j(\omega t - kr)} \tag{5.40}$$

1 M. Abramowitz and A. Stegun, *Handbook of Mathematical Functions*. Dover Publications, New York 1964.

Hence the directional factor of the circular piston reads

$$R(\theta) = \frac{2J_1(ka\sin\theta)}{ka\sin\theta} \tag{5.41}$$

Figure 5.14 is a graphical representation of $|R(\theta)|$, plotted as a function of $ka\sin\theta$. For a given piston and frequency, this quantity cannot exceed ka. Therefore only the central section of this curve as indicated by the bar above it will enter into a polar diagram showing the absolute value of R as function of the angle θ. With increasing frequency this section becomes wider including more and more details. In contrast to the linear array, there is only one direction, namely, that of $\theta = 0$ at which the contributions of all area elements will add in equal phases. Hence, the polar diagram contains just one main lobe. In Figure 5.15 polar directivity diagrams for three values of the frequency parameter ka are represented. At low frequencies (or for small pistons) the strength of radiation is nearly independent of direction. At higher frequencies the sound is increasingly concentrated into the direction of the middle axis. Again, these patterns should be thought as extended in three dimensions by rotation, this time, however, around the middle axis of the piston.

The half-width of the main lobe is approximately:

$$2\Delta\theta \approx \frac{\lambda}{a} \cdot 30° \tag{5.42}$$

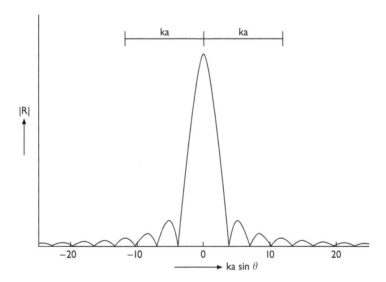

Figure 5.14 Directional factor (magnitude) of a circular piston.

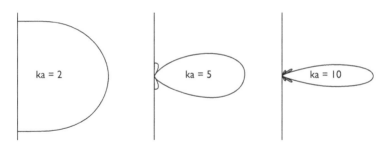

Figure 5.15 Directional diagrams of a circular piston.

5.8.3 Total power radiated and gain

Since it is rather cumbersome to derive an expression for the radiation impedance of the circular piston[2] we present just its result:

$$Z_r = SZ_0 \left(1 - \frac{2J_1(2ka)}{2ka} + j\frac{2H_1(2ka)}{2ka} \right) \tag{5.43}$$

Here H_1 denotes the Struve function of first order.[3] The real and the imaginary part of the radiation impedance is plotted in Figure 5.16 versus ka, that is, the ratio of the circumference of the piston and the wavelength. The real part of the radiation impedance, that is, the radiation resistance is:

$$R_r = SZ_0 \left(1 - \frac{2J_1(2ka)}{2ka} \right) \rightarrow \frac{\rho_0 S^2}{2\pi c} \cdot \omega^2 \quad \text{for ka} \ll 1 \tag{5.44}$$

Combined with eq. (5.19) this equation yields the power output of the piston:

$$P_r = \tfrac{1}{2}\hat{v}_0^2 SZ_0 \left(1 - \frac{2J_1(2ka)}{2ka} \right) \rightarrow \frac{\rho_0 S^2 \hat{v}_0^2}{4\pi c} \cdot \omega^2 \quad \text{for ka} \ll 1 \tag{5.45}$$

In the low-frequency range of up to ka $\approx$ 2 the radiated power grows with the square of the frequency as with the breathing sphere. This is the same range in which the radiated sound energy is more or less uniformly distributed over all directions (see Fig. 5.15); the source radiates like a point source. In this range the imaginary part of the radiation impedance is a mass reactance; the

2 Lord Rayleigh, *The Theory of Sound*, 2nd Edition 1896, Vol II, Ch. XV 1st American edition. Dover Publications, New York 1945.
3 M. Abramowitz and A. Stegun, *Handbook of Mathematical Functions*. Dover publications, New York 1964.

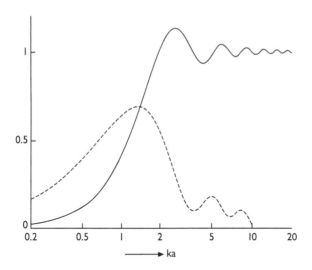

Figure 5.16 Circular piston: normalized radiation impedance (solid line: real part R_r/SZ_0, broken line: imaginary part X_r/SZ_0).

medium mass involved in it, that is, the radiation mass, is approximately

$$m_r \approx \tfrac{8}{3}\rho_0 a^3 \tag{5.46}$$

At high frequencies the radiation resistance approaches the limiting value $\rho_0 c S$, but not asymptotically as in the case of the pulsating sphere but in an oscillating manner. To understand this behaviour we remember that at high ka-values the radiation is mainly concentrated towards the middle axis of the piston. Viewed from a point on this axis the piston's surface is divided into concentric annular zones, so-called Fresnel zones, which contribute to the resulting sound pressure with opposite phases. With increasing frequency additional zones will be created at the rim of the piston, and each new zone will increase or reduce the resulting sound pressure amplitude, depending on the sign of its contribution.

According to eq. (5.40) the sound pressure amplitude on the middle axis ($\theta = 0$) is $\omega \rho_0 \hat{v}_0 S/2\pi r$ from which we obtain the maximum intensity:

$$I_{max} = \frac{\rho_0}{2c} \cdot \left(\frac{\omega \hat{v}_0 S}{2\pi r}\right)^2 \tag{5.47}$$

On the other hand, the average intensity is

$$\langle I \rangle = \frac{P_r}{2\pi r^2} = \frac{\hat{v}_0^2 R_r}{4\pi r^2}$$

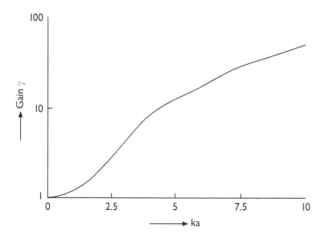

Figure 5.17 Gain of a circular piston 2.5.

both quantities at an arbitrary however very large distance r. Inserting these expressions into eq. (5.16) (first version) yields the gain of the circular piston:

$$\gamma = \frac{\rho_0 \omega^2 S^2}{2\pi c R_r} = \frac{(ka)^2}{2}\left(1 - \frac{2J_1(2ka)}{2ka}\right)^{-1} \qquad (5.48)$$

In Figure 5.17 the ten-fold logarithm of γ is plotted as a function of ka.

The treatment of pistons with different shapes (for instance rectangular or elliptic) is considerably more complicated. More is said on this matter, for instance, in F. Mechel's book on sound absorbers.[4]

Although the circular piston is again an idealised device it can be regarded as a fairly realistic model of many practical sound sources. In particular, it describes the radiation from a circular loudspeaker diaphragm quite well, provided the loudspeaker system is set flush in a very extended rigid baffle. Differences especially at high frequencies are due to the fact that loudspeaker diaphragms are not flat but have conical shape, and that they cannot be regarded any longer as rigid. In particular, they will not vibrate with uniform velocity at elevated frequencies. If the baffle is not very large further deviations from the described behaviour will occur, caused by diffraction of the sound waves around the rim of the baffle. Similar effects occur with loudspeakers built in an enclosure. Nevertheless, the rigid piston as discussed in this section is of great value in the understanding of the loudspeaker function.

4 F. P. Mechel, *Schallabsorber*, Band I. S. Hirzel Verlag, Stuttgart 1989.

Chapter 6

Reflection and refraction

Probably everybody is familiar with the echo produced by shouting or hand-clapping in front of a building facade or a large rock, that is, with the fact that a sound wave is thrown back by an extended surface. A less common experience, however, is that of a sound wave arriving at the boundary between two different media at an oblique angle and changing its direction when it penetrates the interface. The first process is called the reflection of sound waves; it occurs whenever the wave impinges on some wall, even if there is no audible echo. The change of direction which a sound wave undergoes when entering a different medium is named refraction.

Both phenomena – reflection and refraction – will be described at some length in this chapter under the simplifying assumption that the primary sound wave is plane. Then the reflected wave is also a plane wave provided the boundary is plane and of infinite extent (or at least very large in comparison with the acoustic wavelength). The same holds for the refracted wave. Compared with this relatively simple situation the reflection of a spherical wave is generally much more complicated. A comprehensive presentation of this subject can be found in F. P. Mechel's book.[1]

6.1 Angles of reflection and refraction

Let us consider a primary wave arriving from the left as is shown in Figure 6.1a which strikes a plane wall of infinite extension. For the sake of simplicity the waves are represented here just by wave normals, so-to-speak by 'sound rays'. The direction of the impinging wave is characterised by an angle ϑ with respect to the wall normal; it is called the angle of incidence. According to the law of reflection known from optics the reflected wave leaves the boundary under the same angle. Furthermore, the wave normals of both waves and the wall normal are lying in the same plane.

---

1 F. P. Mechel, see footnote on page 93.

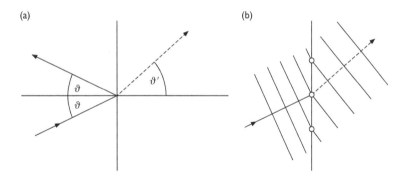

Figure 6.1 Reflection and refraction: (a) definition of the angle of incidence, reflection and refraction, (b) trace fitting.

If the 'wall' is the boundary separating two different fluids then a third wave appears which enters the medium to the right (dotted line in Fig. 6.1a). It is called the refracted wave since generally it proceeds in a different direction characterised by the refraction angle ϑ'. Both angles are related by:

$$\frac{c}{\sin \vartheta} = \frac{c'}{\sin \vartheta'} \qquad\qquad (6.1)$$

In this formula c and c' are the sound velocities in the media on the left and the right of the boundary, respectively. In optics, this relation is named 'Snell's law'. If the sound velocity in the material behind the interface is smaller than that of the medium on the left then the refraction angle is also smaller than the angle of incidence, and the wave will be refracted towards the wall normal. In the reverse case the refraction angle is larger, and the wave will be refracted away from the normal. This latter case is shown in Figure 6.1a.

These laws can be derived most easily by applying the principle of 'trace fitting' which is explained for the case of refraction in Figure 6.1b. For each wave, the incident and the refracted one, a few wavefronts are depicted. If the former travels towards the boundary with speed c, the intersections of the wave surfaces with the boundary move with the 'trace velocity' $c / \sin \vartheta$ along the interface from below to above. The corresponding trace velocity of the refracted wave is $c'/ \sin \vartheta'$. Both velocities must be equal from which eq. (6.1) immediately follows.

If the sound speed in the medium to the right of the interface is larger than that of the left medium ($c' > c$) it may happen that eq. (6.1) cannot be fulfilled since $\sin \vartheta'$ cannot exceed unity. Then no refracted wave will appear; instead, the incident wave is completely reflected from the intersection. This case is referred to a 'total reflection'. It occurs if $\sin \vartheta \geq c / c'$. The angle

$$\vartheta_t = \arcsin \left(\frac{c}{c'} \right) \tag{6.2}$$

is called the critical angle of total reflection.

The law of refraction is easily generalised for two media moving parallel to the boundary with different velocities V und V'. In this case the principle of trace fitting yields at once:

$$\frac{c}{\sin \vartheta} + V = \frac{c'}{\sin \vartheta'} + V' \tag{6.3}$$

In solids not only longitudinal waves but also transverse waves can occur. Accordingly, the reflection and refraction by boundaries turn out to be considerably more complicated. We shall come back to this matter in Chapter 10.

6.2   Sound propagation in the atmosphere

If the sound velocity and the flow velocity of the medium is not constant but varies steadily then the direction of the wave normals (or of the 'sound rays') will also vary continuously, and the sound rays are curved. This fact is of great importance for the sound propagation in the atmosphere since the temperature of the air and hence the sound velocity depends on the height over the ground. The same is true for the speed of the wind which is zero immediately at the ground and grows with increasing height. One can imagine the atmosphere divided into thin horizontal layers with slightly different sound or air speed as sketched in Figure 6.2. Whenever a sound ray

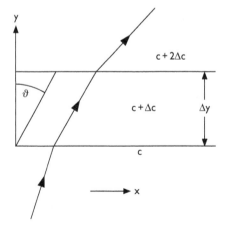

Figure 6.2 Ray path in a layered medium.

crosses the boundary between two layers its direction, that is, the angle ϑ, will be altered by a small amount. In this way the path of a ray will become a polygon which in the limit of vanishing layer thickness will change into a steadily curved line.

For a quantitative treatment we presume that the sound velocity depends on one coordinate only, for instance on the height y over ground. The same holds for the horizontal wind speed V. According to eq. (6.3), along a curved sound ray we have

$$\frac{c}{\sin \vartheta} + V = \text{const.} = c_s$$

Let x_r and y_r denote the rectangular coordinates of a ray point, then $dy_r / dx_r = \cot \vartheta$ with the present meaning of ϑ yields the direction of the tangent of the ray, or, from the equation above:

$$\frac{dy_r}{dx_r} = \cot \vartheta = \pm \frac{1}{c}\sqrt{(c_s - V)^2 - c^2}$$

With this equation the ray path can be stepwise constructed; $c_s = c_0 / \sin \vartheta + V_0$ is determined by the initial data c_0, V_0 of the ray. Vanishing of the quantity under the square root indicates a maximum or minimum of the curve; if this happens the construction must be continued by using the opposite sign. The same holds for the closed representation of the ray path which is immediately obtained from the expression above:

$$x_r = \pm c(y) \int_{y_0}^{y_r} \frac{dy}{\sqrt{(c_s - V)^2 + c^2(y)}} \tag{6.4}$$

As an example Figure 6.3a shows sound ray paths originating from a point source close to the ground. The air is assumed to be still, but with temperature falling or growing linearly with increasing height (left and right half of the figure, respectively). In the first case which might be considered as normal rays running upwards in an oblique direction will be curved upwards. Hence, rays travelling next to the ground will be more 'diluted', and the sound intensity will diminish more rapidly with distance than they would at constant sound velocity. The situation is the opposite when the temperature increases with height as shown in the right half of the figure. This is the case of temperature inversion; it takes place, for instance, when on a clear evening the air at some height above ground is still warm while it has already cooled down below because the ground radiates heat into the space, unimpeded by clouds. Oblique sound rays will now be curved back towards the ground, causing increased density of rays next to the ground. Thus in this case the temperature variation in the air has a similar effect as a collecting lens. This can lead to a quite noticeable intensity increase and to

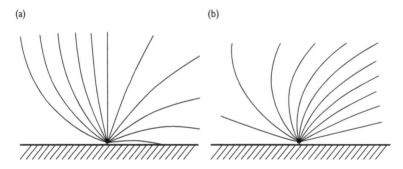

Figure 6.3 Sound rays in the atmosphere: (a) at constant temperature and wind speed linearly increasing with height (direction from the left to the right), (b) at resting air and linearly varying temperature (left side: temperature decreasing; right side: temperature increasing with height).

enhanced distance of sound propagation, sometimes associated with 'dead zones' into which no sound is heard. Everybody knows from experience that at certain evenings the noise from a distant railway or motorway is heard with unexpected loudness.

In Figure 6.3b sound rays at uniform air temperature but with wind speeds linearly increasing with height are represented. All rays are curved towards the right where a higher density of rays will occur, corresponding to an increased sound intensity. On the other, to the left the situation is reversed since here the rays are spread apart. Thus it is the vertical gradient of the wind speed and not the air flow itself which is responsible for the common experience that the sound transmission is better with the wind than against it.

Of course, circumstances are more complicated in the real atmosphere than in the model atmospheres regarded heretofore. In particular, in a thunderstorm the wind speed and the temperature show strong spatial and temporal fluctuations. This is, for instance, the reason why the sharp bang caused by lightning is mostly heard as rolling thunder at some distance and more so the larger the observer's distance from the centre of the thunderstorm.

Curved sound paths occur also in the ocean. They are caused by differences of water temperature, but also by the fact that the sound velocity depends on the salinity of the water and on the static pressure, that is, on the depth below the water surface. We shall come back to these phenomena in Chapter 16.

6.3   Reflection factor and wall impedance

The laws of reflection and refraction as mentioned in Section 6.1 yield information on the directions of the reflected and the refracted waves. In this section we shall determine the relative amplitudes of these waves.

For a quantitative treatment of sound reflection we introduce Cartesian coordinates the x-axis of which is represented by the horizontal line in Figure 6.1, while the y-axis is the vertical line which indicates the projection of the reflecting wall; the intersection of both lines marks the origin of the coordinate system. The z-axis is perpendicular to the plane of the paper.

An expression for the incident sound wave is then obtained from eq. (4.11) by replacing α with our present angle of incidence ϑ. The angle γ is $\pi / 2$ since all waves we consider in this section are travelling within the x-y-plane. Thus $\beta = (\pi / 2) - \alpha$ and therefore $\cos \beta = \sin \alpha$. Then the sound pressure of the incident wave is:

$$p_i(x, y) = \hat{p}e^{-jk(x \cos \vartheta + y \sin \vartheta)} \tag{6.5}$$

For the sake of simplicity the time factor $\exp(j\omega t)$ is suppressed in this and the subsequent expressions.

Perfect reflection will occur when the surface situated at $x = 0$ is completely rigid and non-porous. Such a surface is called 'sound hard' or simply 'hard'. Real walls may come close to this limiting case but never reach it completely. In general, the pressure amplitude will be reduced by some factor $|R| < 1$ when it is reflected, and at the same time its phase will undergo an abrupt change by some angle χ. Therefore the reflected wave can be represented by

$$p_r(x, y) = \hat{p} |R| e^{-jk(-x \cos \vartheta + y \sin \vartheta) + j\chi}$$

The change of the sign of $x \cos \vartheta$ in the exponential indicates that the wave has reversed its direction with respect to the x-coordinate while its y-direction remains unaltered. Both quantities, $|R|$ and $\exp(j\chi)$, can be combined in the complex reflection factor

$$R = |R| e^{j\chi} \tag{6.6}$$

which in general depends on the frequency and the incidence angle ϑ of the primary wave. Thus the sound pressure of the reflected wave is

$$p_r(x, y) = \hat{p}Re^{-jk(-x \cos \vartheta + y \sin \vartheta)} \tag{6.7}$$

The total wave field in front of the reflecting wall is obtained by adding eqs. (6.5) and (6.7):

$$p(x, y) = p_i + p_r = \hat{p}e^{-jky \sin \vartheta} \left(e^{-jkx \cos \vartheta} + Re^{jkx \cos \vartheta} \right) \tag{6.8}$$

The reflection factor contains all acoustical properties of the wall as far as they are relevant for the reflection of sound waves. It can be conceived as the transfer function of a linear transmission system in the sense of

Section 2.10; the sound pressure of the primary wave is its input signal while that of the reflected wave is the output signal. Therefore the following relations are true, according to eq. (2.61):

$$|R(-\omega)| = |R(\omega)| \quad \text{and} \quad \chi(-\omega) = -\chi(\omega) \tag{6.9}$$

The acoustical properties of a wall can be alternatively described by its wall impedance Z. It is defined as the ratio of the sound pressure at the wall surface to the normal component of the particle velocity at the same location. In the present case this component is v_x, hence

$$Z = \left(\frac{p}{v_x}\right)_{x=0} \tag{6.10}$$

In contrast to the mechanical impedance introduced in Section 2.4 this definition contains the force per unit area, hence the wall impedance has the dimension Ns/m^3 as the characteristic impedance Z_0. Generally, it is also complex since the sound pressure and the particle velocity are usually not in phase. The wall impedance divided by the characteristic impedance is called the specific wall impedance:

$$\zeta = \frac{Z}{Z_0} \tag{6.11}$$

Both quantities, the wall impedance and the reflection factor, are closely related. This relationship may be found from eq. (3.15) which reads, after replacing the differentiation with respect to time with a factor $j\omega$:

$$j\omega\rho_0 v_x(x, y) = -\frac{\partial p}{\partial x}$$

or, applied to eq. (6.8) and with $\omega = kc$ and $\rho_0 = Z_0/c$:

$$v_x(x, y) = \frac{\hat{p}}{Z_0} e^{-jky \sin\vartheta} \left(e^{-jkx\cos\vartheta} - Re^{jkx\cos\vartheta}\right)\cos\vartheta \tag{6.12}$$

Finally, we insert eqs. (6.8) and (6.12) with $x = 0$ into the definition (6.10):

$$Z = \frac{Z_0}{\cos\vartheta}\frac{1 + R}{1 - R} \tag{6.13}$$

or in terms of R:

$$R = \frac{Z\cos\vartheta - Z_0}{Z\cos\vartheta + Z_0} = \frac{\zeta\cos\vartheta - 1}{\zeta\cos\vartheta + 1} \tag{6.14}$$

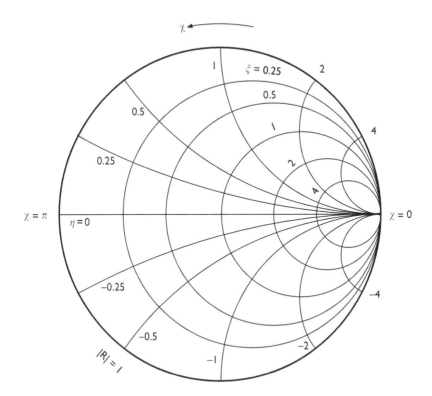

Figure 6.4 Contours of constant real part ξ and imaginary part η of the wall impedance in the complex plane of the reflection factor $|R| \cdot \exp(j\chi)$.

Figure 6.4 shows a graphical representation of this relation for normal sound incidence ($\vartheta = 0$). It shows the complex plane of the reflection factor R, and the outmost circle corresponds to $|R| = 1$. Since the absolute value of R cannot exceed unity only the interior of this circle is of interest. It contains the circles of constant real part $\xi = \text{Re}\{\zeta\}$ and of constant imaginary part $\eta = \text{Im}\{\zeta\}$ of the specific wall impedance. For given values of ξ und η the reflection factor is determined from the the intersection of the corresponding curves; if needed this point must be determined by interpolation. Its distance from the centre of the figure is the absolute value $|R|$, and its angular coordinate measured from the horizontal axis is the phase angle χ of the reflection factor. Conversely, the wall impedance can be determined from the complex reflection factor.

Next we consider the interface between two different fluid media, for instance, two liquids which do not mix. As shown in Figure 6.1, the incident sound wave will be split into a reflected and a refracted wave, the latter entering the medium behind the boundary in which the angular wavenumber is $k' = \omega / c'$. Its sound pressure is represented by

$$p_t(x, y) = T\hat{p}e^{-jk'(x \cos \vartheta' + y \sin \vartheta')} \tag{6.15}$$

The factor T is called the transmission factor of the interface. To determine it we calculate the x-component of the particle velocity, as before, by using eq. (3.15):

$$v_{tx} = \frac{T\hat{p}}{Z_0'}e^{-jk'(x \cos \vartheta' + y \sin \vartheta')} \cos \vartheta' \tag{6.16}$$

$Z_0' = \rho_0'c'$ is the characteristic impedance of the medium on the right. At the interface the sound pressures at both its sides must agree. Equating eqs. (6.8) and (6.15) with $x = 0$ yields the conditions

$$k' \sin \vartheta' = k \sin \vartheta$$

which agrees with eq. (6.1), and

$$\hat{p}(1 + R) = \hat{p}T \tag{6.17}$$

Furthermore, the x-components of the particle velocities must be the same on both sides. From eqs. (6.12) and (6.16), again with $x = 0$ one obtains a further boundary condition:

$$\frac{\hat{p}}{Z_0}(1 - R) \cos \vartheta = \frac{\hat{p}}{Z_0'}T \cos \vartheta' \tag{6.18}$$

From both relations (6.17) and (6.18) the reflection factor and the transmission factor can be calculated; the result is:

$$R = \frac{Z_0' \cos \vartheta - Z_0 \cos \vartheta'}{Z_0' \cos \vartheta + Z_0 \cos \vartheta'} \tag{6.19}$$

and

$$T = \frac{2Z_0' \cos \vartheta}{Z_0' \cos \vartheta + Z_0 \cos \vartheta'} \tag{6.20}$$

Equation (6.19) shows that at a certain incidence angle the reflected wave can completely vanish. A necessary condition for this to happen is that either

$$\frac{\rho_0'}{\rho_0} > \frac{c}{c'} > 1 \quad \text{or} \quad \frac{\rho_0'}{\rho_0} < \frac{c}{c'} < 1 \tag{6.19a}$$

It may be noted that the transmission factor according to eq. (6.20) can exceed unity, that is, the amplitude of the refracted wave may surpass that of the incident wave. This may seem a violation of the energy principle at first sight. This is, however, not so since the factors R and T are related not to energies but to sound pressures.

6.4 Absorption coefficient

A quantity of great practical importance is the absorption coefficient. This is the fraction of the incident sound energy which is not reflected from the wall, either because it is dissipated by losses occurring within the wall, or because it is transmitted through it. After eq. (5.8) the intensities of the incident and the reflected wave are given by

$$I_i = \frac{\hat{p}^2}{2Z_0} \quad \text{and} \quad I_r = \frac{|R|^2 \hat{p}^2}{2Z_0} \tag{6.21}$$

Accordingly, the absorption coefficient is related to the reflection factor by:

$$\alpha = \frac{I_i - I_r}{I_i} = 1 - |R|^2 \tag{6.22}$$

or, after expressing the reflection factor by the specific wall impedance with eq. (6.14):

$$\alpha = \frac{4\xi \cos \vartheta}{|\zeta|^2 \cos^2 \vartheta + 2\xi \cos \vartheta + 1} \tag{6.23}$$

Here ξ is, as earlier, the real part of the specific wall impedance. This expression shows that for $\vartheta \rightarrow 90°$, that is, for grazing sound incidence, the absorption coefficient is zero.

The sound pressure of an incident wave enforces a velocity normal to a boundary (provided the latter is not completely rigid). If this motion depends only on the sound pressure at the considered wall point and not also on the pressure at other points we speak of a 'locally reacting' wall. This holds, for instance, for a thin foil without bending stiffness, or for a rigid, perforated panel. If, on the contrary, the 'wall' consists of a panel with bending stiffness, its motion and hence the normal particle velocity in front of it depends on the pressure distribution over the whole surface and not merely on the pressure at a single surface point. Such a panel would be an example of a 'non-locally reacting' wall. Since a wave which arrives at an oblique angle does not excite all surface elements at equal phases, and since these phase differences depend on the direction of sound incidence, only locally reacting walls show an angle-independent wall impedance. In this case the angle dependence of the reflection factor in eq. (6.14) is just due to the cosine function. The same holds for the absorption coefficient after eq. (6.23).

In Figure 6.5 the content of eq. (6.23) is represented in form of circles of constant absorption coefficient in the plane of the complex wall impedance. This diagram is valid originally for normal incidence of the primary wave. However, if the wall is of the locally reacting type it can be used as well for oblique incidence if the real part ξ of the specific wall impedance is replaced with $\xi \cdot \cos \vartheta$ and the imaginary part η with $\eta \cdot \cos \vartheta$.

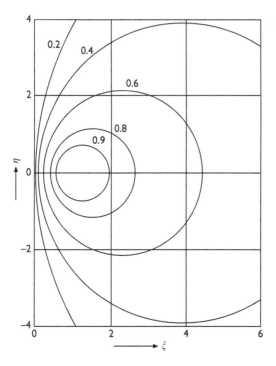

Figure 6.5 Circles of equal absorption coefficient in the complex plane of the specific wall impedance.

Obviously, the absorption coefficient of a wall is the greater the closer the magnitude of the specific wall impedance comes to unity. A wall with the impedance Z_0, that is, with specific impedance 1, does not cause any sound reflection.

6.5   Standing waves

In this section we restrict the discussion to perpendicular sound incidence. Accordingly, we set $\vartheta = 0$ in eqs. (6.8) and (6.12). Then the absolute value of the sound pressure and the particle velocity become (the index x in v_x is omitted):

$$|p(x)| = \hat{p}\left|1 + Re^{j2kx}\right| = \hat{p}\sqrt{1 + |R|^2 + 2\,|R|\cos(2kx + \chi)} \qquad (6.24)$$

$$|v(x)| = \frac{\hat{p}}{Z_0}\sqrt{1 + |R|^2 - 2\,|R|\cos(2kx + \chi)} \qquad (6.25)$$

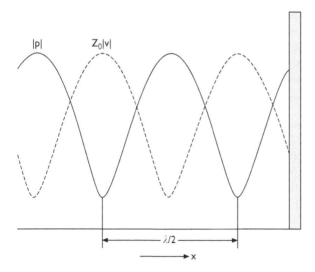

Figure 6.6 Standing wave in front of a plane surface with the reflection factor $R = 0.7 \cdot e^{-j\pi/4}$. Solid line: pressure amplitude, broken line: amplitude of particle velocity.

Both quantities are plotted in Figure 6.6 for $R = 0.7 \cdot \exp(-j\pi/4)$. They vary periodically between maximum and minimum values in such a way that a maximum of the velocity amplitude coincides with a minimum of the pressure amplitude and vice versa. The minima are often referred to as 'nodes'. These amplitude fluctuations are caused by interference of the incident and the reflected wave: in those points where the sound pressures of both waves have equal phases they are added and a maximum will be formed (constructive interference); if they have opposite phase, however, they partially cancel each other and the sound pressure amplitude will be at minimum (destructive interference). The same holds for the particle velocity. Since the cosine function has the period 2π the mutual distance of maxima (or minima) is $\pi/k = \lambda/2$. It should be noted that this pattern of pressure and velocity amplitudes is stationary. Therefore, this kind of a sound field is called a standing wave. It is complete when the reflection factor of the boundary has the absolute value 1; in this case the minima will become sharp tips lying at zero. Equations (6.24) and (6.25) simplify then into

$$|p(x)| = 2\hat{p} \cdot |\cos(kx + \chi/2)| \tag{6.24a}$$

$$|v(x)| = 2\hat{p} \cdot |\sin(kx + \chi/2)| \tag{6.25a}$$

Standing waves are a simple means of determining the magnitude and phase angle of the reflection factor and, by using eq. (6.12), the wall impedance of a sample. For this purpose a tube with a rigid and non-porous wall as in Figure 4.3 is employed in which a sound wave is excited. In order to avoid higher order wave types which would not be plane (see Section 8.5) the tube must not be too wide. Thus the length of the greater side of a tube of rectangular cross section must not exceed half an acoustic wavelength at the highest frequency of interest. If the tube has circular cross section the condition

diameter $< 0.586\,\lambda$

must be fulfilled over the whole frequency range. The primary sound wave is produced by a loudspeaker which is mounted at one end of the tube while the other end is terminated with the test sample. The sound pressure is measured by means of a small microphone which can be moved along the tube axis. The magnitude of the reflection factor is obtained from the maximum and the minimum pressure amplitudes:

$$\frac{|p|_{max}}{|p|_{min}} = \frac{1 + |R|}{1 - |R|} \tag{6.26}$$

from which the absorption coefficient at normal sound incidence is calculated with eq. (6.22). Moreover, the phase angle χ of the reflection factor can be determined from the position of the pressure minimum next to the termination. Let d_{min} denote the distance of this minimum from the surface of the sample then

$$\chi = \pi \left(\frac{4d_{min}}{\lambda} - 1 \right) \tag{6.27}$$

This device is known as Kundt's tube or impedance tube; it is one of the most important devices in acoustical measuring techniques. In particular, it is employed for rapid determination of the absorption coefficient of absorbing materials, linings, etc., of course, at perpendicular sound incidence.

6.6   Sound absorption by walls and linings

In this section the concepts introduced in the preceding sections will be applied to certain types of boundaries which are either of fundamental interest or come close to real walls or linings of walls.

A completely rigid and unporous wall is called 'acoustically hard'. Since the medium particles at its surface cannot carry out vibrations perpendicular to it, that is, since the normal component of the particle velocity is zero, the surface has the impedance $Z = \infty$, according to eq. (6.10). Hence, such a wall reflects the arriving wave perfectly, its reflection factor is 1 for all angles of incidence. Practically, a wall made of concrete or brick can be considered as nearly hard at least in the frequency range of audible sound.

The counterpart of a hard surface is an acoustically soft boundary. Along such a 'wall' it is not the normal component of the particle velocity (v_x in the present notation) which vanishes but the sound pressure. Consequently, its impedance from eq. (6.10) is zero and it has the reflection factor $R = -1$. Again, it is a perfect sound reflector, however, it changes the sign of the sound pressure during reflection corresponding to an abrupt phase shift by 180°. For airborne sound a soft surface can only be realised, if at all, in a limited frequency range. Matters are different for sound waves in a liquid, for instance, in water. If these impinge on a free surface, the adjacent air does not impede the displacement of the water particles to any significant degree. In other words: a free surface is virtually force free; it can be considered as acoustically soft.

In the subsequent examples we restrict discussion to normal incidence of the primary sound wave. Accordingly, we set $\vartheta = 0$ in eqs. (6.13), (6.14) and (6.23).

6.6.1 Impedance of an air layer

If a hard wall is shifted with respect to the reference plane $x = 0$ by a distance d (see Fig. 6.7a), a plane air cushion with the thickness d is formed, so-to-speak. As before, the absorption coefficient of the plane $x = 0$ is zero since a shift of the coordinates leaves the physical situation unaltered. The phase angle of the reflection factor, however, has changed since the wave has to travel twice the distance d before it arrives again at the reference plane $x = 0$ after its reflection. Therefore, the reflection factor of the

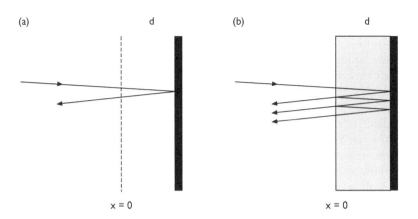

Figure 6.7 (a) Shifting the reference plane $x = 0$ by d, (b) absorbing layer on a rigid wall.

reference plane is $R = \exp(-j2kd)$, and its impedance, with eq. (6.13):

$$Z = Z_0 \frac{1 + e^{-j2kd}}{1 - e^{-j2kd}} = -jZ_0 \cot(kd) \tag{6.28}$$

(see eqs. (2.6a,b)). This is the 'wall' impedance of a layer of air backed with a hard boundary. If the thickness d of the layer is small compared to the acoustic wavelength, that is, for $kd \ll 1$ the cotangent can be approximated by the reciprocal of its argument with the result:

$$Z \approx \frac{Z_0}{jkd} = \frac{\rho_0 c^2}{j\omega d} \tag{6.28a}$$

According to Section 2.5 this is the impedance of a spring. Hence, we learn that a thin air cushion acts like an extended spring with the pressure-related compliance $d/\rho_0 c^2$. This result will be used frequently in the following.

6.6.2   Impedance and absorption of a porous layer backed with a rigid wall

We keep to the air layer considered earlier but imagine that it is filled with some porous material in which sound dissipation occurs. Hence, the angular wave number for waves propagating in this material is complex (see eq. (4.21)), and the same holds for the characteristic impedance $\underline{Z_0'}$ of that material. Replacing k and Z_0 in eq. (6.28) with these quantities yields for the wall impedance:

$$Z = -jZ_0' \cot(\underline{k'}d) \tag{6.29}$$

Viewed from a less formal point of view the following happens in the layer between $x = 0$ and the hard rear wall (see Fig. 6.7b): one part of the arriving wave is immediately reflected from the surface of the layer, another one enters the absorbing material and is repeatedly reflected between the surface and the rear boundary with continuously decreasing amplitude. This attenuation is due to viscous losses within the pores. Each time this internal wave reaches the plane $x = 0$ a portion of it will leave the layer. The sound wave reflected from the surface is formed by superposition of all these portions which have travelled different path lengths within the lossy material and hence cause constructive or destructive interference. Consequently, the reflection factor and the absorption coefficient of the arrangement will show a more or less pronounced frequency dependence, provided the waves reflected from the back wall are still strong enough to cause noticeable interference effects when they reach the surface $x = 0$. High absorption coefficients are observed when the characteristic impedance Z_0' of the layer is not too different from that of air thus allowing the impinging wave to intrude into the porous material, and, at the same time, when the attenuation in it is high enough to extinguish the entered wave before it reappears at the surface.

In acoustics, absorbing wall linings as used in acoustic rooms design or in noise control consist basically of some porous material. For this reason we shall have a closer look at such materials and the way the losses are produced within them. We assume that the cavities inside the material are not closed but that they are interconnected with channels which form a more or less complicated network. If air is pressed through a sample of such a material a pressure difference

$$\Delta p = \Xi \Delta x \cdot v_= \tag{6.30}$$

between its front and rear side must be maintained in order to enforce a flow with the velocity $v_=$ through the sample. In this relation Δx is the thickness of the sample and Ξ is the specific flow resistance of the material.

As an idealised model of a porous material we now consider a rigid body which is traversed by equal and equidistant thin channels perpendicular to its surface (Rayleigh model, see Fig. 6.8). Of course, the whole air flow through the material is carried by these channels. Since the sum of all channel cross sections is smaller by a factor σ than the area of the sample, the flow velocity v_{in} inside a single channel exceeds the gross velocity $v_= : v_{in} = v_= / \sigma$. The quantity σ is called the porosity of the material; it is defined as the volume of all pores or channels divided by the total volume of a sample.

If the fraction $\Delta p / \Delta x$ in eq. (6.30) is replaced with the pressure gradient we obtain the differential relation, valid for one channel:

$$\frac{\partial p}{\partial x} = -\Xi \sigma v_{in} \tag{6.31}$$

We assume that this relation is valid as well for alternating flows, that is, for the particle velocity in a sound wave. In this case, however, additional inertial

Figure 6.8 Rayleigh model of a porous material.

forces must be included. This is achieved by adding a term $(1/\rho_0)(\partial v_{in}/\partial t)$ to the left side of eq. (6.31), according to eq. (3.15). Hence the total pressure gradient is

$$\frac{\partial p}{\partial x} = -\rho_0 \frac{\partial v}{\partial t} - \sigma \,\Xi v \quad \text{with} \ \ v = v_{in} \tag{6.32}$$

The second relation we need is the equation of continuity (3.16):

$$\rho_0 \frac{\partial v}{\partial x} = -\frac{1}{c^2} \frac{\partial p}{\partial t} \tag{6.33}$$

Instead of combining both these equations in a differential equation as in Section 3.4, we try the following solutions:

$$p = A e^{j(\omega t - \underline{k}'x)} \quad \text{and} \quad v = B e^{j(\omega t - \underline{k}'x)}$$

with the goal of determining the complex angular wave number $\underline{k}'$. Inserting both expressions into eqs. (6.32) and (6.33) leads to

$$-j\underline{k}'A = -(j\omega\rho_0 + \sigma\,\Xi)B$$

and

$$-j\underline{k}'\rho_0 B = -\frac{1}{c^2} \cdot j\omega A$$

From these equations the angular wave number

$$k' = \frac{\omega}{c}\sqrt{1 - \frac{j\sigma\,\Xi}{\rho_0\omega}} \tag{6.34}$$

and the ratio of the sound pressure to the particle velocity, that is, the characteristic impedance

$$Z_{0p} = \frac{A}{B} = Z_0\sqrt{1 - \frac{j\sigma\,\Xi}{\rho_0\omega}} \tag{6.35}$$

are readily obtained.

These formulae hold for a sound wave travelling within a pore. To obtain the average characteristic impedance of material, we have again to base it on the average particle velocity which differs from the velocity in the pores by the factor σ. Therefore we have $\underline{Z'_0} = Z_{0p} / \sigma$ or

$$\underline{Z'_0} = \frac{Z_0}{\sigma} \sqrt{1 - \frac{j\sigma\,\Xi}{\rho_0 \omega}} \qquad (6.36)$$

Now we are ready to calculate the wall impedance of the rigidly backed porous layer by inserting the quantities $\underline{k'}$ and $\underline{Z'_0}$ into eq. (6.29); its absorption coefficient for perpendicular sound incidence is then obtained from eq. (6.23) with $\vartheta = 0$.

The result of such a calculation is presented in Figure 6.9 for a material with $\sigma = 0.95$; the parameter of the different curves is the flow resistance of the material divided by the characteristic impedance of air, $\Xi'd / Z_0$ with $\Xi' = \sigma\,\Xi$. The abscissa is the product of the frequency f and the thickness d of the layer in Hz·m. The ascending tendency of these curves finds its explanation in the fact that with increasing frequency the characteristic impedance Z_{0p} of the air in the pores approaches that of free air Z_0 which improves the match. At the same time the imaginary part of k' increases which, after eq. (4.21), characterises the attenuation of the sound waves in the material. The fluctuations of the absorption coefficient occurring at low values of $\Xi'd / Z_0$ level out at random sound incidence and are therefore insignificant from the practical point of view. The absorption coefficient reaches its highest values when the parameter $\Xi'd / Z_0$ is between 1 and 4. In any case, however, the thickness of the layer must not be too small, otherwise

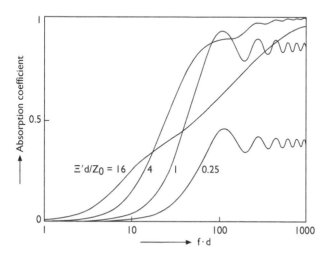

Figure 6.9 Absorption coefficient of a porous layer on a rigid wall (normal sound incidence). Parameter: $\Xi'd / Z_0$.

the absorption will be low. This is easily understood: next to the rigid rear wall the particle velocity is very small, and the same holds, of course, for the losses caused by the motion of the medium particles in the channels. According to Figure 6.9 even in the most favourable case high absorption is reached only if fd exceeds about 25 Hz · m.

The internal structure of practical absorption materials differs considerably from the Rayleigh model as discussed here. This is often accounted for by a so-called structure factor. In this text, however, we refrain from a more detailed treatment of the porous absorber since in practical situations the absorption coefficient of an unknown material must in any case be determined by measurement.

6.6.3   Absorption by a porous sheet

Next we consider the sound absorption by a thin porous sheet, for instance, by a sheet of some textile such as a curtain, or by a thin plate with very fine perforation. At first, it is assumed that this sheet is far from all solid walls, that is, that it is freely suspended in the space. Furthermore, we suppose that it is thin in comparison with all wavelengths of interest, but heavy enough to prevent vibrations of the sheet as a whole. Instead, the sound pressure will force an air flow through the pores or apertures, and consequently, a second wave will originate from the rear side of the sheet. To maintain a mean flow velocity v_s through the layer the pressure difference

$$\Delta p = r_s v_s \tag{6.37}$$

between both sides of the layer is needed. Here the flow resistance r_s applies to the whole layer in contrast to the specific flow resistance Ξ as introduced in eq. (6.30) which is the flow resistance per unit thickness.

On the one hand, the sound pressure of the incident sound wave must overcome the flow resistance of the sheet; additionally, it must balance the sound pressure of the wave radiated from its rear side. Accordingly, the wall impedance of the sheet contains the flow resistance r_s and the characteristic impedance Z_0:

$$Z = Z_0 + r_s \tag{6.38}$$

Inserting this into eq. (6.14) (with $\vartheta = 0$) leads us to the reflection factor of the sheet

$$R = \frac{1}{1 + 2Z_0 / r_s} \tag{6.39}$$

Formally, its absorption coefficient after eq. (6.22) is

$$\alpha = 1 - R^2 = \frac{1 + r_s / Z_0}{(1 + r_s / 2Z_0)^2} \tag{6.40}$$

The latter accounts not only for the viscous losses occurring in the pores of the sheet or fabric but as well for the energy transferred by the wave leaving its rear side. To separate both parts we have to equate the mean flow velocity v_s through the sheet to the particle velocity at its front (for $x = 0$) and both velocities to the particle velocity of the transmitted sound wave:

$$v_s = \frac{(1 - R)\hat{p}}{Z_0} = \frac{T\hat{p}}{Z_0}$$

where we again used the transmission factor T introduced in eq. (6.15). It is obtained from the above equation and eq. (6.39):

$$T = 1 - R = \frac{1}{1 + r_s / 2Z_0} \tag{6.41}$$

Subtracting the intensities of the reflected and the transmitted wave from that of the incident wave leaves the dissipated energy per second and unit area:

$$I_i - I_r - I_d = I_i(1 - R^2 - T^2) = \frac{r_s / Z_0}{(1 + r_s / 2Z_0)^2} I_i$$

The factor of I_i on the right could be named the 'dissipative absorption coefficient'

$$\alpha_d = \frac{r_s / Z_0}{(1 + r_s / 2Z_0)^2} \tag{6.42}$$

from the practical point of view it is more important than the coefficient in eq. (6.40). It reaches its maximum of 0.5 when the flow resistance of the material is equal to twice the characteristic impedance of the air.

Next consider the porous sheet not freely suspended in space but arranged parallel in front of a rigid and non-porous wall at distance d (see Fig. 6.10a). Now the air flow through the pores gives rise to a standing wave in the space behind the sheet. Therefore the characteristic impedance Z_0 in eq. (6.38) must be replaced with the impedance of a air layer according to eq. (6.28). Hence the wall impedance of the whole arrangement is now given by

$$Z = r_s - jZ_0 \cot(kd) \tag{6.43}$$

and the absorption coefficient is:

$$\alpha = \frac{4r_s / Z_0}{(1 + r_s / Z_0)^2 + \cot^2(kd)} \tag{6.44}$$

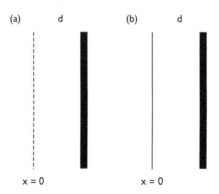

Figure 6.10 (a) Porous layer or (b) non-porous layer in front of a rigid wall.

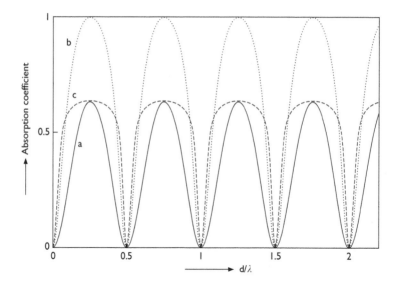

Figure 6.11 Absorption coefficient of a porous sheet (see Fig. 6.10b). Curve a: $r_s = Z_0/4$, curve b: $r_s = Z_0$, curve c: $r_s = 4Z_0$.

For various flow velocities, it is plotted in Figure 6.11 as a function of the ratio d/λ which is proportional to the frequency. Due to the periodicity of the cotangent the absorption coefficient varies periodically between 0 and its maximum value

$$\alpha_{max} = \frac{4r_s/Z_0}{(1 + r_s/Z_0)^2} \tag{6.45}$$

This maximum occurs whenever an odd number of quarter wavelengths fit into the space between the layer and the rear wall. Then the porous sheet coincides with a maximum of the particle velocity of the standing wave. In reality, however, the porous material will be set into vibration because of its finite mass and its absorption will be somewhat lower, at least at low frequencies.

After eq. (6.45) the absorption coefficient can reach unity for certain frequencies, which is the case for $r_s = Z_0$. Hence, a curtain in front of a wall, for instance, in front of a window, may be quite an efficient sound absorber. The strong frequency dependence of the absorption coefficient can be reduced by varying the distance d, for instance, by hanging or stretching the material in deep folds, which may also be desirable from an aesthetic standpoint.

6.6.4   Non-porous sheet

In this section we consider a non-porous panel or foil. It is characterised by its mass m' per unit, often referred to as the specific mass of the layer. A sound wave arriving at the surface will force the panel into vibrations which give rise to a transmitted sound wave travelling into the rearward space. Hence the wall impedance is obtained by replacing the flow resistance in eq. (6.38) with the mass reactance jωm' (per unit area) of the plate (see Section 2.5):

$$Z = Z_0 + j\omega m' \tag{6.46}$$

The reflection factor and the absorption coefficient follow from eqs. (6.14) and (6.23) with $\vartheta = 0$:

$$R = \frac{1}{1 + 2Z_0 / j\omega m'} \tag{6.47}$$

$$\alpha = \frac{1}{1 + (\omega m' / 2Z_0)^2} \tag{6.48}$$

In the present case, the expression 'absorption coefficient' is somewhat misleading in that the reflection loss is not caused by dissipation but by sound transmission through the wall. This process is of great practical importance with respect to the sound insulation of walls and ceilings which will be treated to some more detail in Chapter 14.

Equation (6.48) shows that a light foil is virtually sound transparent at low and even at medium frequencies. A foil with a specific mass of m' = 50 g/m², for example, reflects less than half the arriving sound energy at frequencies below 2.67 kHz.

As before, we imagine that the plate or foil is arranged at a distance d from rigid wall (see Fig. 6.10b). Accordingly, we replace in eq. (6.46) the characteristic impedance Z_0 with the impedance of a air layer backed by a rigid boundary. However, in contrast to the preceding subsection we do not use eq. (6.28). Instead, we suppose that the thickness of the air space is small compared with the wavelength, that is, the air layer is considered as a kind of spring the impedance of which is given by the approximative expression (6.28a). Furthermore, there are inevitably certain losses which will not be specified at this point but will just be taken into account by some frictional resistance r. Then the wall impedance of the arrangement becomes:

$$Z = r + j \left(\omega m' - \frac{\rho_0 c^2}{\omega d} \right)$$

(6.51)

This corresponds to eq. (2.21), and also the consequences drawn from it are the same as in Section 2.5: at the angular frequency

$$\omega_0 = \sqrt{\frac{\rho_0 c^2}{m' d}}$$

(6.52)

the imaginary part of the impedance becomes zero. Then its absolute value is at a minimum, and the oscillations caused by the sound field will attain particularly high amplitudes. This means the arrangement described is an acoustical resonator. The external force exciting the panel or layer must be determined by considering the sound pressures of both the arriving and the reflected sound wave and hence depends on the impedance Z.

In the present context the absorption coefficient of the arrangement is of particular interest. It is obtained from eqs. (6.51) and (6.23) (with $\vartheta = 0$):

$$\alpha = \frac{4 r Z_0}{(r + Z_0)^2 + \left[(m'/\omega) \left(\omega^2 - \omega_0^2 \right) \right]^2}$$

(6.53)

In Figure 6.12 the absorption coefficient is plotted as a function of the frequency for several values of r/Z_0. This diagram shows that the absorption coefficient too shows resonant properties. At the resonance frequency ω_0 it assumes its maximum

$$\alpha_{max} = \frac{4 r Z_0}{(r + Z_0)^2} = \frac{1}{(1/4) \left(\sqrt{r/Z_0} + \sqrt{Z_0/r} \right)^2}$$

(6.54)

which will be 1 for $r = Z_0$, that is, for the case of perfect impedance matching. These curves have some similarity with those of Figure 6.11; when the

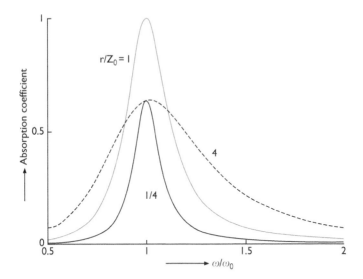

Figure 6.12 Absorption coefficient of a panel in front of a rigid wall (see Fig. 6.10b) (panel resonator) for $m'\omega_0 = 4Z_0$. Parameter: r/Z_0.

loss resistance r exceeds the characteristic impedance they will become flatter and the absorption is less selective at expense of the maximum absorption, however.

On account of its resonant properties, the panel resonator as described here plays an important role in room acoustics and in noise control. More will be said on it in Section 13.5.

Chapter 7

Diffraction and scattering

The preceding chapter was based upon the assumption that all boundaries from which an arriving sound wave is reflected are plane and infinite. If a reflecting boundary, a wall for instance, is limited in extent the described laws still apply more or less provided its dimensions are very large compared with the acoustic wavelength. However, an additional wave will originate from the edge of the surface which distributes its energy over a wide angular range.

This holds even more for limited obstacles of any shape. Let us consider a rigid sphere as sketched in Figure 7.1. If its diameter is much larger than the wavelength of the incident sound, the reflection law according to Section 6.1 applies to each sound ray. Behind the sphere a pronounced shadow will be formed, that is, a region which is virtually free of sound. The situation is different when the dimensions of the sphere are comparable to the wavelength. At first, the concept of a sound ray loses its significance. And second, the sphere – and the same holds for any obstacle – becomes the origin of secondary waves which are superimposed on the original sound field and disturb it more or less. In particular, some sound will reach into the geometric shadow region the limits of which become blurred. For still smaller objects the shadow may entirely disappear. This phenomenon is named diffraction.

Often diffraction is distinguished from the so-called scattering of sound: several authors speak of diffraction when there is still some shadow which is more or less 'brightened' by diffracted sound while scattering denotes the relatively slight modification of a sound field brought about by a small obstacle. In this chapter we use both expressions without making a sharp distinction between both phenomena because they have the same physical basis.

Diffraction occurs with all kinds of waves. It is more pronounced the smaller the obstacle in the way of a wave or an aperture in an otherwise impenetrable surface, compared with the wavelength. Observing diffraction of light waves usually requires special experimental techniques since the wavelengths of visible light are in the range from 0.4 to 0.8 μm. On the contrary, a small object on the surface of a lake, for instance, a rod stuck in

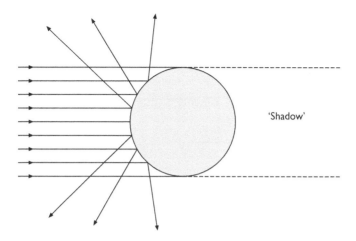

Figure 7.1 Specular sound reflection from a rigid sphere with d ≫ λ (d = diameter).

the bottom of the lake, does not disturb the propagation of surface waves to any noticeable extent; they are almost completely diffracted around it. In acoustics diffraction is an omnipresent phenomenon, since the acoustical wavelengths are comparable to the dimensions of the objects surrounding us in our daily life and – by the way – to those of the human body. For instance, diffraction is responsible for our ability to hear sounds arriving from the side with both our ears and thus determining the direction from which the sounds came; it is responsible for the ease with which we conduct a conversation through an ajar door, or for the limited effect of a screen along an autoroute. Finally, one should keep in mind that the walls of a room are rarely quite plane or quite smooth; in most cases their general shape is interrupted by balconies, pillars, technical installations or plastic decorations, cofferings, etc. All these 'irregularities' diffract or scatter incident sound waves. It would be easy to continue this list of examples.

For a qualitative explanation of diffraction Huygens principle can be brought into play. It states that each point struck by a sound wave becomes the origin of a new or secondary spherical wave. In Figure 7.2 (upper part) such elementary waves are sketched as semicircles for the case of an incident plane wave. Their envelope forms the wavefront of a wave which has travelled a short distance; meanwhile, the wave portions radiated sideways cancel each other. In the lower part of the figure the formation of elementary secondary waves is impeded by a non-transparent wall. In particular, the cancellation of oblique wave portions is incomplete next to the edge of the wall, hence they will reach below the geometric shadow limit, and it is easy

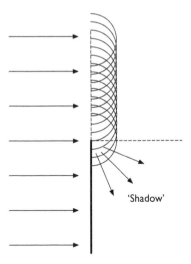

Figure 7.2 Diffraction by the edge of a thin wall.

to imagine that this 'diffraction wave' does not only brighten the shadow behind the wall but modifies all other parts of the sound field as well.

7.1   Exact formulation of diffraction problems

The quantitative treatment of diffraction is considerably more difficult than that of sound reflection from extended plane surfaces. The task is to determine the diffraction wave in such a way that the total sound field consisting of the incident wave and the diffraction wave fulfils the prescribed boundary conditions at the surface of the obstacle. Often these conditions can be expressed by the wall impedance of the surface, that is, after appropriate modification of the definition (6.10), by the ratio

$$Z = \left(\frac{p}{v_n} \right)_{\text{surf}}$$

$$(7.1)$$

Here v_n denotes the normal component of the particle velocity (see Fig. 7.3). If the surface of the scattering object is acoustically hard, this component must vanish. Since it is composed of the contribution of the incident wave $(v_i)_n$, and that of the diffraction wave $(v_d)_n$, the condition to be met at a rigid surface reads:

$$(v_d)_n = -(v_i)_n$$

$$(7.2)$$

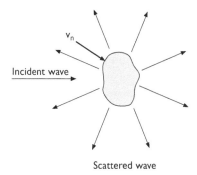

Figure 7.3 Diffraction or scattering by a finite body.

The component $(v_i)_n$ can be calculated from the incident wave, and the remaining task is to determine the scattered wave from $(v_d)_n$. It is obvious that this problem has strong similarity with the problem of sound radiation from vibrating surfaces.

Unfortunately, this method leads to closed solutions only if the surface of the diffracting object is a coordinate surface of a suitable coordinate system. This is true for certain bodies of simple geometry such as spheres or cylinders.

By integrating the intensity of the diffraction wave over all directions the total diffracted or scattered power P_s is obtained. Dividing it by the intensity I_i of the incident wave which is assumed as plane leads to a quantity with the dimension of an area, the so-called scattering cross section of the obstacle:

$$Q_s = \frac{P_s}{I_i} \tag{7.3}$$

7.2   Diffraction by a rigid sphere

In this section we consider as an important example the diffraction (or scattering) of a plane wave by a rigid and resting sphere. Provided its radius a is small in comparison with the wavelength the produced diffraction field can be presented in form of a closed formula. Without the sphere the sound field would alternatively compress and expand the sphere volume $4\pi a^3/3$; moreover, it would set it into vibrations with the velocity $v_i = p_i/Z_0$ with p_i denoting the sound pressure of the incident wave. The sphere must counteract both motions by emission of secondary waves, namely, a spherical wave and a dipole wave. The volume velocity Q of the former is calculated from the relative change in volume $dV/V_0 = \operatorname{div}\vec{s} = -p_i/cZ_0$, the latter expression after eq. (3.24a). By forming the time derivative (i.e. by multiplying with $j\omega$) and after changing the sign we obtain from this with $V_0 = 4\pi a^3/3$:

$$Q = \frac{4\pi a^3}{3} \frac{j\omega p_i}{cZ_0}$$

Thus the secondary spherical wave reads:

$$p_{d1} = -\frac{k^2 a^3 \hat{p}_i}{3r} e^{j(\omega t - kr)}$$

Here we have replaced ω with ck. The sound pressure of the dipole wave is obtained from eq. (5.24) by replacing v_0 with $-p_i/Z_0$:

$$p_{d2} = \frac{k^2 a^3 \hat{p}_i}{2r} \cos\theta \, e^{j(\omega t - kr)}$$

The angle $\theta = 0$ characterises the direction of sound incidence. Adding both contributions yields the total scattering field:

$$p_d(r,\theta,t) = -\frac{k^2 a^3}{r} \left(\frac{1}{3} - \frac{1}{2} \cos\theta \right) \cdot \hat{p}_i e^{j(\omega t - kr)} \tag{7.4}$$

From this we obtain the intensity $I_d = |p_d|^2/Z_0$ of the diffraction wave:

$$I_d(r,\theta) = \left[\frac{k^2 a^3}{r} \left(\frac{1}{3} - \frac{1}{2} \cos\theta \right) \right]^2 \cdot I_i$$

with $I_i = |p_i|^2 / Z_0$. To calculate the total power P_s scattered by the object this expression is integrated over the surface of a sphere with radius r:

$$P_s = \iint_{\text{sphere}} I_d(\theta,\phi) dS$$

In spherical polar coordinates (see Fig. 5.2) the area element on a sphere is $dS = r^2 \sin\theta d\theta d\phi$. The integration over the azimuth reduces to a multiplication with 2π. Then

$$P_s = 2\pi (k^2 a^3)^2 I_i \int_0^\pi \left(\frac{1}{3} - \frac{1}{2} \cos\theta \right)^2 \sin\theta d\theta = \frac{7}{9}\pi k^4 a^6 I_i$$

It is important to observe that P_s increases with the fourth power of k and hence of the frequency. The same holds, of course, for the scattering cross section:

$$Q_s = \frac{7}{9}\pi a^2 (ka)^4 \quad \text{for } a \ll \lambda \tag{7.5}$$

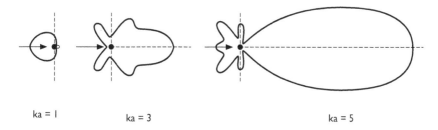

ka = 1   ka = 3   ka = 5

Figure 7.4 Diffraction by a rigid sphere: angular distribution of diffracted sound (sound pressure).

Generally, the diffraction field of a rigid sphere can be represented as an infinite series which will not be shown here. The reader may be referred instead to the book of Morse and Ingard,[1] for instance. Figure 7.4 shows the angular distribution of the sound pressure in the diffracted wave for various values of ka. The primary wave is assumed to be a plane wave arriving from the left. The parameter ka is the perimeter of the sphere divided by the wavelength. Of course, the three-dimensional extension of these diagrams is obtained by rotation around the horizontal axis. For small values of ka, backscattering prevails, that is, the sound portion thrown back towards the sound source. With increasing ka, that is, with increasing frequency if the radius a is given, the forward scattered portion grows, and finally its strength will approach the strength of the primary wave. In fact, the sound shadow behind an object which is large compared to the wavelength is produced by almost complete cancellation of the incident and the scattered waves which have opposite phases.

In Figure 7.5 the scattering cross section of a rigid sphere is plotted as a function of ka. At low values of ka, that is, at low frequencies, one sees the increase with the fourth power of frequency as predicted by eq. (7.5). At higher ka the curve approaches a constant value, the scattering cross section in the limit is $2\pi a^2$ and hence equals twice the visual cross section of the sphere. Then exactly half the incident sound energy is scattered throughout the space whereas the other half forms the shadow. This holds for the scattering cross section of obstacles of any shape.

For our directional hearing the diffraction of the sound signals by a listener's head is of critical importance. It is different for both ears if the sound wave arrives from a lateral direction, and produces different sound signals at both ears. From these 'interaural' differences we infer the direction of sound arrival. In Chapter 12 we shall come back to this matter.

1 P. M. Morse and U. Ingard, *Theoretical Acoustics*, Ch. 8. McGraw-Hill, New York 1968.

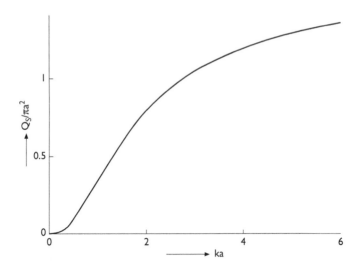

Figure 7.5 Scattering cross section of the rigid sphere, divided by its visual cross section πa^2.

7.3   Sound transmission through apertures

7.3.1   Kirchhoff's integral

Quite a different access to the treatment of diffraction is provided by a diffraction integral which can be derived from Green's theorem in vector analysis. It refers to a region V which is separated from the total space by some boundary with the area S (see Fig. 7.6a). We assume that all sound waves involved are harmonic with angular frequency $\omega = ck$. Then Kirchhoff's diffraction formula states:

$$p(P) = \frac{1}{4\pi} \iint_S \left[p_s \frac{\partial}{\partial n} \left(\frac{e^{-jkr}}{r} \right) - \frac{e^{-jkr}}{r} \frac{\partial p_s}{\partial n} \right] dS \qquad (7.6)$$

It represents the sound pressure $p(P)$ at an arbitrary point P within the region V as expressed by the sound pressure p_s and its derivative $\partial p_s/\partial n$ along the boundary. The distance of P from the area element dS on the boundary is represented by r. The normal vector $\bar{n}$ is supposed to point inward.

Kirchhoff's formula is the mathematical expression of Huygens's principle as explained at the beginning of this chapter. It can be interpreted as follows: The sound pressure at point P is made up of two contributions. The first one consists of spherical waves the sources of which are distributed over the boundary S according to the function $\partial p_s/\partial n$ (second term of the integrand).

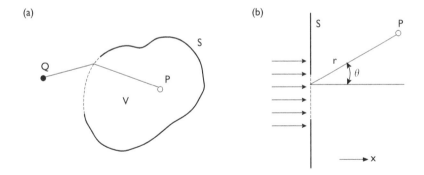

Figure 7.6 Boundary of the Kirchhoff integral (Q = sound source, P = field point): (a) general, (b) diffraction by an aperture in a plane screen.

The second contribution is due to waves of the type

$$\frac{\partial}{\partial n}\left(\frac{e^{-jkr}}{r}\right) = \lim_{d \to 0} \frac{1}{d}\left(\frac{e^{-jkr_1}}{r_1} - \frac{e^{-jkr_2}}{r_2}\right)$$

The expression on the right represents spherical waves originating from two points situated on the normal of the boundary at a small distance d. By comparing this expression with eq. (5.21) we learn that the first term in eq. (5.6) is due to dipoles which are distributed along S according to the function p_s.

In the following it will be supposed that the dashed part of the boundary in Figure 7.6a is acoustically non-transparent; the dotted part is to indicate an opening. The sound source Q is situated outside the screen S. With the help of eq. (5.6) the sound pressure at each point of the region V could be calculated, provided the sound pressure and its normal derivative along the boundary S were known. But these are just the quantities to be calculated. Thus eq. (5.6) turns out to be an integral equation, strictly speaking, which contains the unknown quantity inside the integral. However, it can be used to obtain an approximative solution by assuming that p_s and $\partial p_s / \partial n$ vanish on the inner side of the intransparent screen. In the opening these quantities are set equal to those of the undisturbed primary sound field as produced by the source Q. It is clear that this assumption is justified only if the dimensions of the aperture are large compared with the wavelength. For only then we can expect that the sound field in the opening is essentially undisturbed by the diffraction wave from the rim.

7.3.2   Sound transmission through large apertures

In the following this method will be applied to calculate the sound transmission through an aperture in a rigid plane screen of infinite extension (see Fig. 7.6b) which is struck by a plane wave with the sound pressure $p_i = \hat{p}_i \exp(-jkx)$ arriving from the left. The aperture is assumed large enough to justify Kirchhoff's approximation mentioned earlier.

Since the normal is parallel to the x-axis, $\partial p_s / \partial n$ can be replaced with $-jkp_i$. Furthermore, we get:

$$\frac{\partial}{\partial n}\left(\frac{e^{-jkr}}{r}\right) = -\left(jk + \frac{1}{r}\right)\frac{e^{-jkr}}{r} \cdot \frac{\partial r}{\partial x} = \left(jk + \frac{1}{r}\right)\frac{e^{-jkr}}{r}\cos\theta \qquad (7.7)$$

θ denotes the angle between r and the x-axis. Then eq. (7.6) assumes the form, with $k = \omega/c$:

$$p(P) = \frac{j\omega\hat{p}_i}{4\pi c}\iint_{\text{opening}} \frac{e^{-jkr}}{r}\left[\left(1 + \frac{1}{jkr}\right)\cos\theta + 1\right]dS \qquad (7.8)$$

If Kirchhoff's formula were exact this expression should agree with eq. (5.31) for the sound pressure produced by a piston. After replacing r' with r and $\hat{v}_0$ with $\hat{v}_i = \hat{p}_i/\rho_0 c$ that equation reads:

$$p(P) = \frac{j\omega\hat{p}_i}{2\pi c}\iint_{\text{opening}} \frac{e^{-jkr}}{r}\,dS \qquad (7.9)$$

Obviously, both equations agree if the distance of the point P from the aperture is so large that the second term in the round bracket can be neglected with respect to the first and when point P is so close to the axis that the cosine function can be replaced with unity without much error. This consideration clearly demonstrates the limitations of Kirchhoff's formula.

If the field point is so far from the aperture that r in the denominator can be considered as nearly constant and that the lines connecting it with the points in the aperture are virtually parallel we speak of Fraunhofer diffraction in contrast to Fresnel diffraction where these simplifications are not permitted. This distinction corresponds to that between the near field and the far field in Section 5.8. It indicates once more the close relationship of radiation and diffraction problems. Accordingly, the directional distribution of sound behind a circular aperture with diameter 2a is the same as that shown in the diagrams in Figure 5.15.

7.3.3 Sound transmission through small apertures

As mentioned before the Kirchhoff approximation fails when the opening is small compared with the wavelength, for then the particle velocity in the aperture differs significantly from that of the incident wave.

We assume that the screening wall is acoustically hard and has a very small circular opening. From the left a plane wave with sound pressure p_i is arriving at the screen as is indicated by a few wavefronts (see Fig. 7.7a). It is virtually perfectly reflected; thus on the left side a standing wave will be formed. From the aperture a spherical wave is originating towards both sides. Since this wave is very weak it will cause only a negligible distortion of the wave field on the left. However, on the right, the sound field consists just of this wave.

To calculate the strength of this spherical wave we note that the sound pressure $2p_i$ must overcome a finite mass to set the air in the aperture in motion. This is even so when the thickness of the wall is vanishingly small which may be explained in the following way: each sound wave can be conceived as an alternating air flow. Its flow lines (see Fig. 7.7b) cannot abruptly change their form so as to fit into the small cross section of the opening nor can they assume their original shape immediately after leaving the aperture. Instead, the flow will gradually contract, at the same time as it is accelerated. After passing the opening the same process will take place in the reverse direction. This is taken into account by thinking of the accelerated air as contained in a little tube with the same cross section as the aperture and with length $2\Delta l$, the so-called end correction. If the wall has finite thickness d its 'effective thickness' is

$$d_{eff} = d + 2\Delta l \qquad (7.10)$$

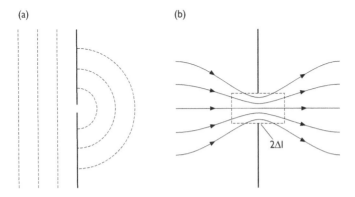

(a)   (b)

$2\Delta l$

Figure 7.7 Sound transmission through a small aperture: (a) wave fronts of the incident and the transmitted wave, (b) stream lines and end corrections.

and the air mass to be accelerated becomes

$$m = S_{op}d_{eff}\rho_0$$

where S_{op} is the area of the opening. For a circular aperture with the radius a S_{op} is πa^2 and[2]

$$2\Delta l = \frac{\pi}{2}a \tag{7.11}$$

Hence the air velocity within the opening enforced by the force $S_{op} \cdot 2p_i$ is

$$v_0 = \frac{S_{op}p_i}{j\omega m} = \frac{2p_i}{j\omega\rho_0 d_{eff}}$$

Therefore, the opening has the same effect as a point source with volume velocity $Q = S_{op}v_0$. According to eq. (5.6) it produces a spherical wave at the rear side of the wall with sound pressure (after replacing the factor $1/4\pi$ with $1/2\pi$)

$$p(r) = \frac{S_{op}\hat{p}_i}{\pi d_{eff}} \frac{e^{-jkr}}{r} \tag{7.12}$$

Another spherical wave with the same pressure, however, with reverse sign is radiated back into the direction of sound incidence.

For a door with the thickness of 2 cm with a circular opening of 1 cm diameter d_{eff} is 0.028 metres and the earlier formula yields a sound pressure level (see eq. (3.34)) which in 1 m distance is

$$\Delta L = 20 \log_{10}\left(\frac{\tilde{p}_i}{\tilde{p}(1\text{ m})}\right) \approx 61\text{ dB}$$

below that of the incident wave. Comparing this figure with the transmission loss of an average door which may be in the range of 20 dB it can be seen that this opening does not significantly deteriorate the sound insulation of the door. It is probably common experience that a conversation behind a closed door can be overheard only by putting the ear close to the keyhole.

Matters are quite different, however, if the aperture is a long, narrow slit with width $b \ll \lambda$. Figure 7.7 applies also to this case, and the wavefronts are those of a cylindrical wave now. Its intensity at the distance r is approximately:[3]

$$I_s \approx \frac{\lambda}{4r[\ln(0.717\lambda/b)]^2}I_i \tag{7.13}$$

however, it is assumed now that the thickness of the wall is vanishingly small.

2 P. M. Morse and H. Feshbach, *Methods of Theoretical Physics*, eq. (10.3.60). McGraw-Hill, New York 1953.
3 See reference above, §11.2, eq. (11.2.106).

As an example, we choose b = 1 cm, a frequency of 500 Hz corresponding to a wavelength of 0.68 m; the distance r is again 1 m. Then the level of the penetrating wave is

$$\Delta L = 10 \log_{10} \left(\frac{I_e}{I_s(1\,\mathrm{m})} \right) \approx 19.5\,\mathrm{dB}$$

below that of the primary wave. This value is comparable with the transmission loss of a door. This example shows clearly that the sound insulation of a wall may be considerably diminished by a slit in it.

7.3.4 Diffraction by a half-plane

In this subsection Kirchhoff's formula simplified according to eq. (7.9) will be applied to the diffraction of a plane wave by a non-transparent half-plane as already mentioned at the beginning of this chapter. The situation is sketched in Figure 7.8. The origin of our coordinate system lies on the edge of the wall which is assumed as very thin. The location of a point in the non-transparent wall x = 0 is given by its coordinates y' and z' (the z'-axis is perpendicular to the plane of the paper). Accordingly, the position of the wall is characterised by y' < 0. The integration is extended over the upper

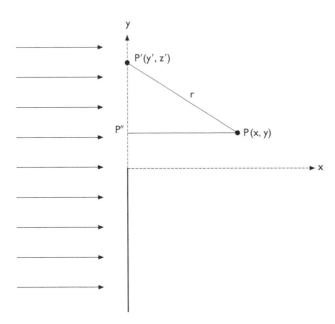

Figure 7.8 Diffraction by a rigid half-plane, coordinates.

half-plane, that is, over the region $y' > 0, -\infty < z' < \infty$. In the integral

$$p(P) = \frac{j\omega\hat{p}_i}{2\pi c} \iint\limits_{\text{Aperture}} \frac{e^{-jkr}}{r} dS = \frac{j\omega\hat{p}_i}{2\pi c} \int_0^\infty dy' \int_{-\infty}^\infty \frac{e^{-jkr}}{r} dz'$$

the variation of the function $1/r$ is relatively slow, hence $1/r$ may be replaced with $1/x$ without too much error. The function $\exp(-jkr)$, however, varies rapidly in general and its contributions to the integral cancel each other to a large extent. An exception is in the vicinity of the point P'' with the coordinates $y' = y, z' = 0$, where r has a minimum. Therefore we use the approximation

$$r = \sqrt{x^2 + (y - y')^2 + z'^2} \approx x + \frac{(y - y')^2}{2x} + \frac{z'^2}{2x}$$

for the r appearing in the exponent. Then the integral above transforms at first into

$$p(P) = \frac{j\omega\hat{p}_i}{2\pi cx} e^{-jkx} \int_0^\infty e^{-jk(y-y')^2/2x} dy' \int_{-\infty}^\infty e^{-jkz'^2/2x} dz' \tag{7.15}$$

With the substitutions

$$y - y' = \sqrt{\frac{\pi x}{k}} s \quad \text{and} \quad z' = \sqrt{\frac{\pi x}{k}} s$$

both integrals in eq. (7.15) can be expressed by the complex function

$$E(z) = \int_0^z e^{-j(\pi/2)s^2} ds = C(z) - jS(z) \tag{7.16}$$

Here $C(z)$ and $S(z)$ are the so-called Fresnel integrals, the values of which are found in relevant tables.[4] Furthermore, $E(\infty) = (1 - j)/2$. With these notations the result can be written in the form

$$p(P) = \frac{1+j}{2} \hat{p}_i e^{-jkx} \left[E\left(y\sqrt{\frac{k}{\pi x}}\right) + \frac{1-j}{2} \right] \tag{7.17}$$

or alternatively

$$p(P) = \tfrac{1}{2}\hat{p}_i [1 + C + S + j(C - S)] \tag{7.18}$$

the functions C and S having the argument $y\sqrt{k/\pi x}$.

---

4 See M. Abramowitz and I. A. Stegun, *Handbook of Mathematical Functions*. Dover Publications, New York 1964.

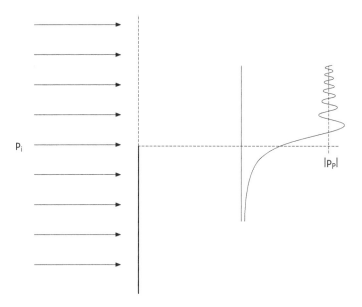

Figure 7.9 Diffraction by a rigid half-plane. Sound pressure amplitude near the shadow limit.

The mean square pressure of the sound field in some plane with $x > 0$ as calculated with eq. (7.18) is shown in Figure 7.9. The limit between the shadow and the 'illuminated' region is blurred now, and the transition from one region to the other is continuous. In the shadow region the amplitude of the diffracted wave decreases monotonically with increasing distance from the geometrical shadow boundary. In the upper half space the diffraction wave originating from the edge of the screen interferes with the incident wave causing the oscillations of the total sound pressure amplitude. These oscillations are very pronounced next to the boundary of both regions and gradually die out with increasing distance from it. Because of the simplifying assumptions which have been made this diagram represents the sound pressure better the larger the distance from the screen, measured in wavelengths. To obtain the true extension of the diffraction pattern one should note that the argument of the function E can be written as well in the form

$$\sqrt{\frac{ky^2}{\pi x}} = \sqrt{\frac{2(y/\lambda)^2}{x/\lambda}} \qquad (7.19)$$

This demonstrates once more that in diffraction problems lengths are only relevant in relation to the wavelength. Furthermore, this formula shows that the extension of the diffraction pattern does not grow in proportion to the observation distance x; to double it the distance must be increased by a factor 4.

The range of the diagram in Figure 7.9 is given by

$$-5 < y\sqrt{\frac{k}{\pi x}} < 5$$

As an example, let us suppose that the wavelength is 1 cm and the observation distance is 2 metres, then $\sqrt{k/\pi x}$ has the value 10 m^{-1} and the diffraction pattern extends over 1 m in total.

Two comments may be added to the preceding discussion:

1   The problem of sound diffraction by an edge can be solved exactly, that is, without any approximations or simplifications. Regarding the method of solution the reader is referred to the literature.[5]
2   If there are two non-transparent planes arranged one above the other in the plane x = 0 in such a way that both edges are running parallel in a distance 2b we obtain a slit. The diffraction field produced by it can be calculated by a simple modification of the above expressions. We have just to insert the limits −b and +b in the first integral of eq. (7.15). Then we obtain instead of eq. (7.17):

$$p(P) = \frac{1+j}{2}\hat{p}_i e^{-jkx} \left[E\left((y+b)\sqrt{\frac{k}{\pi x}}\right) - E\left((y-b)\sqrt{\frac{k}{\pi x}}\right) \right]$$

(7.20)

Again this formula is the more accurate the wider the slit in relation to the acoustical wavelength.

7.4   Babinet's principle

Suppose the roles of both half-plane in Figure 7.8 are interchanged, that is, it is the upper half-plane which is non-transparent, then the position of the 'light' and the 'shadow' region are interchanged; accordingly, the diffraction pattern in Figure 7.9 is turned upside down while remaining unaltered otherwise.

This fact is generalised by Babinet's principle. We suppose that an aperture of arbitrary shape is cut into an obstructing plane (see Fig. 7.10). The diaphragm obtained in this way be called S_1; the part cut out of the plane

---

5 P. M. Morse and U. Ingard, *Theoretical Acoustics*, Ch. 8.4. McGraw-Hill, New York 1968.

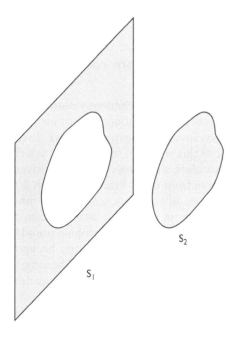

Figure 7.10 Complementary screens.

is called the 'complementary screen' with respect to S_1. We denote it by S_2. Then the sound pressures behind S_1 and S_2 are:

$$p_1 = p_i + p_{d1} \qquad\qquad (7.21a)$$

$$p_2 = p_i + p_{d2} \qquad\qquad (7.21b)$$

As earlier, p_i denotes the sound pressure of the incident wave which may be of arbitrary geometry; p_{d1} and p_{d2} are the sound pressure of the diffraction waves originating from the rims of both screens. Now we imagine S_1 and S_2 reunified to form the original plane. The sound pressure behind it is composed, on the one hand, by the pressure of the incident wave and those of both diffracting waves, on the other; of course, it must be zero:

$$0 = p_i + p_{d1} + p_{d2}$$

From this we obtain immediately with eq. (7.21):

$$p_2 = p_i - p_1 \qquad\qquad (7.22)$$

that is, the sound field behind two complementary screens is the same, apart from the sign of the sound pressure and from the added p_i, the sound pressure of the primary sound field. Therefore, if the diffraction field behind a circular disc is known, we also know the field behind the corresponding circular aperture.

7.5   Multiple scattering, scattering from rough surfaces

Now we consider a medium containing numerous randomly distributed scattering obstacles. Each of them deflects a small amount of the energy out of an incident wave which in a way gets lost from the original wave. Everyone is familiar with the optical analog of this situation, namely, fog, which not only weakens a light beam but also blurs all contours. And car drivers are bothered by the bright milky beams in front of his headlights which are due to light thrown back by the fog droplets. In acoustics, scattering by inhomogeneities becomes noticeable particularly in ultrasonic material testing and medical diagnostics in that it reduces the intensity of the probing sound beam (see also Subsection 4.4.3) and, at the same time, adds a noisy background to the echo signals produced by flaws, tissue boundaries, etc. The same holds more or less for underwater sound where the backscattered energy appears as 'reverberation'.

We assume that the medium under consideration contains N irregularly distributed obstacles per unit volume. For the sake of simplicity we imagine these as small rigid spheres with $ka \ll 1$ (a = radius of a sphere). Figure 7.11 shows a thin layer of this medium. According to our assumption there are $N dx$ scatterers per unit cross section in it. Suppose a sound wave with intensity I arrives from the left at this layer. Each scatterer removes energy $Q_s I$ per second from this sound wave with Q_s denoting the scattering cross section as defined by eq. (7.3). When the wave leaves the layer its intensity has been diminished by

$$dI = -N Q_s I dx$$

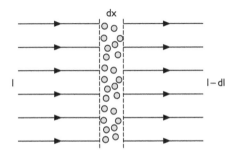

Figure 7.11 Intensity loss in a foggy medium.

Hence scattering causes an exponential intensity decrease with distance:

$$I(x) = I_0 e^{-NQ_s x} \tag{7.23}$$

similar to that effected by dissipation in the medium (see eq. (4.19)). If the scattering obstacles are different in size Q_s may be replaced with the average scattering cross section.

What happens to the scattered sound energy? In any case it will distributed over many directions including that from which the primary wave arrived. It may happen that scattered sound portions will be scattered once more by an obstacle, and this process can even be repeated again and again. In this case we speak of multiple scattering. Then the whole sound field consists of the superposition of all these scattered portions and the primary wave. Neglecting interferences occurring between the various components one can imagine the sound field as being composed of 'sound particles', each of them travelling with constant velocity c and along a straight line. Only when they hit a scattering obstacle will they be deflected into a new direction. This process is sketched in Figure 7.12. Obviously, it has some similarity with diffusion: suppose a droplet of milk is carefully injected into a cup of black coffee. At first the bright spot will remain rather localised. However, after a time it will gradually become broader and, at the same time, more diffuse, and if one waits long enough coffee and milk will have become a uniform mixture even without any stirring. Sound in a strongly scattering medium behaves in a similar way.

Now we turn towards another form of scattering, namely, that resulting from many scattering centres fixed on a wall, for instance, by regular or irregular protrusions on an otherwise plane wall. As an example, we regard a wall with sawtooth-like protrusions as shown in Figure 7.13. The way it reflects a sound wave depends on the size of the surface structure in relation

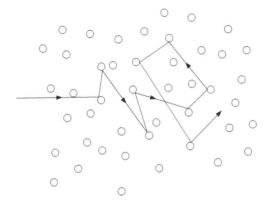

Figure 7.12 Multiple scattering.

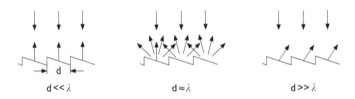

Figure 7.13 Sound scattering from a rough wall for three ranges of wavelengths.

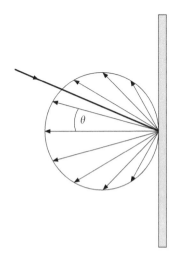

Figure 7.14 Angular distribution of the scattered sound (intensity) according to Lambert' law.

to the wavelength. If the protrusions are small compared to the wavelength (left part of the figure) an incident wave will not 'notice' them, so-to-speak; the wall reflects the sound wave in the same manner as if it were smooth. If, on the contrary, the surface elements are very large in comparison with the wavelength (right part) each element h reflects the arriving wave specularly, that is, according to the rule 'incidence angle = reflection angle'. In the range between these limiting cases each wall element will produce a diffracted wave which distributes the sound more or less in all directions. In this case we speak of a 'diffuse wall reflection' in contrast to the 'specular reflection' occurring at a smooth surface.

If a surface shows very strong scattering the actual angular distribution of the scattered energy is often replaced with a model distribution, expressed by Lambert's cosine law. It states that the intensity of the scattered sound at a distance r from a scattering wall element dS with the absorption coefficient α is

$$I(r, \theta) = B(1 - \alpha) \frac{\cos \theta}{r^2} dS \qquad (7.24)$$

Here B denotes the so-called irradiation density, that is, the energy incident on unit area of the surface per second; θ is the angle subtended by the wall normal and a chosen direction of scattering (see Fig. 7.14).

Real walls will almost never produce completely diffuse reflections according to eq. (5.24). Often, however, the sound is thrown back in a mixture of specularly and diffusely reflected components. Nevertheless, Lambert's law is a very useful concept which describes surface properties often more readily than the law of specular reflections.

Chapter 8

Sound transmission in pipes and horns

In Section 4.1 we found that the field of a plane sound wave in a fluid is not disturbed by rigid surfaces arranged parallel to the direction of sound propagation. The reason for this is the longitudinal character of the sound waves, that is, the fact that the vibrations of fluid particles of the medium are in that direction and hence are not impeded by the walls. This statement remains true as well if such a boundary is bent into the shape of a tube. Accordingly, sound waves can readily travel within rigid pipes of arbitrary cross section. Such 'waveguides' may even be curved. As long as the radius of curvature is large compared with the wavelength the wave follows the course of the pipe.

As we know, pipes or tubes with completely rigid walls do not exist in reality. Practically, however, many pipe walls come fairly close to this ideal if they are sufficiently thick and heavy, if the characteristic impedance of the wall material is very high in comparison with that of the medium inside, and if the inner surfaces are free of pores. This can be achieved easily for gases. Thus sound waves in air can be particularly well guided along metal pipes, but materials like wood or plastics are also suitable. In the well-known stethoscope the sounds picked up from the patient's body are often conducted to the doctor's ears by means of rubber tubes.

The term 'acoustical waveguide' is not restricted to pipes filled with a fluid but applies as well to bars, wires, strips or plates along which elastic waves can be conducted (see Chapter 10).

8.1 Sound attenuation in pipes

Strictly speaking it is not quite correct to say that a sound wave travelling within a pipe filled with air remains completely unaffected by its walls. Even if a wall is entirely rigid and has no pores inside, it will cause a characteristic attenuation of the sound waves in excess of that occurring in the free medium as described in Subsection 4.4.1. As with 'classical attenuation' in gases and liquids, this kind of attenuation is linked to the viscosity and the heat conductivity of the wave medium.

It is evident that the fluid particles immediately on the wall are fixed by it and cannot participate in the vibrations of a sound wave. The transition from the boundary value $v_x = 0$ to the particle velocity of the undisturbed wave at some distance from the wall is gradual and occurs within a thin boundary layer (see Fig. 8.1). In this layer the volume elements experience shear deformations leading to increased frictional losses at the expense of the sound energy conducted in the tube. A similar effect is due to the temperature variations within a sound wave as given by eq. (4.23). They must vanish immediately at the wall since the latter maintains constant temperature on account of its thermal inertia. Therefore a similar boundary layer is formed along the wall in which the transition from the local sound temperature in the wave to the constant wall temperature takes place. The strong temperature gradient within this layer is connected to some heat flow which is irreversible and removes sound energy from the sound wave in the pipe.

The thickness d_{vis} of the viscous boundary, defined as the distance from the wall in which the disturbance of the particle velocity has fallen by a factor $1/e$, is:

$$d_{vis} = \sqrt{\frac{2\eta}{\rho_0 \omega}} \tag{8.1}$$

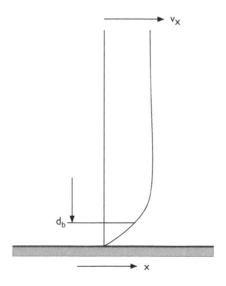

Figure 8.1 Transverse distribution of the particle velocity near the boundary of a pipe (d_b: thickness of the boundary layer).

while the thickness d_{th} of the thermal boundary layer defined in a corresponding way is given by[1]

$$d_{th} = \sqrt{\frac{2\nu}{\rho_0 \omega C_p}} \tag{8.2}$$

As in Subsection 4.4.1 η is the viscosity and ν is the heat conductivity of the gas; C_p is its specific heat at constant pressure related to unit mass. For air under normal conditions inserting the constants $\eta = 1.8 \cdot 10^{-5}$ Ns/m², $\nu/C_p = 1.35\eta$ and $\rho = 1.29$ kg/m³ leads to:

$$d_{vis}/mm = 2.1 \cdot \frac{1}{\sqrt{f/Hz}} \quad \text{and} \quad d_{th}/mm = 2.4 \cdot \frac{1}{\sqrt{f/Hz}} \tag{8.3}$$

Both these values are not very different and have the same frequency dependence which is not too surprising since both processes are related to each other: viscosity is based on momentum exchange, while heat conduction is linked to the exchange of energy between colliding molecules.

Since both boundary layers are quite thin it is only in pipes with small cross section that they occupy a noticeable fraction of the tube volume. In general we can imagine these boundary layers as a sort of skin which covers the inner surface of a wall.

Concerning the somewhat complicated derivation of the attenuation in pipes the reader may be referred to Cremer and Müller (see footnote). The contributions to the attenuation constant due to viscosity and heat conduction are:

$$m_v = \frac{U}{2cS} \cdot \sqrt{\frac{\eta\omega}{2\rho_0}} \tag{8.4}$$

and

$$m_t = \frac{U}{2cS}(\kappa - 1) \cdot \sqrt{\frac{\nu\omega}{2C_p\rho_0}} \tag{8.5}$$

(U = circumference and S = cross-sectional area of the tube.) For air under normal conditions both expressions can be combined in the formula:

$$D = 0.4 \cdot \frac{U/cm}{S/cm^2} \sqrt{f/kHz} \frac{dB}{m} \tag{8.6}$$

According to this expression, the excess attenuation in a circular pipe with a diameter of 5 cm and at a frequency of 10 kHz amounts to about one decibel

1 L. Cremer and H. A. Müller, *Principles and Applications of Room Acoustics*, Vol. 2, Ch. 6. Applied Science Publishers, London 1982.

per metre. However, the measured attenuation turns out to be somewhat higher.

8.2 Basic relations for transmission lines

Although the excess attenuation described in the preceding section may attain noticeable or even considerable values – depending on the actual circumstances – it will be neglected in this and the following sections in order to avoid unnecessary complications. Furthermore, we assume that the plane wave as hitherto considered is the only waveform which can propagate in the tube. In Section 8.5 the justification of this assumption will be discussed.

We consider a section of a pipe of length x and with constant cross section, the area of which is S (see Fig. 8.2). No assumptions are made regarding the total length of the tube, its connection with other tubes, etc. and nothing is said about the way the sound waves are generated. Therefore we expect that in the section under consideration there are two plane waves travelling in opposite directions. Hence the sound pressure in the tube can be represented, omitting the time factor exp(jωt) which is common to both waves, in the following form

$$p(x) = Ae^{-jkx} + Be^{jkx} \qquad (8.7a)$$

with arbitrary constants A and B. The particle velocity $v = v_x$ is related to the pressure by the factor $\pm 1/Z_0$ where the upper sign holds for the first, the lower one for the second part. Therefore we have:

$$Z_0 v(x) = Ae^{-jkx} - Be^{jkx} \qquad (8.7b)$$

Because of the relation $\exp(\pm jkx) = \cos(kx) \pm j\sin(kx)$ the exponentials in these formulae can be expressed by trigonometric functions:

$$p(x) = (A + B)\cos(kx) - j(A - B)\sin(kx) \qquad (8.8a)$$
$$Z_0 v(x) = (A - B)\cos(kx) - j(A + B)\sin(kx) \qquad (8.8b)$$

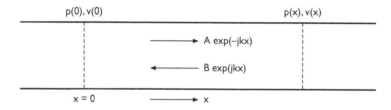

Figure 8.2 Derivation of eqs. (8.9a and b).

Obviously, $p(0) = A + B$ and $Z_0 v(0) = A - B$. Thus we arrive at the final representation:

$$p(x) = p(0) \cos(kx) - jZ_0 v(0) \sin(kx) \tag{8.9a}$$

$$Z_0 v(x) = -jp(0) \sin(kx) + Z_0 v(0) \cos(kx) \tag{8.9b}$$

These are the basic equations of a transmission line, and they relate the sound pressure and the particle velocity at any point x with the corresponding quantities at some other point $(x = 0)$. They are very useful for discussing the sound propagation in acoustical waveguides of any kind. It may be observed that they hold as well for electrical transmission lines; in this case we have to replace the sound pressure and the particle velocity with the electrical voltage and current, respectively.

Since the cross section of the pipe does not appear in eqs. (8.9a and b) and in all other relations of this section they are valid as well for free plane waves.

Now imagine that the pipe is cut open at one side $(x = 0)$ and is terminated with a surface having wall impedance $Z(0) = p(0)/v(0)$ at the other. We call l the distance of the point x from this termination, that is, $x = -l$ in the present notation. We obtain the impedance $Z(l) = p(l)/v(l)$ by dividing eq. (8.9a) by eq. (8.9b):

$$\frac{Z(l)}{Z_0} = \frac{Z(0) + jZ_0 \tan(kl)}{jZ(0) \tan(kl) + Z_0} \tag{8.10}$$

According to eq. (8.10), a tube – or more generally, a waveguide – transforms a given impedance into another one. From eq. (8.10) we can derive some interesting special cases:

1   If l equals a quarter-wavelength, that is, if $kl = \pi/2$, the tangent will grow beyond all limits and we obtain

$$Z(l) = \frac{Z_0^2}{Z(0)} \tag{8.11}$$

This means, a quarter-wavelength pipe transforms a given impedance $Z(0)$ into its reciprocal, apart from the square of the characteristic impedance Z_0.

2   If the pipe is terminated with a rigid end plate at $x = 0$, $Z(0)$ approaches infinity and eq. (8.10) yields $Z(l) = -jZ_0 \cot(kl)$. This expression

agrees – apart from different notations – with eq. (6.28) which was to be expected after what was said earlier. Again,

$$Z(l) \approx \frac{Z_0}{jkl} = \frac{\rho_0 c^2}{j\omega l} \qquad (8.12)$$

for $kl \ll 1$ as in eq. (6.28a): a short and rigidly terminated section of a pipe acts as an air cushion, that is, as a spring.

3   The input impedance of a layer with the thickness l, the characteristic impedance Z_0' and the angular wave number k′ which is embedded into a medium is obtained from eq. (8.10) by replacing Z_0 with Z_0', $Z(0)$ with Z_0 and k with k′. The result is:

$$Z = Z_0 \frac{Z_0' + jZ_0 \tan(k'l)}{jZ_0' \tan(k'l) + Z_0} \qquad (8.13)$$

The reflection factor and the transmission factor of the layer are readily calculated by applying eqs. (6.19) and (6.20) with $\cos\vartheta = \cos\vartheta' = 1$ and replacing Z_0' with Z.

8.3   Pipes with discontinuities in cross section

Now we omit the assumption of constant cross section. Instead, we consider two pipes with different cross sections S_1 and S_2 which are joined to each other at x = 0 (see Fig. 8.3a). At present we leave it open as to which of the pipes is the wider or the narrower one. In any case we assume that the pipe at the right of the junction is infinitely long and that the wave travelling in it is a progressive one. It will turn out that a progressive sound wave arriving from the left will be partially reflected from the junction, and that the wave entering the right part has changed its amplitude. As before it is assumed that all lateral dimensions of the pipes are small compared with the wavelength.

A first requirement is that at both sides of the junction are equal pressures, $p_1 = p_2$. On the other hand, the sound pressure p_1 is composed of the

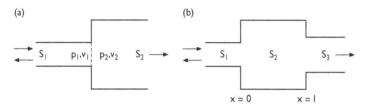

Figure 8.3 Pipes with discontinuities in cross section: (a) simple junction, (b) pipe with two discontinuities.

sound pressure p_i of the incident and that of the reflected wave Rp_i where R is the reflection factor as introduced in Section 6.3. Hence $p_1 = p_i(1 + R)$. Furthermore, we express the amplitude of the transmitted wave by the transmission factor: $p_2 = Tp_i$. Equating both pressures yields as earlier (see eq. (6.17)):

$$1 + R = T \qquad (8.14a)$$

Furthermore, the principle of continuity requires that the volume flow from the left towards the junction must be completely taken up by the right side. Hence $v_1 S_1 = v_2 S_2$ or

$$S_1(1 - R)\frac{p_i}{Z_0} = S_2 T \frac{p_i}{Z_0} \qquad (8.14b)$$

From both eqs. (8.14a and 8.14b) it follows for the reflection factor

$$R = \frac{S_1 - S_2}{S_1 + S_2} \qquad (8.15)$$

and for the transmission factor

$$T = \frac{2S_1}{S_1 + S_2} \qquad (8.16)$$

Here, both factors are real. If the tube on the right is wider than on the left ($S_2 > S_1$), then R is negative and the sound pressure will reverse its sign during reflection. In the reverse case R is positive with the seemingly strange consequence that the transmission factor becomes large than unity. However, this is no contradiction to the energy principle, since the energy passing the junction per second is in any case smaller than the incident energy by a factor $T^2 S_2/S_1$. At the end of Section 6.3 we had a similar discussion concerning the reflection and transmission of free waves.

Finally, the impedances $Z^- = p_1/v_1$ und $Z^+ = p_2/v_2$ at both sides of the junction are related by the equation

$$\frac{Z^-}{S_1} = \frac{Z^+}{S_2} \qquad (8.17)$$

which is an immediate consequence of the continuity equation. It tells us that any abrupt change of a pipe's cross section acts as an impedance transformer.

However, the preceding relations cannot claim strict validity. For a pipe ending in the free space at $x = 0$, for instance, which is tantamount to $S_2 \to \infty$, eq. (8.16) predicts $T = 0$, that is, the incident wave should be totally reflected. This is in glaring contrast to our everyday experience; if it were true we could not hear any noise from the exhaust of a motorbike. In fact, the derivation presented earlier neglects the end correction which has been discussed in Subsection 7.3.3 (see also Fig. 7.7b). Furthermore, a pipe ending in free space may be considered as a sound source and is loaded with a radiation impedance. Therefore an open end just approximates to a 'soft' termination.

Next we consider a pipe with two subsequent discontinuities: the area of the cross section jumps, as shown in Figure 8.3b, from S_1 to S_2 at $x = 0$ and from S_2 to S_3 at $x = l$. This is not just 'more of the same'; we must not expect, for instance, that the reflection factor of the whole arrangement is just the product of the reflection factors of both junctions. Rather, the wave portions reflected from both discontinuities interfere with each other, thus their distance will be of crucial importance.

Again, the primary wave is supposed to arrive from the left side at the first junction. Since the wave leaving both junctions towards the right side was assumed to be progressive the right discontinuity is charged with the impedance Z_0. According to eq. (8.17) it transforms Z_0 into the impedance $Z_0 \cdot (S_2/S_3)$. This value has to replace $Z(0)$ in eq. (8.10) which shows the impedance transformation achieved by the middle section of length l. Finally, to obtain the input impedance of the arrangement we have to multiply the result with the factor S_1/S_2 in order to account for the impedance transformation due to the left discontinuity. By combining these three transformations we arrive at the ultimate result:

$$\frac{Z}{Z_0} = \frac{S_1}{S_2} \cdot \frac{S_2 + jS_3 \tan(kl)}{jS_2 \tan(kl) + S_3} \tag{8.18}$$

Next we assume that the cross section at $x = l$ attains its initial value, that is, we set $S_3 = S_1$. Moreover, l be small compared to the acoustical wavelength, or what means the same, $kl \ll 1$. This enables us to approximate the tangent by its argument:

$$\frac{Z}{Z_0} \approx \frac{S_1}{S_2} \cdot \frac{S_2 + jklS_1}{jklS_2 + S_1} \tag{8.19}$$

Now the following special cases will be considered:

8.3.1   Constriction ($S_2 < S_1$), perforated panel

If $S_2 < S_1$ (see Fig. 8.4a), $jklS_2$ in the denominator of eq. (8.19) can be neglected against S_1:

$$Z \approx Z_0 + j\omega\rho_0 l \frac{S_1}{S_2} \tag{8.20}$$

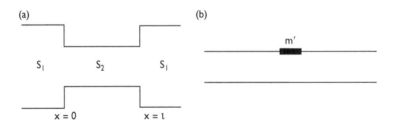

Figure 8.4 Constricted pipe section: (a) longitudinal section, (b) equivalent electrical circuit.

Hence, a constriction of a pipe has the same effect as a foil stretched across the interior of the tube with the specific mass (= mass per unit area)

$$m' = \frac{S_1}{S_2}\rho_0 l \tag{8.21}$$

the inertia of which must be overcome by the incident wave. The electrical equivalent of this arrangement is an electrical transmission line with a series inductance m' (see Fig. 8.4b), where the particle velocity corresponds to the electrical current while the sound pressure is identified with the electrical voltage (see Section 2.7). We should remember that this result is restricted to the condition $kl \ll 1$, otherwise the constricted section cannot be regarded as a lumped element but must be treated as a transmission line.

Now imagine that the constriction of the pipe becomes shorter and shorter ($l \to 0$). In the limit we arrive at a diaphragm of vanishing thickness which is inserted into the pipe. Equation (8.19) or (8.20) would predict no effect whatsoever which is not reasonable from a physical point of view. This discrepancy is solved by adding an end correction Δl to both sides of the aperture, that is, by replacing the geometrical length l with the 'effective length' $l_{eff} = l + 2\Delta l$ (see Subsection 7.3.3). The latter term remains finite even for vanishing l.

If the pipe has rectangular cross section we can imagine an infinite number of pipes of the same kind stacked side by side and on top of each other as indicated in Figure 8.5. Since the particle velocity has no components normal to the pipe walls we can omit them without disturbing the sound field. What remains is a uniformly perforated plate or panel of thickness l with the 'porosity' $\sigma = S_2/S_1$. It has a specific mass according to eq. (8.21) with l_{eff} instead of l. The factor S_1/S_2 accounts for the increase of the flow velocity in the holes and thus for the increased inertial effect of the air in the holes. The reflection factor and the absorption coefficient of the perforated panel – for normal sound incidence – can be calculated with eqs. (6.47) and (6.48). As an example, we consider a metal sheet 1 mm thick, perforated at 20% ($\sigma = 0.2$) with circular holes having a diameter of 3 mm. According to eq. (6.48), its absorption coefficient exceeds 0.9 for frequencies up to

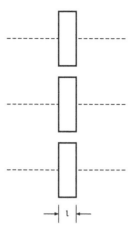

Figure 8.5 Perforated panel.

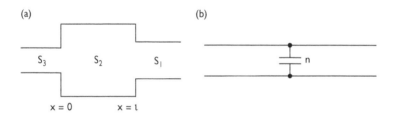

Figure 8.6 Enlarged pipe section: (a) longitudinal section, (b) equivalent electrical circuit.

1.7 kHz, that is, in this range it transmits more than 90% of the incident sound energy. For this reason perforated panels are often used as sound transparent covers to protect or to hide loudspeakers, sound absorbing wall linings, etc.

8.3.2 Enlargement ($S_2 > S_1$)

If, on the contrary, $S_2 > S_1$ (see Fig. 8.6a), the second term in the nominator of eq. (8.19) becomes vanishingly small in comparison with the first one. Then the reciprocal of the input impedance, that is, the input admittance of the arrangement, becomes

$$\frac{1}{Z} = \frac{1}{Z_0} + j\omega n \text{ with } n = \frac{lS_2}{\rho_0 c^2 S_1} \qquad (8.22)$$

The second term at the right side of this equation represents the admittance of a spring and hence is equivalent to a pressure release. Since here the total admittance is represented as the sum of two admittances this arrangement is analogous to an electrical transmission line with a capacitor connected in parallel as shown in Figure 8.6b; its capacity is n if the same correspondence between electrical and acoustical quantities is adopted as in Subsection 8.3.1.

8.3.3 Resonator

In the preceding subsections we got to know two acoustical 'circuit elements', namely, a mass and a spring. By combining them certain acoustical 'networks (circuits)' can be constructed the variety of which, however, is much smaller than the variety of electrical networks made up of inductances and capacitances.

As a simple example, Figure 8.7a shows a rigidly terminated tube connected to a constricted section of length l; the end section has length l'. Another important parameter is the ratio $\sigma = S_2/S_1$. After eq. (8.12) the input impedance of the end section is $\rho_0 c^2/j\omega l'$. Adding the impedance $j\omega m'$ of the constricted section with m' after eq. (8.21) leads us to the input impedance of the arrangement:

$$Z = j\rho_0 S_1 \left(\frac{\omega l}{S_2} - \frac{c^2}{\omega S_1 l'} \right) \tag{8.23}$$

It indicates that this combination is a resonator the impedance Z of which vanishes at the resonance frequency

$$\omega_0 = c \sqrt{\frac{S_2}{l l' S_1}} \tag{8.24}$$

hence the particle velocity would become infinitely large even at finite sound pressure. In reality the unavoidable losses, in particular, frictional losses in

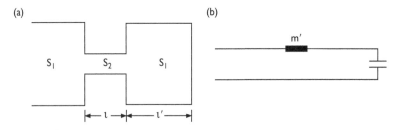

Figure 8.7 Resonator consisting of tube sections: (a) longitudinal section, (b) equivalent electrical circuit.

the constricted section, restrict the particle velocity. In any case, the velocity amplitude for given sound pressure attains a maximum at the resonance frequency. Of course, eq. (8.24) is only meaningful if $\sigma = S_2/S_1$ is relatively small.

This discussion may be followed by two comments: at first, one can apply the same procedure as described in Subsection 8.3.1, namely, arranging numerous resonators near one another as in Figure 8.5. After removing the pipe walls one arrives at a perforated panel with the thickness l, placed in front of a rigid wall in distance l'. This shows that the resonance absorber described in Subsection 6.6.4 can be realised not only with a vibrating foil or panel but as well with a rigid, properly perforated plate. Wall linings of this type are often applied in reverberation or noise control.

Second, the compliance of the end piece in Figure 8.7a depends only on its volume $V = l'S_1$ but not on its shape. Hence it is evident that any cavity with reasonably rigid walls with a narrow neck is an acoustical resonator. The neck of the cavity resonator may be degenerated to just a hole in the wall of the vessel; then most of the mass is contained in the edge correction of this hole. Such resonators have been known for centuries; today they are mostly referred to as 'Helmholtz resonators'. Often the transition from the mass to the spring element is continuous which does not impair its function as a resonator. This is easily demonstrated with an empty beer or wine bottle by softly blowing from the side across its opening. The edge tone produced at the edge of the opening is synchronised by the resonance of the bottle, and with some luck a clear tone will be heard.

8.3.4 Acoustical low-pass filter

Finally, we regard a pipe with periodically alternating constrictions and expansions as shown in Figure 8.8a. The electrical analogue to it is a low-pass filter consisting of series inductors and parallel capacitors (see Fig. 8.8b). It tells us that this arrangement is an acoustical low-pass filter. It transmits sound with frequencies below its cut-off frequency; at higher frequency the acoustical wave is subject to increasing attenuation. The cut-off frequency

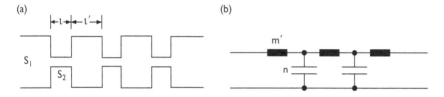

Figure 8.8 Acoustic low-pass filter: (a) longitudinal section, (b) equivalent electrical circuit.

of the filter is given by

$$\omega_c = 2c\sqrt{\frac{S_2}{11'S_1}} \tag{8.25}$$

and its attenuation per section in the range $\omega > \omega_c$ is

$$D' = 8.69 \cdot \operatorname{ar cosh}\left(\frac{\omega}{\omega_c}\right) \text{ decibels} \tag{8.26}$$

8.4   Pipes with continuously changing cross section (horns)

In this section we regard the sound propagation in pipes the cross-sectional area of which show continuous variations according to a given function $S = S(x)$. If $S(x)$ is a monotonically growing function such pipes are usually referred to as horns.

It is well established that the sounds generated by some source can be enhanced by a horn. For this reason this classical acoustical device has found manifold practical applications. One of them is the well-known megaphone, either in its traditional form or in combination with an electroacoustical driver. As everyone knows, many musical instruments are fitted with a horn. Another example is Edison's phonograph and its successors, and in loudspeaker construction horns play an important role.

Of course, a horn does not amplify anything in the modern sense of the word, but it rather improves the impedance match between an original sound source (the human vocal chords, for instance, or the membrane of a loudspeaker) and free space. This holds also in the reverse direction: if a sound wave hits the wide end of a horn, it will produce a considerably higher sound pressure at its narrow end. This fact was the basis of purely mechanical hearing aids as they were widely used in pre-electronic and pre-electroacoustic ages.

Figure 8.9 shows a section of a pipe with variable cross section. To arrive at a quantitative treatment first we proceed as in Section 3.2, namely, by setting up the force and mass balance for a small volume element within the tube bounded by the coordinates x and x + dx. The balance of forces reads:

$$(S)_x - (S)_{x+dx} + [S(x + dx) - S(x)]\,p(x) = \rho_0 \frac{\partial v_x}{\partial t}S dx \tag{8.27}$$

The first and second term of this formula represent forces acting on the left and right boundary of the volume element. The third term is the x-component of reaction force exerted by the wall onto the considered element. These

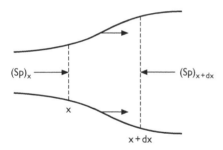

Figure 8.9 Derivation of the horn equation.

forces are balanced by the inertial force on the right. The usual simplifications have already been made. Now we express the differences on the left by derivatives. Then we obtain from the earlier equation, after cancelling dx:

$$-\frac{\partial(Sp)}{\partial x} + p\frac{dS}{dx} = \rho_0 S\frac{\partial v_x}{\partial t}$$

or

$$-S\frac{\partial p}{\partial x} = \rho_0 \frac{\partial(Sv_x)}{\partial t} \tag{8.28}$$

which agrees with eq. (3.15) – apart from a factor S on both sides which could be cancelled in principle. Hence, the pressure gradient is independent of the cross section and its variation.

Now we consider the difference of mass flows passing the left and right cross section per second, assuming that both boundaries are fixed. It is balanced by a change of density of the medium within the element, multiplied by its volume:

$$(\rho_t v_x S)_{x+dx} - (\rho_t v_x S)_x = -Sdx\frac{\partial \rho}{\partial t}$$

or after adequate simplifications and with $\rho = p/c^2$:

$$\rho_0\frac{\partial(Sv_x)}{\partial x} = -\frac{S}{c^2}\frac{\partial p}{\partial t} \tag{8.29}$$

By differentiating eq. (8.28) with respect to x and eq. (8.29) with respect to t, the quantity Sv_x can be eliminated from both equations. The result can be written in the form:

$$\frac{1}{S}\frac{\partial}{\partial x}\left(S\frac{\partial p}{\partial x}\right) = \frac{1}{c^2}\frac{\partial^2 p}{\partial t^2} \tag{8.30a}$$

or alternatively

$$\frac{\partial^2 p}{\partial x^2} + \frac{d(\ln S)}{dx}\frac{\partial p}{\partial x} = \frac{1}{c^2}\frac{\partial^2 p}{\partial t^2} \tag{8.30b}$$

This differential equation is known as Webster's equation or as the horn equation.

Its derivation is based on the assumption that the waves inside the horn are nearly plane. This is true, at best, for slender horns. In contrast, the wavefronts in widely opened horns are nearly spherical and are orthogonal to the horn wall. The consequences drawn from eq. (8.30) will describe the real situation the better the smaller is the slope angle of the horn wall.

A particular horn is specified by its cross-sectional function S(x). Once this is given we can try to solve eq. (8.30). In what follows this will be carried out for two particularly simple horn shapes, namely, the conical and the exponential horn under the assumption that they are infinitely long. Another important type of horns will be briefly described in Subsection 11.5.3.

8.4.1   Conical horn

The simplest horn shape is a cone with the lateral dimensions – for instance, the radius if the cross section is circular – increasing linearly with the length coordinate x as shown in Figure 8.10a. Then the area of the cross section grows proportionally with the square of x:

$$S(x) = Ax^2 \tag{8.31}$$

For small apertures the constant A is roughly equal to the solid angle Ω which the cone subtends. From eq. (8.31) it follows that $d(\ln S)/dx = 2/x$.

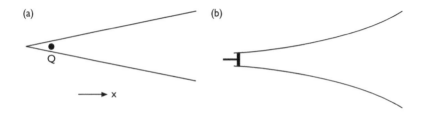

Figure 8.10 Simple horns: (a) conical horn with point source Q, (b) exponential horn with piston.

Hence the horn equation (8.30b) yields immediately

$$\frac{\partial^2 p}{\partial x^2} + \frac{2}{x}\frac{\partial p}{\partial x} = \frac{1}{c^2}\frac{\partial^2 p}{\partial t^2} \tag{8.32}$$

This differential equation agrees with eq. (5.2) with the only difference being that r is now replaced by x. Thus the wave travelling in the horn is a spherical wave the pressure amplitude of which decreases in inverse proportion to the distance x from the tip of the horn at $x = 0$. Hence the results of Section 5.2 can be widely applied to the conical horn.

To demonstrate the amplifying effect of a horn let us imagine a point source located next to its tip as shown in Figure 8.10a. Since its full volume velocity is used now to create a spherical wave in the restricted solid angle Ω the sound pressure within the horn is increased by a factor $4\pi/\Omega$ compared with that in a free spherical wave. Hence we get instead of eq. (5.6):

$$p(r, t) = \frac{j\omega\rho_0 \hat{Q}}{\Omega x} e^{j(\omega t - kx)} \tag{8.33}$$

The intensity which is proportional to the square of the sound pressure amplitude surpasses that of free radiation as given by eq. (5.9) by the factor $(4\pi/\Omega)^2$. This may be a little surprising at first glance. Intuitively, one might explain the amplifying effect by the concentration of the radiated energy into a smaller solid angle. This, however, would only justify one factor $4\pi/\Omega$. Another factor $4\pi/\Omega$ is due to the increased acoustical load which the horn offers to the source. In fact, the total acoustical power delivered by the source in the horn exceeds that of free radiation after eq. (5.10), again by the factor $4\pi/\Omega$

As mentioned earlier these relations hold for the infinitely long horn. When the horn has finite length we expect some sound reflection from its 'mouth' which modifies the sound pressure to some extent. More will be said on this matter in Subsection 19.4.3.

8.4.2 Exponential horn

Significantly different properties are encountered with the exponential horn (see Fig. 8.10b) which will be described in this section. Here, the cross-sectional area varies according to

$$S(x) = S_0 e^{2\varepsilon x} \tag{8.34}$$

with ε denoting the 'flare constant'. Since $\ln(S) = \ln(S_0) + 2\varepsilon x$, eq. (8.30b) yields

$$\frac{\partial^2 p}{\partial x^2} + 2\varepsilon\frac{\partial p}{\partial x} = \frac{1}{c^2}\frac{\partial^2 p}{\partial t^2} \tag{8.35}$$

Into this equation we insert as a tentative solution:

$$p(x, t) = \hat{p}_0 e^{j(\omega t - \underline{k}x)}$$

Then we arrive at a quadratic equation for the unknown $\underline{k}$:

$$\underline{k}^2 + j2\varepsilon\underline{k} - \frac{\omega^2}{c^2} = 0$$

From both possible solutions we select that corresponding to a wave progressing in positive x-direction:

$$\underline{k} = k' - j\varepsilon \tag{8.36}$$

with

$$k' = \frac{1}{c}\sqrt{\omega^2 - c^2\varepsilon^2} \tag{8.37}$$

Then the sound pressure in the horn wave is

$$p(x, t) = \hat{p}_0 e^{-\varepsilon x} \cdot e^{j(\omega t - k'x)} \tag{8.38}$$

The first factor expresses the 'dilution' of the wave, which must fill increasingly larger cross sections in the course of its propagation. However, eq. (8.38) represents a wave only if the angular wave number k' is real, that is, if $\omega \geq c\varepsilon$. In the other case k' and hence $\underline{k}$ will become imaginary and there will be no wave propagation at all. This shows us that the exponential horn is a high-pass line with the lower cut-off frequency

$$\omega_c = c\varepsilon \tag{8.39}$$

The wave velocity in the horn is $c' = \omega/k'$ or

$$c' = \frac{c}{\sqrt{1 - (\omega_c/\omega)^2}} \tag{8.40}$$

Hence it exceeds the free field velocity c. Another remarkable fact is that the wave velocity depends on the sound frequency ω. This phenomenon is called 'dispersion'. We shall come back to it in Section 8.6.

From eq. (3.15) we calculate the particle velocity using eq. (8.38):

$$v(x, t) = -\frac{1}{j\rho_0\omega}\frac{\partial p}{\partial x} = \frac{p}{Z_0}\left(\sqrt{1 - (\omega_c/\omega)^2} - j\omega_c/\omega\right) \tag{8.41}$$

Hence the ratio of the sound pressure and the particle velocity, the characteristic impedance within the horn, is

$$Z_0' = Z_0\left(\sqrt{1 - (\omega_c/\omega)^2} + j\omega_c/\omega\right) \tag{8.42}$$

To explain the 'amplifying' effect of the horn we again suppose that the sound wave is excited by a sound source in the narrow part of the horn at $x = 0$ – the so-called throat – this time, however, in the form of an oscillating piston as depicted in Figure 8.10b. It is loaded with the radiation resistance:

$$R_r = S_0\text{Re}\{Z_0'\} = Z_0S_0\sqrt{1 - (\omega_c/\omega)^2} \tag{8.43}$$

which is plotted in Figure 8.11b as a function of frequency (solid curve). At high frequencies, it approaches asymptotically to the value S_0Z_0 which would hold for a tube of constant cross section. For comparison, the diagram shows as a dotted line the radiation resistance of a piston of the same size without a horn, but inserted into an infinite rigid plane wall (compare Subsection 5.8.3). It is assumed that the piston is circular and has a diameter

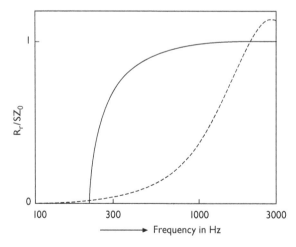

Figure 8.11 Radiation resistance of a circular piston of 10 cm in diameter in an exponential horn with the flare constant $\varepsilon = 4$ m^{-1} (solid line) and in an infinite boundary (dotted line).

of 10 cm, the lower limiting frequency $f_c = \omega_c/2\pi$ of the horn is 100 Hz, and the medium is air. Obviously, the horn significantly improves the impedance match of the sound source to the medium. The radiation resistance of the piston fitted with the horn grows relatively fast when the frequency is above the cut-off frequency, and it exceeds that of the free piston in a wide frequency range. This advantage is paid for by zero radiation when driven below the cut-off frequency.

Again, it should be noted that these properties are strictly valid for the infinitely long horn only. Even with this limitation the advantages of horns are so prominent that they find wide technical application.

8.5   Higher order wave types

Up to this point it has been supposed that the lateral dimensions of pipes are significantly smaller than the acoustic wavelength – an assumption which may be rather questionable for the horn as has been described in the preceding section.

Now we omit this presupposition and investigate whether there are other waveforms – apart from the plane wave – which can be propagated in a pipe or – more generally – in an acoustical waveguide. We restrict this discussion to pipes of constant cross section.

At first, let us consider two plane harmonic waves of equal frequency propagating in directions which are inclined by the arbitrary angle $\pm\varphi$ to the x-axis. Their wave normals are represented in Figure 8.12 as arrows.

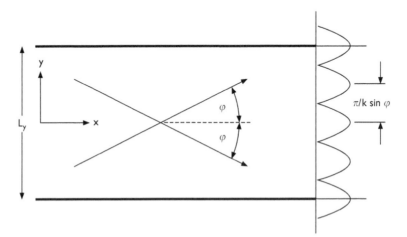

Figure 8.12 Two crossed plane waves. Right side: transverse distribution of the resulting sound pressure amplitude.

According to eq. (4.11) the sound pressures associated with them are, with $\alpha = \varphi$, $\cos \beta = \sin \varphi$, $\cos \gamma = 0$:

$$p_{1,2}(x, y) = \hat{p} e^{jk(-x \cos \varphi \pm y \sin \varphi)} \tag{8.44}$$

By adding p_1 and p_2 and using Euler's formula we obtain for the total sound pressure:

$$p(x, y) = 2\hat{p} \cos(ky \sin \varphi) \cdot e^{-jkx \cos \varphi} \tag{8.45}$$

Here as well as in the following expressions the exponential factor $\exp(j\omega t)$ containing the time dependence has been omitted. Equation (8.45) represents a wave progressing in the positive x-direction with angular wavenumber $k' = k \cos \varphi$. Obviously, this is not a plane wave since its amplitude is modulated with respect to the y-coordinate in the same way in a standing wave. Maximum sound pressure amplitudes occur whenever $ky \cdot \sin \varphi$ is an integral multiple of π, that is, in the planes $y = m\pi/(k \sin \varphi)$ where m is an integer. At these positions the vertical component of the particle velocity is zero. Accordingly, the sound field would not be disturbed by replacing two of this planes with rigid surfaces. Conversely, if the distance L_y of two parallel and rigid surfaces is given, a sound wave given by eq. (8.45) can propagate between them if $kL_y \cdot \sin \varphi = m\pi$; this relation defines the angle φ in eq. (8.45). From this condition we obtain the angular wavenumber relevant for the propagation in the x- direction:

$$k' = k \cos \varphi = \frac{\omega}{c} \sqrt{1 - \left(\frac{m\pi c}{\omega L_y}\right)^2} \tag{8.46}$$

It is real if the angular frequency ω exceeds a certain cut-off frequency ω_m which depends on the order m and distance L_y of both surfaces, that is, on the height of the 'channel':

$$\omega_m = \frac{m\pi c}{L_y} \tag{8.47}$$

The wave represented by eq. (8.45) is called a mth order wave type or wave mode. The wave characterised by $m = 0$ is the fundamental wave, and it is identical with the plane wave of our earlier sections and has a cut-off frequency zero. The wave velocity $c' = \omega/k'$ of the wave type of order m is

$$c' = \frac{c}{\sqrt{1 - (\omega_m/\omega)^2}} \tag{8.48}$$

This formula agrees with eq. (8.40) for the exponential horn and shows the same frequency dependence. Figure 8.13a plots the wave speed (which

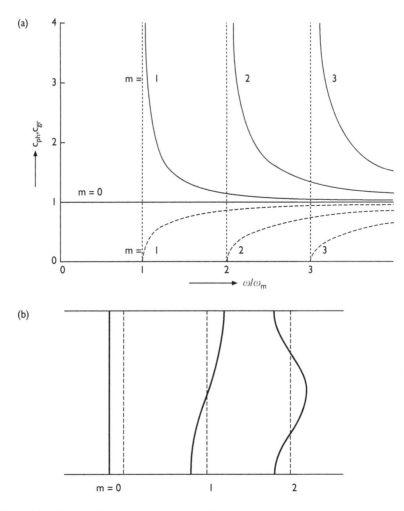

Figure 8.13 Higher order wave types in a two-dimensional waveguide: (a) speed of propagation (solid lines: phase velocity, broken lines: group velocity) as a function of the
angular frequency, (b) transverse distribution of the sound pressure amplitude.

should be more correctly referred to as the 'phase velocity' as will be
explained in the following section) as a function of the frequency (solid
lines). For m > 0 it is always larger than the free field sound velocity c which
it approaches asymptotically at very high frequencies.

Now we can represent the sound pressure in the mth mode by

$$p(x, y) = 2\hat{p} \cos \left(\frac{m\pi y}{L_y} \right) \cdot e^{-jk'x} \tag{8.49}$$

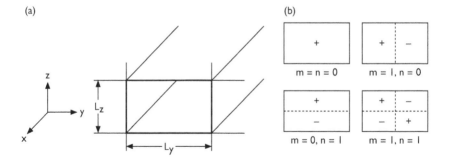

Figure 8.14 Sound propagation in a duct with rectangular cross section: (a) sketch of the duct, (b) nodal planes (section) of a few wave types (broken lines), m, n: integers in eq. (8.51a).

The lateral distribution of the sound pressure amplitude is shown in Figure 8.13b.

The preceding considerations can be extended without difficulties to a one-dimensional waveguide, that is, to a rigid-walled rectangular duct. For this purpose we imagine a second pair of rigid planes with a mutual distance L_z arranged perpendicular to the z-axis (see Fig. 8.14a).

As before we compose the sound field of two waves after eq. (8.49), how-ever not travelling into the x-direction. Instead, they propagate in directions which are parallel to the x-z-plane but subtend the angles $\pm\varphi'$ with the x-axis. Hence, analogous to eq. (8.44):

$$p_{1,2}(x, y, z) = 2\hat{p} \cos\left(\frac{m\pi y}{L_y}\right) \cdot e^{-jk'(x\cos\varphi' \pm z\sin\varphi')} \tag{8.50}$$

The same considerations as in the derivation of eq. (8.49) lead to the expression

$$p(x, y, z) = 4\hat{p} \cos\left(\frac{m\pi y}{L_y}\right) \cos\left(\frac{n\pi z}{L_z}\right) \cdot e^{-jk''x\cos\varphi'} \tag{8.51a}$$

for the sound pressure, with n denoting a second integer. Now the wave field consists of standing waves with respect to the y- and the z-direction and it travels along the x-direction with the angular wave number:

$$k'' = \frac{\omega}{c}\sqrt{1 - (\omega_{mn}/\omega)^2} \tag{8.52}$$

Here we introduced the cut-off frequency:

$$\omega_{mn} = \sqrt{\left(\frac{m\pi c}{L_y}\right)^2 + \left(\frac{n\pi c}{L_z}\right)^2} \tag{8.53}$$

A particular wave mode is characterised by two integers m and n; it is a progressive wave only if the driving frequency ω is above this cut-off frequency. For $\omega < \omega_{nm}$ the wavenumber k'' becomes imaginary; according to eq. (8.51a) this corresponds to a pressure oscillation with constant phase and with an amplitude decaying (or growing) exponentially with x. Below the lowest non-zero cut-off frequency the fundamental wave with m = n = 0 is the only propagating wave mode. Depending on whether L_y or L_z is the larger lateral dimension this cut-off frequency is either ω_{10} or ω_{01}. Or expressed in terms of the free field wavelength λ: the range where only the fundamental wave can be propagated is given by:

$$\lambda > 2 \cdot \text{Max}\{L_y, L_z\} \tag{8.54}$$

Figure 8.14b presents an overview of the pressure distributions within the channel associated with some modes. The dotted lines indicate nodal planes where the pressure amplitude is zero at any time. They separate regions with opposite phase as marked by the signs. Both indices m and n indicate the number of nodal planes in the respective direction. Again, the wave velocity is given by eq. (8.48) after replacing ω_m with ω_{nm}. The same holds for Figure 8.13a, however, the limiting frequencies are no longer equidistant in this case.

The formulae derived earlier are valid in a slightly modified form for channels with soft boundaries the wall impedance of which is zero. In eqs. (8.49) and (8.51) we have just to replace the cosine functions with sine functions:

$$p(x, y, z) = 4\hat{p} \sin\left(\frac{m\pi y}{L_y}\right) \sin\left(\frac{n\pi z}{L_z}\right) \cdot e^{-jk''x \cos \varphi'} \tag{8.51b}$$

because this time it is the sound pressure which has to vanish along the walls of the channel and not the normal components of the particle velocity. However, there exists no fundamental wave in a channel with soft walls since; according to eq. (8.51b) the sound pressure vanishes everywhere if m or n is zero. Waveguides with soft boundaries cannot be realised for gaseous media, but for liquid ones. Thus the surface of water, viewed from inside the water, has nearly zero impedance; furthermore, almost soft boundaries can be made with porous plastics with the pores filled with air.

Now we return to sound propagation in gas-filled waveguides. Most important are pipes with circular cross section for the transport of air or other gases. To calculate the sound propagation in such pipes one has to apply an

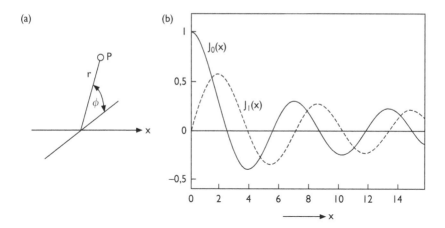

Figure 8.15 (a) Cylindrical coordinates, (b) the lowest order Bessel functions J_0 and J_1.

approach which we could have chosen as well for rectangular waveguides: it starts with the wave equation (3.21) expressed in coordinates which are appropriate to the geometry of the tube. In the present case these are cylindrical coordinates (see Fig. 8.15a). One coordinate axis coincides with the pipe axis, and we denote it as earlier with x. Furthermore, the position of a point P is characterised by its perpendicular distance r from the axis and by the angle between r and a fixed reference line which is also perpendicular to the axis. After specifying the wave equation in this way, the next step is to adapt its general solution to the boundary condition prescribed at the wall. As a result one arrives at the following expression for the sound pressure in a rigidly walled pipe:

$$p(x, r, \phi, t) = A J_m \left(v_{mn} \frac{r}{a} \right) \cdot \cos(m\phi) \cdot e^{j(\omega t - k''x)} \tag{8.55}$$

Here J_m is the Bessel function of mth order which we encountered already in Subsection 5.8.2 (see eq. (5.39)). The first two of these functions, namely, J_0 and J_1, are plotted in Figure 8.15b. Each of them – and the same holds for higher order Bessel functions – has infinitely many maxima and minima. One of these must coincide with the pipe wall since in these points the derivative of the Bessel function and hence the radial component of the particle velocity vanishes. To achieve this, r/a in the argument of the Bessel function is multiplied with a number v_{mn} which is the nth zero (starting with n = 0) of the derivative of the mth order Bessel function. Table 8.1 lists some of these zeros. Again, the angular wavenumber with respect to propagation along the

Table 8.1 Characteristic values v_{mn} in eq. (8.55)

Order m of Bessel function	n = 0	n = 1	n = 2
0	0	3.832	7.015
1	1.841	5.331	8.526
2	3.054	6.706	9.970

axis is given by eq. (8.52); the cut-off frequency of the corresponding wave mode is

$$\omega_{mn} = v_{mn} \frac{c}{a} \tag{8.56}$$

Figure 8.16 shows the nodal surfaces for the lowest wave modes after eq. (8.55); the presentations are ordered after increasing cut-off frequencies. The nodal surfaces associated with the number m are concentric cylinders, and the other ones are planes containing the axis of the tube. The lowest cut-off frequency is that with m = 1 and n = 0, hence the frequency range in which the fundamental wave is the only one which can be propagated is characterised by

$$\lambda \geq 3.41 \cdot a \tag{8.57}$$

The earlier discussion provided some insight into the structure and properties of possible wave modes in a pipe. Whether these modes are actually excited by a certain sound source and participate in the transport of sound energy is a different question which cannot be answered without knowing the kind and position of the sound source. If the source consists of a rigid oscillating piston forming the termination of the pipe, we expect that the fundamental wave will be generated almost exclusively. However, if it is a point source located on the axis of a cylindrical pipe, only modes with m = 0 will be produced, since these are the only ones without any nodal surfaces containing the axis. If the point source is in an asymmetric position we have to reckon with the excitation of all wave types the cut-off frequencies of which are below the driving frequency, and their relative strengths depend on the position of the source.

8.6 Dispersion

In this chapter we have encountered two cases in which the wave velocity depends on the sound frequency, namely, the exponential horn and the higher order wave types in pipes or ducts. This phenomenon which is not restricted to acoustical waves is known as 'dispersion'. We shall have a somewhat closer look at it in this section.

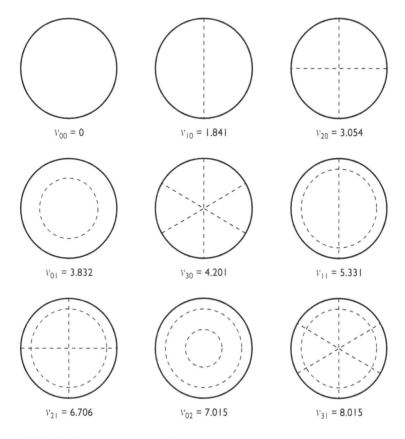

Figure 8.16 Nodal planes (section) of a few wave types in the cylindrical tube (ν_{mn}: see eq. (8.55)).

First we try to find a justification for the word dispersion. According to the Fourier theorem as introduced in Section 2.8 a short impulse consists of numerous harmonic vibrations. The standard example is a signal described by the delta function after eq. (2.55):

$$\delta(t) = \lim_{\Omega \to \infty} \frac{1}{2\pi} \int_{-\Omega}^{\Omega} e^{j\omega t} d\omega$$

which represents an impulse of infinite height and of infinitely short duration. If this signal travels a distance x in the form of a wave, each of its Fourier components will be delayed by the time x/c' with c' denoting the wave velocity. Accordingly, we have to replace in the exponent of the above equation t with $t - x/c'$:

$$s(t) = \lim_{\Omega \to \infty} \frac{1}{2\pi} \int_{-\Omega}^{\Omega} e^{j\omega(t-x/c')} d\omega$$

If the wave velocity is independent of ω, we have $s(t) = \delta(t - x/c')$; the original impulse is delayed as a whole while its shape remains unaltered. If, on the other hand, c' is a function of the frequency, the Fourier components will be delayed by different amounts and will not fit together any more after having travelled the distance x. The result is a different signal which in any case will be longer than the original impulse.

Further conclusions can be drawn if we consider a signal consisting of two harmonic waves with slightly different angular frequencies and wavenumbers:

$$s(x, t) = \hat{s} \left[e^{j(\omega_1 t - k_1 x)} + e^{j(\omega_2 t - k_2 x)} \right]$$

with $\omega_{1,2} = \omega \pm \Delta\omega$ and $k_{1,2} = k \pm \Delta k$. It represents beats of the kind as described in Figure 2.5 (see Section 2.3), which, however, are not fixed to a certain location but travels as a wave. The earlier expression may be written in the form

$$s(x, t) = 2\hat{s} \cos(t\Delta\omega - x\Delta k)e^{j(\omega t - kx)} \tag{8.58}$$

The exponential function represents the rapid oscillations of the signal. They travel with the speed

$$c_{ph} = \frac{\omega}{k} \tag{8.59}$$

called the 'phase velocity' of the medium because it concerns the propagation of phases, for instance, of a particular zero. It is identical with the quantity c' used in eqs. (8.40) and (8.48) which was vaguely referred to as 'wave velocity'.

Now we consider the envelope, that is, the slow variation of the instantaneous amplitude represented by the cosine function in eq. (8.58). Obviously, it propagates with the velocity $\Delta\omega/\Delta k$, or, in the limit of vanishing frequency difference, with speed

$$c_{gr} = \frac{d\omega}{dk} \tag{8.60}$$

This quantity is called the 'group velocity' since it refers to the propagation of a wave group as considered in this example. If the angular frequency ω

is proportional to the angular wavenumber k the phase velocity and group velocity are equal and there is no dispersion.

For the higher order wave modes between two parallel rigid plates the group velocity can be calculated easily by differentiating $k' = \sqrt{\omega^2 - \omega_n^2}/c$ (see eq. (8.46) with eq. (8.47)) with respect to ω. Then the group velocity is the reciprocal of the derivative $dk'/d\omega$:

$$c_{gr} = c\sqrt{1 - (\omega_m/\omega)^2} \tag{8.61}$$

Its frequency dependence is shown in Figure (8.13a) for a few wave modes as dotted lines. Obviously, the group velocity is always smaller than the free field velocity c, and for high frequencies it approaches c asymptotically. Moreover, it follows from eq. (8.48) (with $c' = c_{ph}$):

$$c_{ph} \cdot c_{gr} = c^2 \tag{8.62}$$

Equations (8.61) and (8.62) hold for all rigid channels and also for the exponential horn where we have to replace, of course, ω_n with ω_c. It may be noted, however, that there exist quite different dispersion laws for which eq. (8.61) is not valid (see Chapter 10).

Chapter 9

Sound in closed spaces

The preceding chapters demonstrated that increasing confinement of the sound field is accompanied with an increase in complexity. Thus a reflecting plane hit by a plane wave at oblique incidence creates a wave field which is progressive with respect to one coordinate while it appears as a more or less pronounced standing wave with respect to another one. If the range of propagation is further restricted by a pipe or channel a sound source will generally excite a variety of discrete wave types or wave modes. Proceeding along this line leads us to sound in a completely closed space which will be discussed in this chapter. It will turn out that in this case the sound field is composed of discrete wave patterns which make the idea of sound propagation somewhat questionable at first glance. This fact is the physical basis of room acoustics although the room acoustical practitioner will usually prefer simpler and less formal ways of describing sound in enclosures.

9.1 Normal modes in a one-dimensional space

As a preparation for the main content of this chapter we begin with a one-dimensional enclosure, that is, with a pipe of finite length with rigid walls. It is assumed that the lateral dimensions are small enough to guarantee that in the frequency range considered only the fundamental wave can exist. Any loss processes as may occur in the medium and at the walls according to Sections 4.4 and 8.1 are neglected. The relevant coordinate is an x-axis coinciding with the axis of the pipe.

At first we consider the case that both ends of the pipe are closed with a rigid plate or lid. Any harmonic sound field within the pipe consists of a standing wave with maximum pressure amplitude occurring at its terminations. This is only possible if an integral number of half-wavelengths fits into the tube length L (see Fig. 9.1a). This requirement defines the allowed frequencies which will be called 'eigenfrequencies' in the following:

$$f_n = n\frac{c}{2L} \qquad (n = 0, 1, 2, \ldots) \tag{9.1}$$

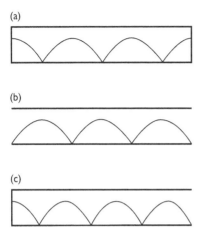

Figure 9.1 Normal modes in a rigid tube of finite length: (a) rigid terminations at both ends, (b) open at both ends and (c) one end rigidly terminated, the other one is open.

The standing wave associated with a particular eigenfrequency is given by

$$p_n(x, t) = \hat{p} \cos\left(\frac{n\pi x}{L}\right) e^{j\omega_n t} \tag{9.2}$$

with $\omega_n = 2\pi f_n$. Here the left end of the pipe is located at $x = 0$. These characteristic amplitudes are called normal modes of the enclosure.

Now it is assumed that, in contrast to the case mentioned earlier, the pipe is terminated with zero impedance at both its ends. This condition can be approximated by leaving the pipe open. Again, the sound field with harmonic pressure variation consists of a standing wave, but now the pressure amplitude at $x = 0$ and $x = L$ must be zero. As with rigid terminations, this is achieved when the pipe length equals an integral number of half-wavelengths, hence the eigenfrequencies of the pipe are given by eq. (9.1) as before. However, the distribution of sound pressure amplitudes is different in that the cosine in eq. (9.2) is replaced with a sine function with the same argument. Obviously, no sound field can exist for $n = 0$, in contrast to the case considered earlier.

As a third case we consider a pipe with one open end at $x = L$ while the other end at $x = 0$ is terminated with a rigid cap (see Fig. 9.1c). Since the resulting standing wave has a pressure maximum at its left side and a pressure node at the right end, the pipe length must be equal to an integral number of half-wavelengths plus one quarter-wavelength, or, in other words, to an odd multiple of a quarter-wavelength. It follows that

$$f_n = (2n - 1) \cdot \frac{c}{4L} \quad (n = 1, 2, 3, \ldots) \tag{9.3}$$

Again, the eigenfrequencies are equidistant along the frequency axis; however, the lowest one is only half as high as that of a pipe with equal terminations. Now the sound pressure is represented by

$$p_n(x, t) = \hat{p} \cos\left[\left(n - \frac{1}{2}\right) \frac{\pi x}{L}\right] e^{j\omega_n t} \tag{9.4}$$

We conclude this section with an example where the eigenfrequencies cannot be expressed by a closed formula, namely, a conical horn which is rigidly closed at its narrow end located at $x = x_1$, while the wide end at $x = x_1 + L$ is open (see Fig. 9.2a).

As we have seen in Subsection 8.4.1, waves in a conical horn are spherical waves travelling either towards the tip or away from it. In the general case both waves are present:

$$p(x, t) = \frac{1}{x}(Ae^{-jkx} + Be^{jkx})$$

with constants A and B still to be determined. Differentiation with respect to x yields

$$\frac{\partial p}{\partial x} = \frac{A}{x} e^{-jkx} \left(-jk - \frac{1}{x}\right) + \frac{B}{x} e^{jkx} \left(jk - \frac{1}{x}\right)$$

The boundary conditions are $\partial p / \partial x = 0$ at $x = x_1$ and $p = 0$ at $x_1 + L$ at x_1 at this place (closed end), while the wide end located at $x =$ is open ($p = 0$). From both the earlier equations we find

$$Ae^{-jkx_1} \left(jk + \frac{1}{x_1}\right) = Be^{jkx_1} \left(jk - \frac{1}{x_1}\right)$$

$$Ae^{-jk(x_1+L)} = -Be^{jk(x_1+L)}$$

Dividing the first of these equation by the second one yields

$$e^{jkL} \left(jk + \frac{1}{x_1}\right) = -e^{-jkL} \left(jk - \frac{1}{x_1}\right)$$

or, by using eqs. (2.6a) and (2.6b):

$$\tan(kL) = -kx_1 = -\frac{x_1}{L}(kL) \tag{9.5}$$

This transcendental equation is graphically represented in Figure 9.2b, and the solutions $k_n L = 2\pi L f_n / c$ correspond to the intersections of the falling

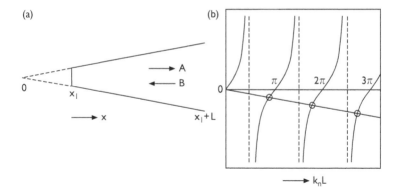

Figure 9.2 Eigenfrequencies of a conical horn with rigid terminations: (a) longitudinal section of the horn, (b) illustration of eq. (9.5).

straight line with the tangent function. The limit $x_1 \rightarrow \infty$ characterises the cylindrical pipe for which the solutions are $k_n L = (2n-1) \cdot (\pi/2)(n = 1, 2, \ldots)$ in agreement with eq. (9.3) (dotted vertical lines in the figure). Compared with these lines the solutions of eq. (9.6) are shifted towards higher kL, that is, towards higher frequencies. Particularly, the lowest eigenfrequencies are significantly higher than those after eq. (9.3).

The contents of this section can be summarised as follows: provided all losses are neglected, a sound field in a pipe of finite length can exist only at certain discrete frequencies, called eigenfrequencies. Each of them is associated with a characteristic standing wave which is known as a normal mode.

9.2   Normal modes in a rectangular room with rigid walls

After this preparation it is not difficult to find the eigenfrequencies and normal modes of a rectangular room with the dimensions L_x, L_y and L_z all walls of which are rigid (see Fig. 9.3). For this purpose we replace the 'thin' pipe considered in the preceding section with a channel with arbitrary lateral dimensions L_y and L_z. Hence we admit that higher order wave types as described in Section 8.5 propagate in this channel each of them characterised by two integers m and n. The angular wavenumber k'' of any such wave type is given by eqs. (8.52) and (8.53); the wavelength associated with it is $\lambda'' = 2\pi/k''$. The requirement that an integral number of half-wavelengths fit into the length L_x of the channel is equivalent to $L_x = 1 \cdot (\pi/k'')$ with

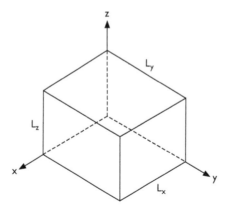

Figure 9.3 Rectangular room.

$l = 0, 1, 2$, etc. Combined with eq. (8.52) this yields the allowed angular frequencies

$$\omega^2_{lmn} = \left(\frac{l\pi c}{L_x}\right)^2 + \omega^2_{mn} \tag{9.6}$$

The eigenfrequencies $f_{lmn} = \omega_{lmn}/2\pi$ of the room are found by inserting ω_{mn} from eq. (8.53):

$$f_{lmn} = \frac{c}{2}\sqrt{\left(\frac{1}{L_x}\right)^2 + \left(\frac{m}{L_y}\right)^2 + \left(\frac{n}{L_z}\right)^2} \tag{9.7}$$

The normal modes associated with these eigenfrequencies are given, in principle, by eq. (9.2), after replacing n with l and L with L_x. However, the amplitude is no longer independent of y and z but shows a lateral distribution according to the cosine functions in eq. (8.51). Hence the sound pressure of a normal mode marked by the integers l, m and n reads:

$$p_{lmn}(x, y, z, t) = \hat{p}\cos\left(\frac{l\pi x}{L_x}\right)\cos\left(\frac{m\pi y}{L_y}\right)\cos\left(\frac{n\pi z}{L_z}\right)\cdot e^{j\omega_{lmn}t} \tag{9.8}$$

This is the three-dimensional extension of the standing wave as treated in Section 6.5 with $R = 1$. A given mode has l nodal planes perpendicular to the x-axis, m nodal planes perpendicular to the y-axis and n nodal planes perpendicular to the z-axis. Along these planes the sound pressure is always zero. Figure 9.4 shows for a normal mode with $l = 3$ and $m = 2$ the amplitude distribution over the ground plane $z = 0$ in the form of contours of equal sound pressure amplitude ($|p_{lmn}/\hat{p}| = 0.25, 0.5$ and 0.75). The nodal

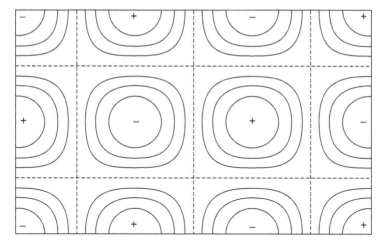

Figure 9.4 Contours of equal sound pressure amplitude in the plane z = 0 for the normal mode l = 3 and m = 2 of a rectangular room.

Table 9.1 The twenty lowest eigenfrequencies of a rectangular room with dimensions 4.7 × 4.1 × 3.1 m³

Eigenfrequency (Hz)	l	m	n	Eigenfrequency (Hz)	l	m	n
36.17	1	0	0	90.47	1	2	0
41.46	0	1	0	90.97	2	0	1
54.84	0	0	1	99.42	0	2	1
55.02	1	1	0	99.80	2	1	1
65.69	1	0	1	105.79	1	2	1
68.55	0	1	1	108.51	3	0	0
72.34	2	0	0	109.68	0	0	2
77.68	1	1	1	110.05	2	2	0
82.93	0	2	0	115.49	1	0	2
83.38	2	1	0	116.16	3	1	0

planes are indicated by dotted lines. Each of them separates two regions with opposite signs of the instantaneous sound pressure.

Table 9.1 lists the first twenty eigenfrequencies of a rectangular room with dimensions $4.7 \times 4.1 \times 3.1$ m³. Of course, they are not equidistant along the frequency axis; apparently, their density increases with frequency.

To get an overview of the number and density of eigenfrequencies we imagine a 'frequency space' with cartesian coordinates f_x, f_y and f_z. It is sufficient to restrict oneself to the octant in which all numbers l, m and n are positive or 0, since changing the sign does not alter eqs. (9.7) and (9.8).

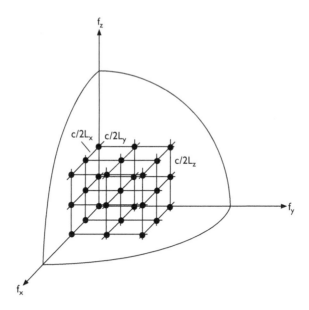

Figure 9.5 Eigenfrequency lattice of a rectangular room.

In this space a particular eigenfrequency corresponds to a point with the coordinates

$$f_x = \frac{lc}{2L_x}, \; f_y = \frac{mc}{2L_y} \quad \text{and} \quad f_z = \frac{nc}{2L_z}$$

the totality of all eigenfrequencies forms a regular lattice as shown in Figure 9.5. The distance of a particular lattice point from the origin is, from eq. (9.7):

$$\sqrt{f_x^2 + f_y^2 + f_z^2} = f_{lmn}$$

The number of eigenfrequencies within the frequency interval from 0 to some frequency f can be estimated as follows: we consider a sphere with radius f described around the origin. The 'frequency volume' of the octant $f_x \geq 0$, $f_y \geq 0$ and $f_z \geq 0$ is $V^{(f)} = (4\pi/3) \cdot f^3/8$; it contains all lattice points falling into the said frequency interval. To each of them one lattice cell with the 'frequency volume'

$$\frac{c}{2L_x} \cdot \frac{c}{2L_y} \cdot \frac{c}{2L_z} = \frac{c^3}{8V}$$

is attributed where V is the geometric volume of the room. The number N_f of eigenfrequencies up to the limit f is obtained by dividing $V^{(f)}$ by the cell volume $c^3/8V$ with the result

$$N_f \approx \frac{4\pi}{3} V \left(\frac{f}{c}\right)^3 \tag{9.9a}$$

Applied to the room specified in Table 9.1 this formula predicts only 10 eigenfrequencies in the range from 0 to f = 116.5 Hz while the correct number is 20. The reason of this discrepancy lies in the eigenfrequency points situated on the coordinate planes; they belong not only to the considered octant but also to the adjacent ones, therefore only half of them are accounted for in eq. (9.9a) although they represent full eigenfrequencies. Likewise, only one quarter of the points lying along the coordinate axes are taken into account in eq. (9.9a) since each of them belongs to four octants. After adding corresponding correction terms we obtain the improved formula

$$N_f \approx \frac{4\pi}{3} V \left(\frac{f}{c}\right)^3 + \frac{\pi}{4} S \left(\frac{f}{c}\right)^2 + \frac{L}{8} \left(\frac{f}{c}\right) \tag{9.9b}$$

Of course, at higher frequencies these additional terms may be neglected. Equation (9.9b) is valid only for a rectangular room while eq. (9.9a), as is not shown here, can be applied to enclosures of any shape.

By differentiating eq. (9.9a) with respect to the frequency f one obtains the number of eigenfrequencies per Hertz, that is, the density of eigenfrequencies at frequency f:

$$\frac{dN_f}{df} \approx 4\pi V \frac{f^2}{c^3} \tag{9.10}$$

As an example we regard again the rectangular room with the dimensions 4.7 m × 4.1 m × 3.1 m. According to eq. (9.9a) we have to reckon with more than 6 millions eigenfrequencies in the range from 0 to 10 000 Hz. At 1000 Hz the average number of eigenfrequencies per Hertz after eq. (9.10) is about 19, hence the mean spacing between eigenfrequencies is as small as about 0.05 Hz.

9.3 Normal modes in cylindrical and spherical cavities

The normal modes of a cylindrical cavity can be calculated in the same way as those of a rectangular room. Our starting point is eq. (8.55) which represents the wave modes in a rigidly walled cylindrical tube. We assume now that the pipe is terminated at both sides with a rigid plate. As explained in Section 9.1 a wave field can only exist in the pipe if its length L_x is equal

Table 9.2 Characteristic values χ_{mn} in eq. (9.13) (nth zero of the derivative $j_{m'}(x)$)

Order m of spherical Bessel function	$n = 1$	$n = 2$	$n = 3$
0	0	4.493	7.725
1	2.082	5.940	9.206
2	3.342	7.290	10.614

to an integer number of half-wavelengths the wavelength being $\lambda'' = 2\pi/k''$. The angular wavenumber k'' is given by eq. (8.52) together with eq. (8.56). Again, the angular eigenfrequencies are obtained from eq. (9.6), so we arrive at the final result

$$f_{lmn} = \frac{c}{2}\sqrt{\left(\frac{1}{L_x}\right)^2 + \left(\frac{\nu_{mn}}{\pi a}\right)^2} \tag{9.11}$$

as earlier a denotes the radius of the tube and l is an integer. The sound pressure of a normal mode is

$$p_{lmn}(x, y, z, t) = \hat{p} J_m\left(\nu_{mn}\frac{r}{a}\right) \cdot \cos(m\phi) \cdot \cos\left(\frac{l\pi x}{L_x}\right) \cdot e^{j\omega_{lmn}t} \tag{9.12}$$

The numbers ν_{nm} are explained in Section 8.5 (see Table 8.1).

For the sake of completeness we present here the eigenfrequencies of a spherical cavity with rigid walls. They are characterised by two subscripts only:

$$f_{mn} = \chi_{mn}\frac{c}{2\pi a} \tag{9.13}$$

In this formula χ_{mn} is the nth zero of the derivative of the spherical Bessel function of order m, j_m.[1] In Table 9.2 some of these numbers are listed.[2]

9.4   Forced vibrations in a one-dimensional enclosure

So far it was assumed that all boundaries of an enclosure are rigid and hence free of losses. Therefore the question of how the normal modes are generated

1 M. Abramowitz and A. Stegun, *Handbook of Mathematical Functions*. Dover Publications, New York 1964.
2 P. M. Morse and H. Feshbach, Methods of Theoretical Physics, §11.3. McGraw-Hill, New York 1953.

Figure 9.6 Pipe with reciprocating piston as sound source, any termination at the right side.

did not arise. Once they are excited they will persist forever without any energy supply. This concept is very useful since it yields reasonable results even for real cavities as long as the losses occurring in them or at their boundary are not too high. However, to get a more realistic picture we must discuss the influence at least of wall losses. We expect that a stationary sound field can only be maintained if there is a sound source which continuously compensates for the energy lost at the boundaries of an enclosure waveguide. We restrict the discussion to one-dimensional space, that is, to the fundamental wave in a rigid-walled pipe as in Section 9.1. The losses are introduced by its right termination at $x = L$ which may be thought of as a plate with some reflection factor $R = |R| \exp(j\chi)$ (see Fig. 9.6). Furthermore, its left termination consists of a rigid moveable piston that vibrates with the velocity $v_0 \exp(j\omega t)$ and makes good the end losses. In other words: we now discuss forced vibrations of the pipe at a given angular frequency ω.

Basically, the sound pressure and particle velocity in the pipe are given by eqs. (6.8) and (6.12) with $\vartheta = 0$. Since it is more practical to have the left end of the pipe at $x = 0$ and the right one at $x = L$ we shift the coordinate axis by L which means that in these equations x is replaced with $x - L$. Furthermore, for the sake of clarity we show here the time factor $\exp(j\omega t)$ which was omitted in the equations of Section 6.3. With this in mind, we obtain from eq. (6.8)

$$p(x) = \hat{p} \left(e^{-jk(x-L)} + R e^{jk(x-L)} \right) e^{j\omega t} \tag{9.14}$$

and from eq. (6.12):

$$v_x(x) = \frac{\hat{p}}{Z_0} \left(e^{-jk(x-L)} - R e^{jk(x-L)} \right) e^{j\omega t} \tag{9.15}$$

For $x = 0$ the particle velocity v_x must equal the velocity $v_0 \cdot \exp(j\omega t)$ of the piston from which

$$\hat{p} = \frac{v_0 Z_0}{e^{jkL} - R e^{-jkL}} \tag{9.16}$$

follows. Then the expression in eq. (9.14) reads:

$$p_\omega(x) = v_0 Z_0 \frac{e^{jk(L-x)} + Re^{-jk(L-x)}}{e^{jkL} - Re^{-jkL}} e^{j\omega t} \tag{9.17}$$

(The index ω of p is to underline the dependence of the sound pressure amplitude on the frequency $\omega = ck$.)

Viewed as a function of x the earlier expression represents a standing wave (see Section 6.5). Its magnitude at the piston's surface (x = 0) is

$$\left|p_\omega(0)\right| = v_0 Z_0 \sqrt{\frac{1 + |R|^2 + 2|R|\cos(2kL - \chi)}{1 + |R|^2 - 2|R|\cos(2kL - \chi)}} \tag{9.18}$$

The sound pressure amplitude assumes particularly high values if the argument of the cosine functions are integral multiples of 2π, that is, if k assumes one of the values $k_n = (2n\pi + \chi)/2L$ with integer n. Hence, the angular eigenfrequencies of the pipe are obtained as solutions $\omega_n = ck_n$ of the equation

$$\omega_n = [2\pi n + \chi(\omega_n)] \frac{c}{2L} \tag{9.19}$$

For $\chi = 0$ the eigenfrequencies f_n agree with those given in eq. (9.1).

In Figure 9.7 the absolute value of the sound pressure at x = 0 after eq. (9.18) is plotted as a function of the frequency parameter $\omega L/c$. For the sake of simplicity $|R| = 0.7$ and $\chi = 45°$ was chosen for the reflection factor.

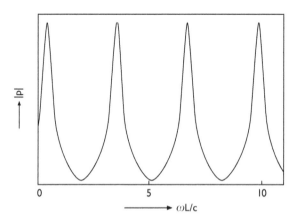

Figure 9.7 Frequency dependence of sound pressure amplitude in a pipe after Figure (9.6), reflection factor of termination is $R = 0.7 \cdot \exp(\pi/4)$.

It should be noted that in general both the magnitude and the phase angle of the reflection factor are frequency dependent. Therefore the resonances of a real pipe are not equidistant along the frequency axis. Nevertheless, the diagram shows the essential fact: a pipe of finite length shows infinitely many more or less pronounced resonances, and the resonance frequencies are identical with the eigenfrequencies of the pipe.

The resonant properties of the pipe can be demonstrated even more readily if eq. (9.17) is split in partial fractions. This is possible since this function has infinitely many single poles which of course coincide with the zeros of its denominator which we denote with

$$k_n = \frac{\omega_n}{c} + j\frac{\delta_n}{c} \tag{9.20}$$

The angular eigenfrequencies ω_n are the solutions of eq. (9.19) while the imaginary parts of k_n are given by

$$\delta_n = -\frac{c}{2L} \ln |R(\omega_n)| \tag{9.21}$$

Their physical meaning will become more obvious in Section 9.6.

With these quantities the expansion of eq. (9.17) in partial fractions reads:

$$\frac{e^{jk(L-x)} + Re^{-jk(L-x)}}{e^{jkL} - Re^{-jkL}} = \frac{1}{jL}\sum_{n=-\infty}^{\infty}\frac{\cos(k_n x)}{k - k_n} = \frac{c}{jL}\sum_{n=-\infty}^{\infty}\frac{\cos(k_n x)}{\omega - \omega_n - j\delta_n}$$

(In the latter expression the angular wavenumbers have been replaced with angular frequencies $\omega = ck$.) Hence eq. (9.17) can be written in the form

$$p_\omega(x) = \frac{v_0 Z_0 c}{jL}\sum_{n=-\infty}^{\infty}\frac{\cos(k_n x)}{\omega - \omega_n - j\delta_n}e^{j\omega t} \tag{9.22}$$

The cosine in the nominator of each term represents the normal mode associated with the eigenfrequency ω_n. Obviously, the contribution of a particular term to the total pressure $p_\omega(x)$ is the larger the closer the driving frequency of the piston in Figure 9.6 is to ω_n.

Since $|R(\omega)|$ is an even and $\chi(\omega)$ is an odd function (see eq. (6.9) we learn from eqs. (9.19) and (9.21) that

$$\omega_{-n} = -\omega_n \quad \text{and} \quad \delta_{-n} = \delta_n \tag{9.23a}$$

furthermore,

$$k_{-n} = -k_n^* \quad \text{and} \quad \cos(k_{-n}x) = \left[\cos(k_n x)\right]^* \tag{9.23b}$$

Hence the terms with subscripts $\pm n$ in eq. (9.22) may be combined. Then we obtain after some simplifications justified if $\delta_n \ll \omega_n$:

$$p_\omega(x) = \frac{2v_0 Z_0 c\omega}{jL} \sum_{n=0}^{\infty} \frac{\cos(k_n x)}{\omega^2 - \omega_n^2 - j2\omega\delta_n} e^{j\omega t} \tag{9.24}$$

The relevant frequency dependence in these terms is that of the denominator. Accordingly, each term of this sum represents a resonance curve. Its half-width (see Section 2.5) expressed by the angular frequency is $2(\Delta\omega)_n = 2\delta_n$ or, alternatively,

$$2(\Delta f)_n = \frac{\delta_n}{\pi} \tag{9.25}$$

The discussions in this section were related in a particular way to sound production in the pipe, namely, by an oscillating piston at one of its ends. Instead the wave could be generated as well by a point source immediately in front of a rigid termination. If the sound source is situated at any position x_0, an additional factor $\cos(k_n x_0)$ appears in the nominator of each sum term. In this case not all normal modes will be excited at equal strength.

The representation of the sound field by eq. (9.24) could suggest the idea that no sound propagation whatsoever occurs in the considered cavity. This, however, is not so. For if the magnitude of the reflection factor is smaller than unity the right termination of the pipe dissipates energy which must be supplied by the sound source. In fact eq. (9.14) can be transformed into

$$p(x) = 2\hat{p}R \cos k(L - x)e^{j\omega t} + \hat{p}(1 - R)e^{j[\omega t + k(L-x)]}$$

The first term of this formula represents a standing wave without any energy transport while the wave represented by the second one is purely progressive and hence continuously transfers energy from the source towards the lossy termination. In eq. (9.24) this fact finds its expression in the complex nature of $\cos(k_n x)$. A corresponding statement holds for the sound fields in three-dimensional cavities to be discussed in the next section.

9.5   Forced vibrations in enclosures of any shape

The detailed treatment of forced vibrations in a closed pipe is justified by the fact that many of the results can be transferred to three-dimensional cavities of any shape. As before we consider harmonic vibrations with the angular frequency ω. Then the wave equation (3.25) is converted into the so-called Helmholtz equation (with $k = \omega/c$):

$$\Delta p + k^2 p = 0 \tag{9.26}$$

We are looking for such solutions of this equation which are adapted to the shape of the cavity and the acoustical properties of its boundary.

Suppose the latter are expressed by the wall impedance of the boundary defined by

$$Z = \left(\frac{p}{v_n}\right)_{\text{boundary}} \tag{9.27}$$

(see eq. (6.10)). The normal component v_n of the particle velocity is related to the sound pressure, according to eq. (3.22), by

$$v_n = -\frac{1}{j\omega\rho_0}\frac{\partial p}{\partial n} \tag{9.28}$$

The derivative $\partial p/\partial n$ is the normal component of grad p as in eq. (7.6). Then the boundary condition can be written in the form

$$Z\frac{\partial p}{\partial n} + j\omega\rho_0 p = 0 \tag{9.29}$$

It can be shown that solutions of eq. (9.26) which satisfy the boundary condition (9.29) exist for certain discrete values of $\underline{k_n}$ only. These values are complex in general and are called the eigenvalues or characteristic values of the enclosure. Their real parts are related to the characteristic angular frequencies $\omega_n = c\,\mathrm{Re}\{\underline{k_n}\}$, and the eigenfrequencies are $f_n = \omega_n/2\pi$. The solutions associated with these eigenvalues are called eigenfunctions, and they are the mathematical expressions of the normal modes $p_n(r)$ of the cavity. The symbol r characterises the position of a point expressed by three suitably selected spatial coordinates. Accordingly, n stands for three integers, for instance, for l, m and n as in Section 9.2. Again, the normal modes can be thought of as three-dimensional standing waves. However, their nodal surfaces are generally not plane, in contrast to those of a rectangular room.

A closed solution of the boundary problem as outlined here can only be worked out for simple room shapes and simple distributions of the wall impedance. One of these cases is the rectangular room with rigid walls which was treated in Section 9.2 in a somewhat different way. For enclosures with more general geometry and boundary conditions the normal modes and eigenfrequencies must be calculated numerically, for instance, with the methods of finite elements (FEM) or boundary elements (BEM), methods which will not be described here.

In any case the forced sound field can be imagined as being composed of normal modes – similar as in eq. (9.24).[3] If the room is excited by a point source with the volume velocity $\hat{Q}_0 \exp(j\omega t)$, the sound pressure in a point situated at $\mathbf{r}$ is:

$$p_\omega(\mathbf{r}) = \hat{Q}_0 \sum_{n=0}^{\infty} \frac{\omega C_n p_n(\mathbf{r})}{\omega^2 - \omega_n^2 - j2\omega\delta_n} e^{j\omega t} \tag{9.30}$$

The coefficients C_n depend on the position of the sound source. Again, it is assumed that

$$\delta_n \ll \omega_n \tag{9.30a}$$

This expression corresponds completely to eq. (9.24). However, the angular eigenfrequencies ω_n or eigenfrequencies f_n are now not regularly arranged along the frequency axis as in the one-dimensional case (see, for instance, Table 9.1). Likewise, the constants δ_n and hence the half-widths $2(\Delta f)_n$ from eq. (9.25) may assume quite different values.

Now one has to distinguish two limiting cases:

1   The spacings between adjacent eigenfrequencies along the frequency axis are significantly larger than the half-widths of the resonances. Then the function $p_\omega(\mathbf{r})$ describes a succession of clearly separated resonance curves. This case is shown in Figure 9.8a; each peak of this curve corresponds to one eigenfrequency. In the vicinity of the resonance frequency ω_n the nth term of the sum yields by far the predominant contribution to $p_\omega(\mathbf{r})$, and therefore the normal modes can be excited and observed virtually independently from each other by choosing the driving frequency equal to the corresponding eigenfrequency.

2   The eigenfrequencies are so close to each other that several or many of them are located within the half-widths of a resonance, hence the resonance curves show strong overlap (see Fig. 9.8b). Then at any driving frequency several or even many terms of the sum in eq. (9.30) have significant values and contribute with quite different phases to the total sound pressure $p_\omega(\mathbf{r})$. Figure 9.9 illustrates this case. Here each phasor represents one term of eq. (9.30) in the complex plane; its length is proportional to its magnitude, and its direction corresponds to its phase. The diagram holds for one particular frequency; if the frequency – or the point of observation – is changed, its general character would be the

---

3 A rigorous derivation of eq. (9.29) can be found, for instance, in P. M. Morse and U. Ingard, *Theoretical Acoustics*, Ch. 9.4. McGraw-Hill, New York 1968.

(a)   (b)

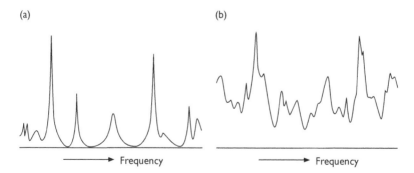

Frequency   Frequency

Figure 9.8 Frequency dependence of sound pressure amplitude after eq. (9.30): (a) overlap of resonances negligible, (b) heavy overlap.

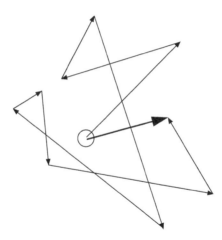

Figure 9.9 Phasor diagram showing the contributions of various room modes at a particular driving frequency, case 2. Solid phasor: resulting pressure amplitude.

same, but its details would be completely different, and the same holds, of course, for the resulting phasor. A maximum of the magnitude of p_ω occurs when many terms of the sum eq. (9.30) happen to contribute with similar phases which in Figure 9.9 corresponds to many phasors pointing in about the same direction. On the contrary, a minimum of the sound pressure amplitude comes about when the contributing terms cancel each other more or less leading to a short resulting arrow in Figure 9.9. In this case of strong modal overlap, the normal modes cannot be separately excited.

Now we know from eq. (9.10) that the average density of eigenfrequencies dN_f/df increases with the square of the frequency. Therefore one should expect that case 1 occurs at very low frequencies, while at sufficiently high frequencies we have case 2. To find a limiting frequency which separates both cases we will speak of significant modal overlap when, on the average, at least three eigenfrequencies fall into the frequency interval $\langle 2\Delta f \rangle$, the mean half-width of the resonances. This is true if $dN_f/df \geq 3/\langle 2\Delta f \rangle$, or, with eqs. (9.10) and (9.25)

$$f \geq \sqrt{\frac{3c^3}{4V\langle\delta\rangle}} \tag{9.31}$$

$\langle\delta\rangle$ is the mean of the constants δ_n.

The sum in eq. (9.30) is the transfer function of the enclosure cavity if the latter is conceived as a linear system in the sense of Section 2.10:

$$\underline{G}(\omega) = \sum_{n=0}^{\infty} \frac{\omega C_n p_n(\mathbf{r})}{\omega^2 - \omega_n^2 - j2\omega\delta_n} \tag{9.32}$$

In Section 9.7 some general properties of this transfer function will be presented.

9.6   Free vibrations

In room acoustics one is concerned with signals which are variable in time. For this reason we shall discuss now the transient behaviour of an enclosure, that is, its response to variable excitations. The prototype of a transient excitation signal is a very short impulse represented by a Dirac function $\delta(t)$, and the output signal of the system, that is, the sound pressure received at some observation point is its impulse response. Formally, it can be calculated – according to Section 2.9 – by performing a Fourier transformation on the transfer function given by eq. (9.32). The result has the form

$$g(t) = \sum_{n=0}^{\infty} B_n \cdot \cos(\omega_n t + \varphi_n) \cdot e^{-\delta_n t} \qquad \text{for } t \geq 0 \tag{9.33}$$

which can be verified by back transformation into the frequency domain. It is composed of many damped harmonic vibrations with different frequencies and phases, and the constants δ_n as introduced by eq. (9.20) turn out to be the decay constants of these components.

As an example of such an 'impulse response' Figure 9.10 shows the super-position of three sum terms with different frequencies, amplitude factors B_n, phase angles and decay constants. The partial vibrations with the largest

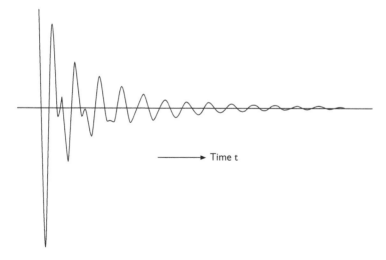

Figure 9.10 Impulse response of a room consisting of three decaying modes.

decay constant vanish first from the mixture of vibrations, and the tail of the response consists mainly of the component with the smallest decay constant.

Matters become more transparent if not the decay of the sound pressure is considered but that of the energy density which is proportional to the square of the pressure. First, we find by squaring eq. (9.33):

$$[g(t)]^2 = \sum_{n=0}^{\infty} \sum_{m=0}^{\infty} B_n B_m \cdot \cos(\omega_n t + \varphi_n)$$

$$\times \cos(\omega_m t + \varphi_m) \cdot e^{-(\delta_n + \delta_m)t} \quad \text{for } t \geq 0 \qquad (9.34)$$

Next some short-time averaging of this expression is carried out with an averaging time which is significantly longer than the periods of the cosine functions but noticeably smaller than $1/\delta_n$ or $1/\delta_m$ which is possible because of the condition (9.30a). The product of the cosine functions can be written as

$$\tfrac{1}{2} \cdot \cos\left[(\omega_n + \omega_m)t + \varphi_n + \varphi_m\right] + \tfrac{1}{2}\cos\left[(\omega_n - \omega_m)t + \varphi_n - \varphi_m\right]$$

For $m \neq n$ each of these term represents a rapidly varying function with the mean value 0, hence it vanishes by the averaging process. The only exception is the second term for m $=$ n which becomes $\frac{1}{2}$. Hence the short-time averaging of eq. (9.34) results in a much simpler expression:

$$\overline{[g(t)]^2} = \frac{1}{2} \sum_{n=0}^{\infty} B_n^2 \cdot e^{-2\delta_n t} \quad \text{for } t \geq 0 \tag{9.35}$$

If the decay constants are not too different they can be replaced by their average $\langle\delta\rangle$. Then the energy density in the decaying sound field which is proportional to the squared sound pressure becomes

$$w(t) = w_0 e^{-2\langle\delta\rangle t} \quad \text{for } t \geq 0 \tag{9.36}$$

In room acoustics such decay processes play an important role. They are named reverberation, and quite often they follow more or less an exponential law as shown in eq. (9.36). Usually, the duration of the decay is not characterised by $1/\langle\delta\rangle$ but by the so-called reverberation time or decay time. This is the time in which the energy density drops to one millionth of its initial value (see Fig. 9.11). From the equation

$$10^{-6} = e^{-2\langle\delta\rangle T}$$

one obtains

$$T = \frac{3 \cdot \ln 10}{\langle\delta\rangle} \approx \frac{6.9}{\langle\delta\rangle} \tag{9.37}$$

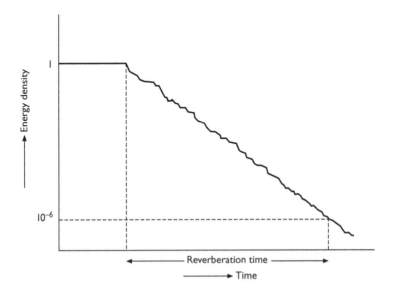

Figure 9.11 Definition of the reverberation time.

With this relation and the numerical value of the sound velocity of air the important condition eq. (9.31) reads

$$f > f_s = 2000\sqrt{\frac{T}{V}} \tag{9.38}$$

It is often called the 'condition of large rooms' and f_s is known as 'Schroeder frequency'.

9.7   Statistical properties of the transfer function

For somewhat larger rooms the condition eq. (9.38) is fulfilled within the whole frequency range of interest since f_s is significantly below 100 Hz. Hence the limiting case 2 in Section 9.5 can be considered as typical for most auditoria, theatres, lecture rooms, etc. Then eq. (9.32) may contain hundreds of significant terms at a given exciting frequency which ought to be taken into account for a correct calculation of the sound field. From a practical standpoint this is just impossible; moreover, the exact knowledge of the transfer function of a room is of little practical use. Instead, we shall restrict the discussion to certain general statistical properties of such transfer functions, following the ideas of M. R. Schroeder.

At first we consider the range in which the absolute value of the transfer function $\underline{G}(\omega)$ varies. Suppose the room is excited with a sine signal the frequency of which is varied slowly enough to ensure steady-state conditions. Then the sound pressure level recorded simultaneously in some observation yields what is called a 'frequency response curve'. A typical example of such a curve is shown in Figure 9.12. Its details, but not its general appearance, depend on the enclosure and on the position of the sound source and the receiver. It is the logarithmic representation of the absolute value of the room transfer function $\underline{G}(\omega)$.

As was pointed out in Section 9.5 the real part G_1 as well as the imaginary part G_2 of a room transfer function is composed of a large number of components which can be considered as virtually independent from each other. Under these circumstances the central limit theorem of probability theory can be applied according to which G_1 and G_2 obey a normal distribution (Gauß distribution). Then the absolute value $|\underline{G}| = (G_1^2 + G_2^2)^{1/2}$ is also a random variable, however, following the Rayleigh distribution. If we denote with z the absolute value of the sound pressure divided by its average then the probability to encounter at some frequency (or at some room point) a value between z and z + dz is:

$$W(z)dz = \frac{\pi}{2}e^{-\pi z^2/4}zdz \quad \text{with } z = \frac{|p_\omega|}{\overline{|p_\omega|}} \tag{9.39}$$

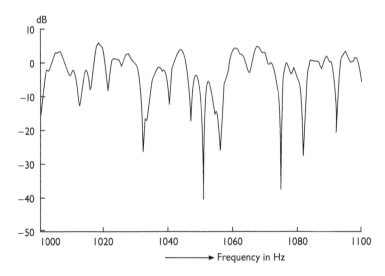

Figure 9.12 Typical frequency response curve of a room (section).

This distribution is shown in Figure 9.13. Its mean standard deviation is

$$\sigma_z = \sqrt{\langle z^2 \rangle - \langle z \rangle^2} = \sqrt{\frac{4}{\pi} - 1} = 0.523$$

About 67% of all values are lying within the range from $1 - \sigma_z$ to $1 + \sigma_z$ (see dotted lines). This is tantamount to the statement that about 67% of the ordinate values of a frequency curve are contained in a band of width

$$20 \cdot \log_{10}\left(\frac{1 + 0.523}{1 - 0.523}\right) \approx 10\,\text{dB}$$

Furthermore, certain statements can be made on the succession of maxima along a frequency curve. Thus the mean distance of two adjacent maxima (or minima) is:

$$(\Delta f)_{max} \approx \frac{4}{T} \tag{9.40}$$

where T is the reverberation time introduced in the preceding section.

According to the distribution (9.39) there is no upper limit to levels occurring in a frequency curve, although the probability of encountering very high values is very small. Now not all of the values of a frequency

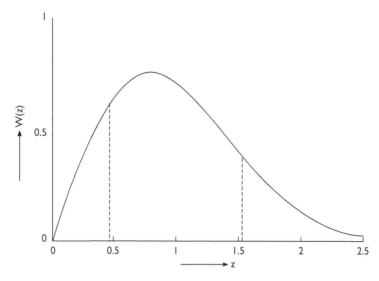

Figure 9.13 Rayleigh distribution $\left(z = |p_\omega| \big/ \overline{|p_\omega|} \right)$.

curve are independent of each other; in fact, the resonance denominator in eq. (9.30) or (9.32) is a strong link which connects the function values at neighbouring frequencies to each other. Hence a finite section of a frequency curve can be represented by a finite number of samples taken at equidistant frequencies. Under these circumstances there is an absolute level maximum occurring in the considered section, characterised by maximum probability of its occurrence. Its height over the quadratic mean value of the frequency curve is

$$\Delta L_{max} = 4.34 \ln \left[\ln(BT) \right] \; dB \tag{9.41}$$

with B denoting the bandwidth of the considered section in Hz. According to this formulae the frequency curve of a room with a reverberation time of 2 seconds shows each 2 Hz a maximum on average. Furthermore, its absolute maximum in a frequency range of 10 000 Hz exceeds the average value by nearly 10 dB. This value is of significance for the performance of sound systems in closed rooms.

It is remarkable that the statistical properties of frequency curves as out-lined earlier are the same for all sorts of rooms, that is, that they do not reflect individual peculiarities of a room – apart from its reverberation time.

This seems to contradict our earlier statement that the transfer function of a linear system contains all its properties. In fact, this is true for a room as well, of course. However, those acoustical properties of a room which are significant for what we hear in it do not show up in obvious features of its frequency curve, but rather in its impulse response as we shall see in Chapter 13.

Chapter 10

Sound waves in isotropic solids

It seems that sound in solids play only a minor role compared to that of sound in fluids and, in particular, in air. This impression may arise because we cannot hear sound waves in solids. But it is wrong: a good deal of the noise from which we suffer in our everyday life is due to sound waves which are primarily excited in solid bodies, for instance, in machinery components, and are radiated afterwards into the environment from some covering, etc. In this context we often speak of 'solid-borne sound' or 'structure-borne sound'. This type of sound is also encountered in buildings; it propagates in walls and ceilings and is responsible for our experience that we are not completely isolated against sound intruding from outside or from noise sources within the building.

Less common in everyday life, but no less important, is the role which solid-borne sound plays in ultrasound technology. Here non-destructive material testing by ultrasound should be mentioned in the first place which will be described in more detail in Chapter 16.

10.1 Sound waves in unbounded solids

At first we consider an isotropic solid of uniform composition which is unbounded in all directions. As explained in Section 3.1 the relevant variables are the elastic stresses (see eqs. (3.3) and (3.4)) and the Cartesian components ξ, η and ζ of the particle displacement. The latter obey three wave equations (3.27) which are coupled to each other. That each of them contains all of the three components makes wave propagation in the solid considerably more complicated than that in a fluid.

However, we can get some idea on possible wave types by restricting the discussion to plane waves which propagate, say, in the x-direction. Then all partial derivatives of the displacement components with respect to y and z become zero in these wave equations. What remains from the Laplace operators on the left is just a second order differentiation with respect to x; likewise div($\vec{s}$) reduces to $\partial\xi/\partial x$. Hence the second term of eq. (3.27a) becomes $\partial^2\xi/\partial x^2$ while the second terms of eqs. (3.27b and c) are zero. In this way a set of mutually independent wave equations for the three components of the displacement vector are obtained:

$$(2\mu + \lambda)\frac{\partial^2 \xi}{\partial x^2} = \rho_0 \frac{\partial^2 \xi}{\partial t^2} \tag{10.1a}$$

$$\mu \frac{\partial^2 \eta}{\partial x^2} = \rho_0 \frac{\partial^2 \eta}{\partial t^2} \tag{10.1b}$$

$$\mu \frac{\partial^2 \zeta}{\partial x^2} = \rho_0 \frac{\partial^2 \zeta}{\partial t^2} \tag{10.1c}$$

The first one refers to a wave with particle vibration in the direction of sound propagation – as in sound waves in gases and liquids. This is a longitudinal wave in which the only non-zero stress component is σ_{xx} because of eq. (3.18). In contrast, in the waves described by eqs. (10.1b) and (10.1c) the medium particles move perpendicularly to the direction of propagation. These waves are called transverse, and the medium undergoes only shear deformations which would be impossible in a non-viscous fluid. So we can state that three independent wave types can exist in an isotropic solid, namely, one longitudinal wave and two transverse waves with particle vibrations perpendicular to each other. Which of these waves are actually present in a particular situation and which is the ratio of their amplitudes depends on the method of their excitation.

By comparing the earlier equations with eq. (3.21) it is obvious that the velocity of the longitudinal wave is given by

$$c_L = \sqrt{\frac{2\mu + \lambda}{\rho_0}} \tag{10.2}$$

The transverse waves travel with a smaller speed:

$$c_T = \sqrt{\frac{\mu}{\rho_0}} \tag{10.3}$$

In Table 10.1 the wave velocities of both wave types are listed for a number of materials.

Figure 10.1a and b shows the deformations of a medium caused by a longitudinal and a transverse wave in the form of a lattice consisting of cells which are squares (or rather cubes) when the medium is at rest. Under the influence of a longitudinal wave the volume elements are stretched or compressed in the direction of propagation. These deformations alter, of course, the density of the medium. Therefore the longitudinal wave is also called a compressional wave or density wave. In contrast, a transverse wave leaves the volume of an elementary cell unaltered; what is changed is just its

Table 10.1 Sound velocity of solids

Material	Density (kg/m^3)	Sound velocity (m/s)		Characteristic impedance (longitudinal) (Ns/m^3)
		Longitudinal	Transverse	
Metals				
Aluminium (rolled)	2700	6420	3040	17.3
Lead (rolled)	11 400	2160	700	24.6
Gold	19 700	3240	1200	63.8
Silver	10 400	3640	1610	37.9
Copper (rolled)	8930	5010	2270	44.7
Copper (annealed)	8930	4760	2325	42.5
Magnesium	1740	5770	3050	10.0
Brass (70% Cu, 30% Zn)	8600	4700	2110	40.4
Steel (stainless)	7900	5790	3100	45.7
Steel (1% C)	7840	5940	3220	46.6
Zinc (rolled)	7100	4210	2440	29.9
Tin (rolled)	7300	3320	1670	24.2
Nonmetals				
Glass (Flint)	3600	4260	2552	15.3
Glass (Crown)	2500	5660	3391	14.2
Quartz, fused	2200	5968	3764	13.1
Plexiglas	1180	2680	1100	3.16
Polyethylene	900	1950	540	1.76
Polystyrene	1060	2350	1120	2.49

shape: the elementary cells undergo a shear deformation. Therefore the transverse wave is also known as a shear wave.

Of course, the particle vibration in a transverse wave is not necessarily parallel to the y- or the z-axis; by linear combination of both displacement components one can arrive at an infinity of possibilities. Let the displacement be described by

$$\eta(x,t) = \hat{\eta}\cos(\omega t - k_T x - \varphi_1) \quad \text{and} \quad \zeta(x,t) = \hat{\zeta}\cos(\omega t - k_T x - \varphi_2) \tag{10.4}$$

with $k_T = \omega/c_T$. Then the particle motion is along a line which subtends the angle $\varepsilon = \arctan(\hat{\zeta}/\hat{\eta})$ with the y-direction if $\varphi_1 = \varphi_2$. In all these cases we speak of 'linearly polarised waves'. If, on the other hand, both phase angles are different the a particle moves on an elliptic orbit around its resting position (elliptic polarisation); their angular velocity is ω. A special case of elliptically polarised waves occurs when the amplitude waves of both displacement components are equal ($\hat{\zeta} = \hat{\eta}$) and when $\varphi_2 - \varphi_1 = \pm\pi/2$. Then the orbit of the particles becomes a circle because $\zeta^2 + \eta^2 = \text{const.}$ along which the particle travels either clockwise or counterclockwise, depending on the sign

(a)

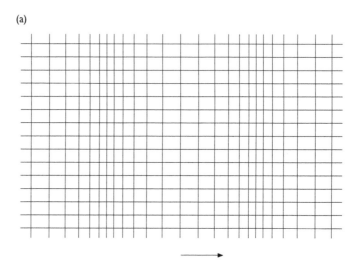

(b)

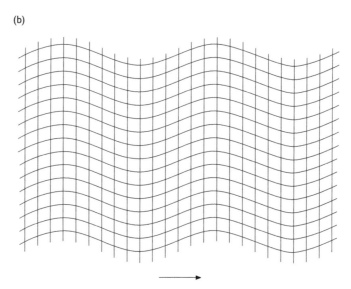

Figure 10.1 Plane waves in an isotropic solid: (a) longitudinal wave, (b) transverse wave.

of phase difference. This is the case of right or left circular polarisation. Figure 10.2 presents different sorts of polarisation of transverse waves.

We come once more back to the deformation pattern of a transverse wave shown in Figure 10.1b. Let $z = 0$ denote the plane of the paper which also contains the direction of motion. According to eqs. (3.18) and (3.19) σ_{xy} is the only non-vanishing elastic stress component, that is, there are no forces perpendicular to the plane of the paper. Hence the illustrated wave field will not be influenced by free surfaces parallel to the plane z = 0. We conclude from this fact that plane transverse waves can propagate in plates of any thickness and that they travel with the same velocity as in the unbounded body.

Furthermore, purely transverse waves can propagate in rods with circular cross section or in circular tubes. Here adjacent cross sections are rotated with respect to each other. They are also known as torsional waves. Figure 10.3 represents a torsional wave on a cylindrical rod. Obviously, there is no radial or axial displacement and the wave velocity is again given by eq. (10.3).

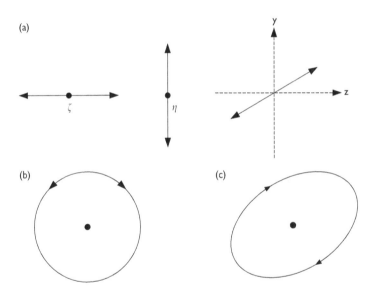

Figure 10.2 Polarisation of transverse waves: (a) linear polarisation, (b) circular polarisation, clockwise or counterclockwise and (c) elliptical polarisation.

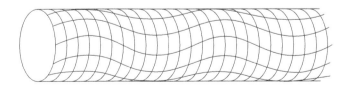

Figure 10.3 Torsional wave on a cylindrical rod.

In the course of time, numerous precision methods for measuring the speed of the various types of sound waves have been developed. They can be used to determine the Lamé constants λ and μ as well as other elastic constants related to them (see Subsection 10.3.1) even from small samples. Likewise, the elastic constants of anisotropic solids such as crystals can be exactly determined in this way.

10.2   Reflection and refraction, Rayleigh wave

Now we consider two different solids which are fixed to each other as depicted in Figure 10.4. Suppose a plane sound wave is impinging on the boundary. If both media were fluids the only conditions to be fulfilled at the boundary were just that of equal sound pressures and of equal normal displacements at both sides (see Section 6.3). However, in the present case a requirement is that not only all components of the particle velocity are continuous at the plane x = 0, but also the normal stress σ_{xx} and the shear stress σ_{xy} (σ_{xz} is anyway zero).

To meet all these conditions a wave field of increased complexity is required. Therefore the arriving sound wave will produce not just one reflected and one refracted wave in general, but two of each sort, namely, longitudinal and transverse, the latter with particles moving parallel to the plane of the paper. The only exception occurs when the primary sound wave arrives perpendicularly at the boundary. When the arriving sound wave is longitudinal, the reflection and refraction angles of the secondary waves are related by a generalised Snell's law (compare eq. (6.1)):

$$\frac{c_{1L}}{\sin \vartheta_{1L}} = \frac{c_{1L}}{\sin \vartheta'_{1L}} = \frac{c_{1T}}{\sin \vartheta'_{1T}} = \frac{c_{2L}}{\sin \vartheta_{2L}} = \frac{c_{2T}}{\sin \vartheta_{2T}} \qquad (10.5)$$

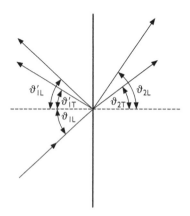

Figure 10.4 Reflection and refraction at the interface between two solids.

The dashed symbol refers to the reflected waves; c_{iL} and c_{iT} ($i = 1$ or 2) are the velocities of the longitudinal and the transverse waves in both materials. If the incident wave is transverse with particle motion parallel to the plane of the paper then the first fraction in eq. (10.5) has to be replaced with $c_{1T}/\sin\vartheta_{1T}$. When the primary wave is transverse, however, with the particles vibrating perpendicular to that plane no longitudinal waves are formed at the interface and the second and the fourth term are missing. Apart from this latter case, there will always occur some wave type conversion at a boundary.

It may happen that one or other of the partial equations in eq. (10.5) cannot be satisfied since the absolute value of the sine function cannot exceed unity. Then one of the secondary waves (reflected or refracted) will vanish. If, for instance, $c_{2L} > c_{1L}$, then the fourth term in eq. (10.5) must vanish for all angles of incidence $\vartheta_{1L} > \arcsin(c_{1L}/c_{2L})$; there will be no refracted longitudinal wave. If, additionally, $c_{2T} > c_{1L}$, then for $\vartheta_{1L} > \arcsin(c_{1L}/c_{2T})$ there will be no refracted wave whatsoever. Then the incident wave will be totally reflected (see Section 6.1).

If one of both media is fluid, that is, a liquid or a gas, then no transverse wave can exist in it. This corresponds to a reduced number of boundary conditions since the shear stress σ_{xy} is zero at the boundary; likewise, the displacement components parallel to the boundary must not have the same value in both materials. If the solid body fills the half space only, that is, if it has a free surface at $x = 0$, then all stress components including σ_{xx} are zero along the boundary. An incident longitudinal or transverse wave with particles vibrating in the x-y-plane will give rise to two reflected waves in general, namely, a longitudinal and a transverse one. Again, the reflection angles can be found from eq. (10.5). This does not hold, however, if the incident wave is transverse with an angle of incidence $\vartheta_{1T} > \arcsin(c_{1T}/c_{1L})$; in this case only a transverse wave will be reflected.

Along a free solid surface another wave type can propagate, the surface or Rayleigh wave. It is similar to the surface waves on a water surface with the difference that the restoring force is not due to gravitation or to the surface tension but to the elasticity of the solid. The Rayleigh wave can be conceived as a particular combination of longitudinal and transverse wave components (see Fig. 10.5); particles close to the surface move on elliptical orbits. The most important feature is, however, that the wave motion is restricted to a region next to the surface; with increasing depth the displacement diminishes exponentially. At the depth of two Rayleigh wavelengths the displacement is nearly zero.

The velocity c_R of the Rayleigh wave is a little smaller than that of the transverse wave. It cannot be represented in a closed formula. Figure 10.6 shows it as a function of the Poisson's ratio ν (see eq. (10.7)).

The Rayleigh wave plays a particularly important role in seismics since among all waves generated in an earthquake the Rayleigh wave has the

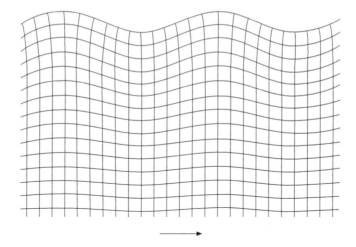

Figure 10.5 Surface or Rayleigh wave.

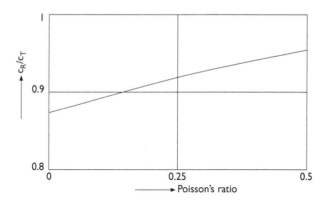

Figure 10.6 Velocity of Rayleigh waves as a function of Poisson's ratio.

largest amplitudes and hence is primarily responsible for the destructions. Furthermore, it has important technical applications in signal processing: high-frequency Rayleigh waves on piezoelectric substrates can be produced and received with highly frequency-selective transducer structures. In this way very small electrical filters, delay lines and other components with prescribed properties can be realised.

10.3   Waves in plates and bars

The volume waves discussed in Section 10.1, that is, the compressional wave, and the shear wave, are of particular interest in ultrasound technology. At ultrasonic frequencies the wavelengths are often in the range of millimetres, therefore even a moderately sized body, for instance, a piece of material or a machine component, can be considered as practically unbounded. On the contrary, in the range of audible sound, wave propagation in bars or plates is of more practical interest. Examples are sound waves in walls and ceilings of buildings, or in machinery components and also in certain musical instruments. One kind of such waves have already been mentioned, namely, pure shear or transverse waves. Further wave types will now be described in the rest of this chapter.

If not stated otherwise we assume that the plates and rods in question consist of some isotropic material, that their extension is infinite and that no exterior forces are acting on them. This condition means in particular that all tensile or shear stresses directed perpendicular to the surface are zero.

10.3.1   Extension and bending

As a preparation for the following discussions this section treats a few facts of elasticity which, although elementary, may not be familiar to every reader.

If a bar is stretched by a tensile force F, then its length l will increase by a certain amount δl (see Fig. 10.7a). Within certain limits the relative change of length is proportional to the force per unit area (Hooke's law):

$$\frac{\delta l}{l} = \frac{1}{Y} \cdot \frac{F}{S} \tag{10.6}$$

where S is the cross-sectional area of the rod; Y is a material constant and is called Young's modulus. Along with extension, the bar undergoes a reduction of all its lateral dimensions, that is, it becomes a little thinner with the relative thickness being a certain fraction of the relative increase in length. This change of the lateral dimensions is called lateral contraction. For a cylindrical bar with the radius a, for instance, we have:

$$\frac{\delta a}{a} = -\nu \frac{\delta l}{l} \tag{10.7}$$

The constant ν is named Poisson's ratio, it depends on the kind of material and lies in the range from 0 to 0.5.

Let the axis of the rod coincide with the x-axis of a rectangular coordinate system; accordingly, the axial tensile stress is denoted by σ_{xx} (see Section 3.1). Then eq. (10.6) is equivalent to

$$\sigma_{xx} = Y \frac{\partial \xi}{\partial x} \tag{10.8}$$

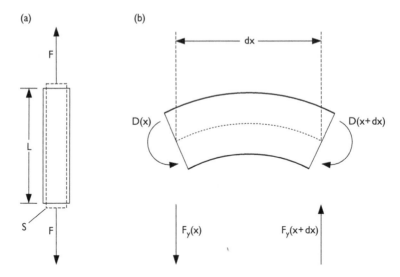

Figure 10.7 Elastic deformations: (a) extension and lateral contraction, (b) bending.

The derivative on the right is the so-called strain, that is, the differential expression for the relative change in length of the rod.

Another elementary deformation of a bar or a plate is bending as depicted in Figure 10.7b. The material layer in the middle remains unaltered when the bar or plate is bent; it is the so-called 'neutral fiber'. Underneath that the material is compressed while it is stretched in the upper half or vice versa. The axial stresses combine in a moment D acting on both cross sections with distance dx. This moment depends on the degree of bending, or more precisely, it is proportional to the radius of curvature which in turn approximately equals the second derivative of the displacement η with regard to x:

$$D = -B\frac{\partial^2\eta}{\partial x^2} \tag{10.9}$$

The proportionality factor B is the bending stiffness of the bar or the plate, and it depends on its dimensions as well as on the elastic properties of the material. In the first place we are interested in the bending of plates, hence it is useful to refer both the moment and the bending stiffness to unit width of the plate. Then the bending stiffness of the plate is given by:

$$B = \frac{d^3}{12} \cdot \frac{Y}{1 - v^2} \tag{10.10}$$

with d denoting its thickness.

When the degree of bending varies with x the same is true of the moment D. Hence the moment $D(x + dx)$ at the right cross section in Figure 10.7b may differ from that at the left side right, $D(x)$, the difference

$$D(x + dx) - D(x) = \frac{\partial D}{\partial x} dx$$

must be kept in equilibrium by a pair of forces consisting of two lateral forces $\pm F_y$ at distance dx:

$$-F_y dx = \frac{\partial D}{\partial x} dx \tag{10.11}$$

When the lateral force F_y is also a function of x, each length element is associated with a difference force

$$dF_y = F_y(x + dx) - F_y(x) = \frac{\partial F_y}{\partial x} dx$$

Combining this equation with eqs. (10.11) and (10.9) gives the result:

$$dF_y = B \frac{\partial^4 \eta}{\partial x^4} dx \tag{10.12}$$

which must be somehow balanced, for instance, by exterior forces (which we excluded) or by inertial forces as will be detailed in Subsection 10.3.3.

The elasticity constants Y and v are related to the Lamé constants which were introduced already in Section 3.3. These relations are:

$$\mu = \frac{Y}{2(1 + v)} \quad \text{and} \quad \lambda = \frac{vY}{(1 + v)(1 - 2v)} \tag{10.13}$$

The constant μ is identical with the shear modulus or torsion modulus G often used in technical elasticity. Inserting these relations into eqs. (10.2) and (10.3) shows that the ratio of c_L und c_T depends only on Poisson's ratio:

$$\left(\frac{c_L}{c_T}\right)^2 = 2\frac{1 - v}{1 - 2v} \tag{10.14}$$

Table 10.2 lists Young's modulus and Poisson's ratio of a few materials.

10.3.2 Extensional waves

The preceding subsection dealt with static or quasistatic elastic deformations of a straight bar or a plate. If, on the contrary, deformations take place at finite speed then not only the elasticity of the material determines what happens but its inertia becomes noticeable as well. To account for it we set

Table 10.2 Young's modulus and Poisson's ratio of solids

Material	Density (kg/m³)	Young's modulus (10^{10} N/m²)	Poisson's ratio
Aluminium	2700	6.765	0.36
Brass (70% Cu, 30% Zn)	8600	10.520	0.37
Steel	7900	19.725	0.30
Glass (Flint)	3600	5.739	0.22
Glass (Crown)	2500	7.060	0.22
Plexiglas	1180	0.3994	0.40
Polyethylene	900	0.0764	0.45

up a force balance similar to that of eq. (3.5). The result can be immediately taken on by replacing the sound pressure p with the (negative) tensile stress σ_{xx}. Furthermore; we carry out the same linearisations as in Section 3.2; in particular, we replace the total acceleration by the local one and the total density ρ_t with its average value ρ_0. Then we arrive at:

$$\frac{\partial \sigma_{xx}}{\partial x} = \rho_0 \frac{\partial v_x}{\partial t} = \rho_0 \frac{\partial^2 \xi}{\partial t^2} \tag{10.15}$$

Combining this relation with eq. (10.8) leads to the following wave equation:

$$\frac{\partial^2 \xi}{\partial x^2} = \frac{\rho_0}{Y} \frac{\partial^2 \xi}{\partial t^2} \tag{10.16}$$

By comparing this equation with earlier wave equations, for instance, with eq. (3.21), we see that the wave velocity of extensional waves on a bar is

$$c_{E1} = \sqrt{\frac{Y}{\rho_0}} \tag{10.17}$$

The general solution corresponds to eq. (4.2). In a similar way the propagation of extensional waves in plates with parallel boundaries is derived. Their wave velocity is found to be

$$c_{E2} = \sqrt{\frac{Y}{\rho_0(1 - v^2)}} \tag{10.18}$$

It is slightly higher than c_{E1} due to the fact that the elastic constraint in a bar is lesser than that in a plate where stress relief due to lateral contraction can only occur in one direction, namely, perpendicular to the plate surfaces. By employing eq. (10.13) it is easily verified that

$$c_L > c_{E2} > c_{E1} > c_T$$

(a)

(b)

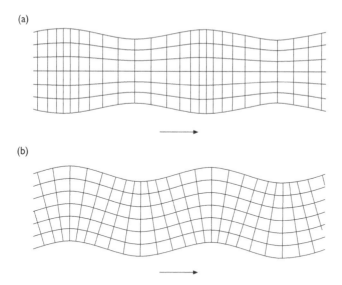

Figure 10.8 (a) Extensional wave (quasi-longitudinal wave), (b) bending wave.

Figure 10.8a depicts the deformations associated with an extensional wave travelling horizontally. Because of the lateral contraction the motion of material particles is not purely longitudinal but there are also displacement components perpendicular to the surface. The bar or plate is thickest where the longitudinal compression of its material is at maximum which agrees with our intuition. Hence, extensional waves are not purely longitudinal waves although the longitudinal displacement prevails. Accordingly, they are often referred to as quasi-longitudinal.

Equations (10.17) and (10.18) are only true as long as the thickness of the plate or the rod is small compared with the extensional wavelength. If this is not the case, the velocity of the extensional waves depends on the thickness of the bar or the plate and also on the frequency, that is, a wave will be subject to dispersion. Moreover, higher order wave types can occur similar to those described in Section 8.5 for gas-filled pipes. Generally, the variety of possible wave types in a solid 'waveguide' is considerably higher than that in a tube filled with a fluid.

10.3.3 Bending waves

The excitation of purely transverse or extensional waves in a plate requires particular precautions which guarantee that exactly the sort of vibrations is induced into the plate which corresponds to the desired wave type. The bending wave, however, is the plate wave per se; if one knocks with a hammer against a panel almost pure bending waves are generated.

The transition from the static bending deformations as described in Subsection 10.3.1 to bending waves requires the consideration of inertial forces. They must compensate for the transverse force dF_y in Subsection 10.3.1. By using the specific mass $m' = \rho_0 d$ of the plate as introduced in Subsection 6.6.4 the force balance reads:

$$m'dx\frac{\partial^2\eta}{\partial t^2} + \frac{\partial F_y}{\partial x}dx = 0 \tag{10.19}$$

or with the use of eq. (10.12):

$$\frac{\partial^4\eta}{\partial x^4} + \frac{m'}{B}\frac{\partial^2\eta}{\partial t^2} = 0 \tag{10.20}$$

Its extension in two dimensions reads:

$$\Delta\Delta\,\eta + \frac{m'}{B}\frac{\partial^2\eta}{\partial t^2} = 0 \tag{10.20a}$$

where $\Delta\Delta\eta$ means $\Delta(\Delta\eta)$. It will turn out that both eqs. (10.20) and (10.20a), have solutions with wave character although they are of fourth order with regard to the space variable(s), in contrast to the wave equations which we encountered so far. Consequently, we expect a greater variety of possible solutions and also of independent variables: while in extensional waves there are only two independent variables, a displacement component, say ξ (or its time derivative, the particle velocity v_x), and the stress σ_{xx}, there are four of them in a bending wave, namely, the displacement η perpendicular to the plate (or the corresponding particle velocity v_y), its spatial derivative and, furthermore, two force-related quantities, the bending moment D and the transverse force F_y. This large number of variables correspond to a greater variety of boundary conditions, which, however, we shall not discuss here in detail.

To keep the mathematics simple we look for a plane, harmonic bending wave propagating in the x-direction with unknown angular wavenumber k_B:

$$\eta(x, t) = \hat{\eta} \cdot e^{j(\omega t - k_B x)} \tag{10.21}$$

Before inserting this expression into eq. (10.20) we note that each time derivative is tantamount to multiplying the variable η with a factor $j\omega$ while each spatial differentiation corresponds to a factor $-jk_B$. This leads us immediately to $k_B^4 = \omega^2 m'/B$ or $k_B^2 = \pm\omega\sqrt{m'/B}$. For the upper sign we get:

$$(k_B)_{1,2} = \pm\sqrt{\omega} \cdot \sqrt[4]{\frac{m'}{B}} \tag{10.22}$$

while the lower one yields:

$$(k_B)_{3,4} = \pm j\sqrt{\omega} \cdot \sqrt[4]{\frac{m'}{B}} \tag{10.23}$$

The angular wavenumbers from eq. (10.22) correspond to waves travelling in the positive or negative x-direction. They are not proportional to the angular frequency ω; accordingly, the wave speed is frequency dependent:

$$c_B = \frac{\omega}{k_B} = \sqrt{\omega} \cdot \sqrt[4]{\frac{B}{m'}} \tag{10.24}$$

These are the characteristics of dispersion. Consequently, the shape of a wave will not be preserved in the course of propagation, or, in other words: the general solution is not of the type in eq. (4.2). Correctly, the wave speed in eq. (10.24) is the phase velocity while the group velocity of the bending wave is:

$$c_B' = \frac{d\omega}{dk_B} = 2\sqrt{\omega} \cdot \sqrt[4]{\frac{B}{m'}} = 2c_B \tag{10.25}$$

However, eqs. (10.22) to (10.25) are only valid for sufficiently low frequencies at which the bending wavelength is large in comparison with the thickness of the plate.

Figure 10.8b shows the deformation pattern of a bending wave. Just as the extensional wave is not purely longitudinal but contains transverse displacement components as well, the bending wave is not purely transverse, that is, it is associated with relatively small displacements parallel to the direction of propagation.

Now we consider the solutions $\eta(x,t)$ which belong to the imaginary wavenumbers $(k_B)_{3,4}$ in eq. (10.23). By inserting them into eq. (10.21) one obtains

$$\eta(x,t) = \hat{\eta} \cdot e^{\pm x\sqrt{\omega^2 m'/B}} \cdot e^{j\omega t}$$

which describes vibrations at equal phase everywhere while their amplitudes increase or decrease exponentially with the distance x. When considering the propagation of free bending waves in an infinite plate we can safely neglect such 'near field' solutions. They are needed, however, if there are edges at which certain boundary conditions (free, clamped, etc.) must be fulfilled which will not be discussed further. Likewise, in the vicinity of a source or of inhomogeneities they must be taken into account.

10.3.4   Sound radiation from a vibrating plate

As was mentioned at the beginning of Section 10.3 our treatment of extensional and bending waves was based on the assumption that the surface of the bar or plate is free of forces. Strictly speaking, this means that these solid bodies are not surrounded by a medium which may interact with the plate vibration. This assumption can be omitted without any harm if transverse waves are considered of the kind discussed in Section 10.1, since these waves are not associated with any displacements perpendicular to the surface. Likewise, the normal displacements occurring in extensional waves as discussed in Subsection 10.3.2 are so small that any significant interaction can be excluded, at least if the bar or plate is embedded in a gas. This is different with bending waves where the lateral displacement is the dominant one. Even if the influence of the surrounding gas on the propagation of bending waves is negligible, the lateral displacements of the plate lead to significant sound radiation into the adjacent medium, at least under certain circumstances.

To find these circumstances let us have a look at Figure 10.9a. It shows a plate carrying a bending wave; the medium around it is assumed to be air. Any sound wave emitted by the plate must be a plane wave. Furthermore, the principle of 'trace fitting' as explained in Section 6.1 applies also to this case, that is, the periodicity of the sound wave in air must agree with that of the bending wave at the plate surface. This agreement is also referred to as 'coincidence'. Let ϑ be the angle between the direction of radiation and the plate normal. Then we see immediately from the figure that, with $\lambda_B = 2\pi/k_B$ denoting the wavelength of the bending wave:

$$\sin \vartheta = \frac{\lambda}{\lambda_B} = \frac{c}{c_B} \tag{10.26}$$

The latter equation holds because both the radiated wave and the bending wave have the same frequency. This equation is only meaningful if the phase velocity c_B of the bending wave is greater than the sound velocity in air. Since, according to eq. (10.24), the former grows with the square root of the

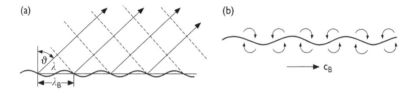

Figure 10.9 Reaction of the adjacent air to a bending wave travelling on a plate: (a) above the critical frequency: radiation of a sound wave, (b) below the critical frequency: local air flows.

frequency there must be critical frequency ω_c below which the plate cannot emit a sound wave. It is found by setting $c_B = c$ in eq. (10.24) and solving for the angular frequency:

$$\omega_c = c^2 \sqrt{\frac{m'}{B}} \tag{10.27}$$

or, after dividing this equation by 2π and inserting the bending stiffness from eq. (10.10):

$$f_c = \frac{c^2}{\pi} \sqrt{\frac{3(1 - v^2)m'}{Yd^3}} \tag{10.28}$$

Thus, the critical frequency is particularly high for heavy and thin plates made of a material with low Young's modulus.

After eliminating B/m' from eqs. (10.24) and (10.27), the phase velocity of the bending wave can also be represented as:

$$c_B = c\sqrt{\frac{\omega}{\omega_c}} = c\sqrt{\frac{f}{f_c}} \tag{10.29}$$

The intensity of the wave radiated into the air can be derived from the requirement that the velocity $j\omega\eta$ of the plate equals the normal component of the air velocity at the plate surface:

$$j\omega\eta = v_y(y = 0) = \frac{p}{Z_0} \cdot \cos \vartheta \tag{10.30}$$

By combining eqs. (10.26), (10.29) and (10.30) we obtain for the sound pressure in the radiated wave:

$$p = \frac{j\omega Z_0 \eta}{\sqrt{1 - \omega_c/\omega}}$$

and for the sound intensity of the airborne wave:

$$I = \frac{\hat{\eta}^2}{2} \cdot \frac{\omega^3 Z_0}{\omega - \omega_c} \tag{10.31}$$

From this equation it becomes evident once more that a plate of infinite extension does not radiate any sound below its critical frequency. Instead, the pressure and density differences produced by displacement of the plate will immediately level out in local air flows as sketched in Figure 10.9b. This phenomenon is another example of an 'acoustic short-circuit' which was mentioned already in Section 5.5.

The critical frequencies of several sorts of plates vary over a wide range, mostly, within the range of audio frequencies. Thus, a massive brick wall with a thickness of 24 cm, for instance, has a critical frequency of about 100 Hz, while it lies at about 12 kHz for a 1 mm thick steel plate. (Further values of the critical frequency can be found in Table 14.1.) The latter example seems to contradict every experience since a thin steel plate produces loud sounds with frequencies covering the whole audio range when knocked with a hammer. This contradiction is due to the fact that the laws derived above hold strictly for infinitely extended plates only. Real plates often have free boundaries. Here the term 'free' means not only absence of external forces but also of moments acting on the boundary. The latter condition means that, according to eq. (10.9), the second derivative of the elongation η be zero along the boundary while the former one is tantamount to the requirement that the third derivative vanishes too, according to eqs. (10.11) and (10.9). These conditions cannot be satisfied by just two bending waves, one running in the positive x-direction and the other in the negative one. Instead, additional solutions representing near fields, that is, solutions with $(k_B)_{3,2}$ after eq. (10.23) are required, and it is the latter which are responsible for the radiation of audible sounds even at frequencies below the critical frequency. Nevertheless, even with bounded plates the sound radiation above the critical frequency is considerably stronger than in the frequency range below.

10.3.5   Internal losses

If a solid body is deformed it will store elastic energy. When this process is reversed not all of this energy can be regained as mechanical energy, instead, a certain fraction of it will be lost, that is, it will be transformed into heat. In elastic waves these energy losses will take place periodically and lead to an attenuation of the wave. Some of the causes to which this attenuation is attributed have been described already in Subsection 4.4.3. At present, the formal treatment of deformation losses is in the foreground. They can be accounted for by introducing a complex Young's modulus:

$$\underline{Y} = Y' + jY'' = Y'(1 + j\eta) \tag{10.32}$$

A similar procedure can be applied to the other elastic constants and quantities related to them including the bending stiffness since all of them are linked to the Young's modulus by linear relations. The constant η which is frequency dependent in general is named the 'loss factor'. Its significance for the propagation of extensional waves becomes clear if the angular wavenumber is expressed by the wave velocity using eqs. (10.17) or (10.18) and is inserted into eq. (10.32):

$$\underline{k_E} = \frac{\omega}{\underline{c_E}} = \frac{\omega}{\sqrt{\underline{Y}/\rho_0}} = \frac{\omega}{\sqrt{Y'(1 + j\eta)/\rho_0}}$$

Now we assume that $\eta \ll 1$. Then the square root can be expanded into a power series which is truncated after the second term:

$$\underline{k_E} \approx \frac{\omega}{\sqrt{Y'/\rho_0}} \left(1 - j\frac{\eta}{2}\right) = k_E \left(1 - j\frac{\eta}{2}\right) \tag{10.33}$$

Then the 'wave factor' $\exp(-jk_E x)$ reads:

$$e^{-j\underline{k_E}x} = e^{-k_E\eta x/2} \cdot e^{-jk_E x}$$

Comparing it with eq. (4.20) shows that the intensity-related attenuation constant is

$$m_E = k_E\eta = \frac{2\pi}{\lambda_E}\eta \tag{10.34}$$

This relation holds as well for other wave types such as, for instance, the torsional wave; we just have to replace the subscript E with T. For bending waves, however, the angular wavenumber is inversely proportional to the fourth root of the bending stiffness (see eq. (10.22)) and hence of the Young's modulus. Therefore we have instead of eq. (10.33):

$$\underline{k_B} = k_B \left(1 - j\frac{\eta}{4}\right) \tag{10.35}$$

and the attenuation constant is obtained as

$$m_B = \frac{1}{2}k_B\eta = \frac{\pi}{\lambda_B}\eta \tag{10.36}$$

In spite of the difference by a factor 2 this attenuation constant is greater than that of the extensional wave since its λ_B is much smaller than λ_E.

Experimentally, the loss factor is mostly determined by setting up a resonance system with a mass and a spring which is made of the material under test. In principle, it does not matter whether the spring responds to bending, torsion or to a change in length. In any case its compliance is inversely proportional to the Young's modulus and so is complex:

$$\frac{1}{\underline{n}} = a \cdot \underline{Y} = a \cdot Y'(1 + j\eta)$$

with some constant a. Then the resonance frequency is complex as well:

$$\underline{\omega_0} = \frac{1}{\sqrt{m\underline{n}}} = \sqrt{\frac{aY'}{m}} \left(1 + j\frac{\eta}{2}\right) = \omega_0 \left(1 + j\frac{\eta}{2}\right)$$

Table 10.3 Loss factor of some materials

Material	Loss factor η
Aluminium	$<10^{-4}$
Brass (70% Cu, 30% Zn)	$<10^{-3}$
Steel	10^{-4}
Glass	10^{-3}
Dense concrete	$4-8 \cdot 10^{-3}$
Lightweight concrete, brick	10^{-2}
Wood, plastics, rubber	$0.1-0.5$

where the square root of $1+j\eta$ has been approximated by $1+j\eta/2$, as earlier. Inserting into the exponential typical for harmonic vibration yields:

$$e^{-j\underline{\omega}_0 t} = e^{-\omega_0 \eta t/2} \cdot e^{-j\omega_0 t}$$

that is, a decaying vibration with the decay constant

$$\delta = \omega_0 \eta/2 \tag{10.37}$$

which is easily observed and can be used to determine η. Alternatively, the loss factor can be obtained from the Q-factor of the resonance. According to eq. (2.31) both quantities are related by:

$$Q = \frac{\omega_0}{2\delta} = \frac{1}{\eta} \tag{10.38}$$

In Table 10.3 the loss factors of some materials are listed. However, these data must be considered as approximate only. In particular, the loss factors of high polymers depend often strongly on the temperature and also on the frequency.

Chapter 11

Music and speech

In Chapter 5 the sound radiation from idealised sound sources was described, for instance, from point sources, or dipoles or from rigid surfaces. Sometimes, real sound sources come rather close to such simple types of sources, regarding their radiative properties, at least in limited frequency ranges. And sometimes a real source can be modelled by a composition of several idealised sources. Very often, however, the sound fields produced by real sound sources are much more complicated than those generated by the simple types described in Chapter 5, either because the radiating surfaces are of more complex shapes and are possibly not excited uniformly, or because there are diffracting bodies in their vicinity such as boxes, screens, etc. which deform the sound waves.

In any case, the prerequisite of any sound production is the generation of vibrations which excite a more or less extended surface or are radiated from a suitably shaped opening of a pipe, that is, from a horn. We shall encounter this question in several places in this book. The present chapter deals with certain 'natural' sound sources where vibrations are produced mechanically in contrast to electrically driven sound generators such as loudspeakers or ultrasound generators. Likewise, the treatment of noise and its control will be postponed until Chapter 15. Accordingly, this chapter is devoted to sound generation by the human speech organ and by conventional musical instruments.

11.1 Simple and complex tones, noise

We are used to calling a sound a tone if a definite pitch can be ascribed to it. For instance, we perceive sounds which are produced with a clarinet or a piano as a sequence of tones. A tone which is associated with a harmonic oscillation of the air pressure is called a simple or pure tone, and its pitch is closely related to the frequency of the oscillation. Several simple tones with different frequencies sounding simultaneously form what is called a complex tone.

Virtually all tones produced by musical instruments are complex tones consisting of several or many simple tones the relative strengths and pitches of which determine their timbre. Many musical instruments as for instance string instruments or wood winds can produce steady tones associated with periodic pressure signals; the pitch of such a tone is determined by the reciprocal of the period. According to Subsection 2.9.1 such a periodic signal can be split into a series of harmonic partials each of them corresponding to a pure tone (fundamental tone and overtones).

If, on the other hand, the spectrum of a sound signal is continuous, that is, if it does not contain discrete partials, the signal is aperiodic, and usually no clearly defined pitch can be attributed to it. This is a typical feature of noise. Examples of noise are bangs, or the rolling of thunder, the sounds produced by drums, by wind or other air flows, by flowing water in a river or in a water pipe. The latter sorts of noise are more or less stationary, and their aperiodic character is due to random processes involved in their generation (see Subsection 15.2.2). We can observe and analyse such sorts of noise but we cannot predict it in detail but only characterise them by average values as, for instance, their power spectrum. (The term noise has a two-fold meaning: it denotes an aperiodic sound signal, or it means just any undesired sound. More will be said about the latter aspect of noise in Chapter 15.)

Finally, periodic and aperiodic sound signals may occur in mixed form. The aperiodic component may be, for instance, the bowing noise of string instruments, or the air noise accompanied with the tone production in wind instruments, in particular, of flutes. Basically, this part of the sound signal is an unwanted side effect of tone production; nevertheless, it contributes often to the characteristics of the instrument's sound.

Figure 11.1 shows the time dependence of the sound pressure for a few sound signals. Speech consists of a succession of phonemes which often merge with each other. All vowels (/a/, /e/, /i/, /o/, /u/, etc.) are complex tones in our present terminology, and their spectra consist of discrete 'spectral lines'. However, due to the speech intonation, their fundamental frequency and hence the frequency distance of overtones is continuously changing. Voiced consonants such as the voiced /s/ or /ch/ represent mixtures of both spectral types. Purely aperiodic phonemes are the unvoiced fricatives as well as the plosives (/p/, /k/ or /t/). According to what has been said earlier the unvoiced consonant /s/ could well be regarded as noise.

The reader may have noted that it is difficult to arrive at a clear delimitation between tones and noise. This is even more so when it comes to distinguishing between desired and undesired sounds, that is, to define noise in the second meaning of the word. It is evident that here the subjective attitude of the listener plays an important role. Thus a motorbike freak will not assess the sound of his own vehicle as noise. On the other hand the most wonderful

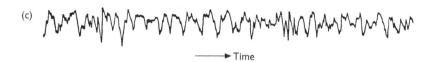

$\longrightarrow$ Time

Figure 11.1 Time function of a few sound signals: (a) vowel /a/, (b) vowel /i/ and (c) noise (road traffic).

Source: H. Fletcher, *Speech and Hearing in Communication*. Van Nostrand, New York 1958.

concert produced by a neighbour's radio or CD player may become noise for the unwilling listener.

11.2   Pitch, intervals and scales

One of the most characteristic features of a pure or complex tone – apart from its loudness – is its pitch. After all, the melodic content of any music is based on a sequence of tones with different pitches.

When the unbiased reader hears the term pitch he may perhaps imagine the keyboard of a piano as shown in Figure 11.2. Here every key belongs to a different tone, and all tones are arranged in the order of pitch ascending from the left to the right. The tone interval between adjacent keys is one semitone (see later). A certain key in the middle of the keyboard is associated with a tone named A_4 in international notation, and the corresponding frequency represents the internationally accepted reference pitch:

$$f(A_4) = 440\,\text{Hz} \tag{11.1}$$

In musical practice, however, instruments are often tuned to a slightly different (usually higher) pitch.

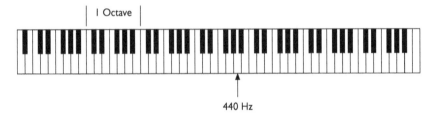

I Octave

440 Hz

Figure 11.2 Keyboard of a piano.

As already mentioned in the preceding section, stationary sounds produced by musical instruments and also by the human voice consist of a fundamental tone and numerous overtones the frequencies of which are integral multiples of the fundamental tone's frequency. In this case the perceived pitch is given by the frequency of the fundamental tone.

The basis of any musical scale system is consonance. By this term the following is meant: if two tones of different pitch are simultaneously presented to a listener, the combination sounds more or less pleasant depending on the pitch interval between both tones which is given by the frequency ratio of their fundamentals. This phenomenon is particularly pronounced if both tones are complex tones containing many overtones.

Generally, the degree of consonance is higher the smaller the integers expressing the frequency ratio. Table 11.1 lists a few tone intervals in the order of decreasing consonance. According to *Helmholtz*, consonance can be explained by more or less complete coincidence of the overtones or partials of both components. For two tones with the fundamental frequencies 100 Hz and 150 Hz which correspond to a fifth, the successions of partials read (all figures in Hz):

100	200	300	400	500	600	700	800	900	1000
	150		300		450		600		750		900

As is easily seen, every third overtone of the upper series coincides with every second overtone of the lower one. However, if both tones include an interval of a major third (100 Hz and 125 Hz), the frequencies corresponding to the overtones of both tones are

100	200	300	400	500	600	700	800	900	1000
	125	250	375	500	625	750	875	1000

In this case only every fifth overtone of the upper series has the same frequency as every fourth overtone of the lower one; the number of coincidences and hence the degree of consonance is smaller than in the first example.

Table 11.1 Frequency ratios and number of semitones for some intervals of high consonance

Name	Number of semitones	Frequency ratio (pure intervals)	Frequency ratio (equally tempered scale)
Octave	12	2:1 = 2	2
Fifth	7	3:2 = 1.5	1.4983
Fourth	5	4:3 = 1.3333...	1.3348
Major third	4	5:4 = 1.25	1.2599
Minor third	3	6:5 = 1.2	1.1892

By combining intervals of the kind shown in Table 11.1 so-called pure tone scales can be generated. In this process the octave showing the highest degree of consonance plays an outstanding role, since two tones in octave distance are perceived as 'equal' in a way and hence are denoted by the same letter. If we, for instance, combine two-fifth intervals, we arrive at an interval with the frequency ratio

$$\frac{3}{2} \cdot \frac{3}{2} = \frac{9}{4} = \frac{2}{1} \cdot \frac{9}{8}$$

which is composed of an octave and an interval corresponding to the frequency ratio 9/8. The latter is called a (Pythagorean) whole tone.

However, continuing this process leads to strange discrepancies: for instance, if a fifth is combined with a fourth, the resulting interval should be an octave. In fact,

$$\frac{3}{2} \cdot \frac{4}{3} = \frac{12}{6} = \frac{2}{1} \tag{11.2}$$

A musically educated person would expect the same result if he combines three major thirds to a common interval. In reality he finds:

$$\frac{5}{4} \cdot \frac{5}{4} \cdot \frac{5}{4} = \frac{125}{64} \neq \frac{2}{1} \tag{11.3}$$

which is a deviation 125/128 from the octave. Furthermore, twelve-fifth intervals should be equivalent to seven octaves which is easily checked with a piano. This process which generates all semitones is called the circle of fifths. In reality, however, $(3/2)^{12} = 531441/4096$, while $2^7 = 128/1 = 524288/4096$. Hence the circle of fifths does not end exactly at its starting point (plus seven octaves). The difference of 531441/524288 is not very large, nevertheless, it is audible. (Strictly speaking, it amounts to about 23.5 cents, see eq. (11.5).) Similar inconsistencies are encountered with every scale which is based upon pure tone intervals represented by simple frequency intervals. Finally, after many attempts to construct more satisfactory tone scales a compromise has been found: the equally tempered tuning. Here the deviation occurring in eq. (11.3) is equally distributed over all semitones within an octave with the octave being the only 'pure' interval. Accordingly, the frequency ratio of a semitone is

$$\sqrt[12]{2} : 1 = 1.0594\ldots \tag{11.4}$$

If one semitone is subdivided into 100 equal intervals one arrives at a tiny interval called 1 cent. It is represented by the frequency ratio:

$$\sqrt[1200]{2} : 1 = 1.000577 \tag{11.5}$$

The last column of Table 11.1 shows the frequency ratios associated with several intervals calculated on the basis of the equally tempered tuning.

Nevertheless, the equally tempered scale does not solve all tuning problems. A violinist, for instance, usually tunes his instrument in pure fifths, which necessarily leads to problems when he plays together with an instrument with fixed tuning such as a piano or an organ.

11.3   General remark on the function of musical instruments

The function of most musical instruments may be illustrated by the schematic representation of Figure 11.3. At first, it shows on the left a primary generator of mechanical oscillations G, next to it an element which determines the fundamental frequency f_0, for instance, a resonator. Mostly, both these parts are in close interaction which is indicated by a feedback loop; often they cannot be distinguished from each other. At the right side, the signal will be radiated into the surrounding air. In some wind instruments this is achieved by a horn which improves the efficiency of radiation. In other types of instruments the radiation is effected by a plate, a 'sound board' or 'resonance box' on which bending waves are excited by the primary oscillator. On account of its numerous vibrational modes it acts as a filter F. Its function is to reduce the strength of some overtones while that of others

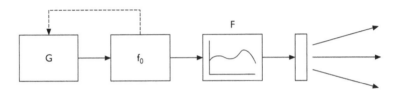

Figure 11.3 Functional scheme of a musical instrument (G: primary generator of mechanical vibrations, f_0: frequency-determining element (resonator), F: filter).

is enhanced. A similar effect is brought about by the resonances of the air column in wind instruments. It is evident that this filter is decisive for the instruments' timbre.

In most musical instruments the fundamental frequency of the tone produced is controlled by a linear resonator, namely, a string or a pipe. In the latter case the pipe is open at one or both ends. Hence its eigenfrequencies, apart from end corrections (see Subsection 7.3.3), can be calculated by eq. (9.1) or (9.2). Organ pipes have a fixed length L which cannot be varied; in woodwinds such as the flute the effective length can be changed by opening side holes, which are normally closed by the player's fingers or by mechanical valves. An open side hole acts as a pressure release; hence its position determines the effective length L of the resonator. In most brass instruments the length of the resonator can be changed by mechanical devices.

The eigenfrequencies of a string stretched between two supports are also given by eq. (9.1) where the sound velocity c of air has to be replaced with the velocity of transverse waves on a string:

$$c_s = \sqrt{\frac{F_s}{\mu}} \tag{11.6}$$

Here F_s is the tension in the string and μ its mass per unit length. While the piano has strings with fixed lengths, with bowed string instruments such as the violin the player varies the effective string length L and hence the pitch of a tone by pressing his finger at the desired position against the finger board. In contrast, the finger board of a guitar is equipped with a number of metal bridges, called 'frets', arranged at right angles to the string. The effective length of the string is varied by pressing it against one of them.

The following description of some groups of instruments is to shed more light on the process of sound generation. The bowed string instruments will be dealt with in more detail in order to give an impression of the complexity of mechanical sound generation at least with one example.

11.4   String instruments

The various string instruments differ in the way the strings are excited into vibration. This can be effected either by bowing them, which results in a stationary tone, or by plucking or striking them. In the latter case the tone will gradually die out since the vibrational losses are not compensated for by a continuous energy supply. In any case it is the free length of the string which determines the fundamental and hence the pitch of the produced tone.

11.4.1   Bowed string instruments

The bowed string instruments which are in use today are the violin, the viola, violoncello and the double bass. The process of tone generation is the same

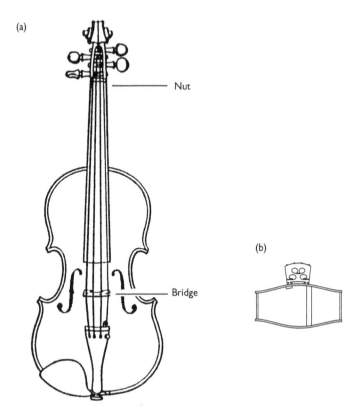

Figure 11.4 Bowed string instrument (violin): (a) top view of a violin, (b) cross section of its corpus.

in all these instruments; the differences are in their sizes and hence in their tonal range.

A bowed string instrument consists at first of a hollow wooden body, the so-called corpus. Figure 11.4a presents the top view of a violin, Figure 11.4b shows a cross section of its corpus. Over this body and the attached finger board four strings (some double basses have five of them) are stretched passing over a bridge which is supported on the top of the corpus. The freely vibrating length of a string is the portion between the bridge and the nut, that is, the far end of the finger board, or, alternatively, a players's finger pressed upon the finger board.

The primary oscillation is created by means of the bow, a thin wooden rod which stretches a bundle of horsehair. The latter has been made sticky by rubbing it with some rosin. The bow is placed on a string at right angles at some distance from the bridge which is about one-tenth of the free string length. When the bow is drawn across the string, and when the normal force and the bowing speed are properly chosen a complicated interplay between the bow and the string will take place, often referred to as stick-slip motion. At first, the bow adheres to the string and hence pulls it away from its equilibrium until the friction cannot overcome the restoring force. Then the string will slip back due to its inertia even beyond its equilibrium position. When its kinetic energy is exhausted it will be caught again by the hair of the bow and the process is repeated. The resulting motion of the string at the point where it is in contact with the bow is shown in Figure 11.5a. It consists of the relatively short slipping time t_g and the longer sticking phase with the duration t_s.

The accompanying waveform of the string must be of the kind that any deflection emerging from the bow must travel along the string with the wave velocity c_s after eq. (11.6), and that this disturbance changes its sign when it is being reflected from the bridge and the nut (or a finger set on the fingerboard) since at these points the string remains at rest. And finally, the waveform must be compatible with the sawtooth oscillation of the string at the point of bow contact (see Fig. 11.5a). All these conditions are fulfilled by the waveform sketched in Figure 11.5b. At any instant, the string has the shape of two straight sections separated by a sharp bend. This bend travels with the speed c_s along the upper portion of the dashed curve from the nut towards the bridge. There it will be reflected and start to travel back towards the right end of the string, this time along the lower portion of the dashed curve. This process is periodically repeated; the period is the duration of one round trip, $T = 2L/c_s$. The dashed curves are parabolas given by the equation:

$$y_b = \pm 4 y_{max} \frac{x_b}{L} \cdot \left(1 - \frac{x_b}{L}\right) \tag{11.7}$$

Here x_b and y_b are the instantaneous coordinates of the bend, y_{max} is the maximum deviation of the string. It depends on the bowing speed and the force with which the bow is pressed onto the string. It is easily seen from the figure that the ratio of the sticking time t_s to the time t_g during which the string moves almost freely is

$$\frac{t_s}{t_g} = \frac{L - x_0}{x_0} \tag{11.8}$$

with x_0 denoting the distance of the bowing spot from the bridge. Of course, this motion cannot be observed without some technical means; to the naked eye only the parabolic curves and a somewhat diffuse area between them are visible.

The excitation of a bowed string into vibrations has been described here in a much simplified manner, leaving out many subtleties. In particular, all losses have been neglected. Such losses are caused by imperfect reflections

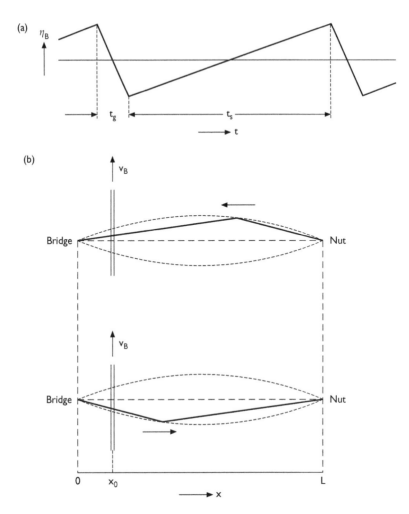

Figure 11.5 (a) Displacement of a bowed string at the point of bow contact, as a function of time, t_s: sticking time, t_g: slipping time, (b) displacement of the whole string (*Helmholtz* motion) at two different instants (v_B = bow speed).

of the string wave, furthermore, by the fact that during the slipping phase the string is still in loose contact with the bow hairs without adhering to them. These losses round off the sharp bend in Figure 11.5b more or less depending on the bowing force and the distance x_0 of the bowing point from the bridge. Viewed in this way the effect of the bow is to sharpen and thus to restore the bend at the beginning and the end of each sticking phase.

A string has a very small radiation impedance, therefore it cannot itself radiate sound to any significant extent. A clearly audible tone is produced by the lateral force F_y which the string applies to the bridge. This force is the transverse component of the string tension F_s:

$$F_y = F_s \frac{y_b}{x_b} = \pm \frac{4y_{max}}{L} \cdot \left(1 - \frac{x_b}{L}\right) \cdot F_s \tag{11.9}$$

the latter according to eq. (11.7). We assume that the starts round trip at $x_b = 0$; accordingly, the lateral force is $-4y_{max}F_s/L$. While the bend travels towards the nut on the lower parabola in Figure 11.5b and back to the bridge on the upper one, F_y grows linearly to $+4y_{max}F_s/L$, then the force jumps abruptly back to its initial value $-4y_{max}F_s/L$, etc. The result is a 'sawtooth', that is, a time function of the force as shown in the example of Figure 2.10a with a frequency spectrum as represented in Figure 2.10b.

As may be seen from Figure 11.4b the top plate and the back plate of the corpus are connected by a vertical post next to the right foot of the bridge, the so-called 'sound post'. Hence the corpus underneath the right foot is stiffer than under the left one. Consequently, under the action of the transverse string forces, the bridge will perform some kind of rotational motion, and it is mainly the left foot of the bridge which excites the corpus to bending waves. The corpus itself acts as a 'resonance box' in the way mentioned earlier: due to its resonances the unspecific spectrum in Figure 2.10b undergoes significant changes which are responsible for the individual timbre of the instrument. Furthermore, the corpus is responsible for efficient radiation into the surrounding air.

11.4.2 Instruments with plucked or struck strings

Another way to excite a string is by plucking, that is, by lifting the string from its rest position with the finger or a mechanical device and then releasing it. Let us denote the displacement of the string by $\eta(x, t)$; it must be a solution of the wave equation (3.21) with c replaced with c_s. Accordingly, $\eta_0(x) = \eta(x, 0)$ is the initial displacement at time $t = 0$. This is shown in the upper part of Figure 11.6 as a fat triangle. Since η_0 is only defined in the interval $0 \leq x \leq L$ it must be continued as an uneven function beyond $x = 0$ and $x = L$ as shown by the thin lines in Figure 11.6. It is evident that now η_0 is a periodic function with period 2L. As in eq. (4.2) we combine the solution of two functions $f(x - c_s t)$ and $g(x + c_s t)$ where f and g must be determined from the initial conditions of the string. One of these conditions is

$$\eta_0(x) = \eta(x, 0) = f(x) + g(x) \tag{11.10}$$

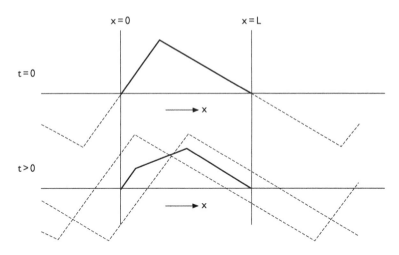

Figure 11.6 Constructing the shape of a plucked string.

The second one requires that the string is at rest at the time $t = 0$:

$$\left(\frac{\partial \eta}{\partial t}\right)_{t=0} = -c_s f'(x) + c_s g'(x) = 0 \tag{11.11}$$

From both equations it follows that $f(x) = g(x) = \eta_0(x)/2$, and eq. (4.2), adapted to the present problem, reads

$$\eta(x, t) = \tfrac{1}{2} \cdot [\eta_0(x - c_s t) + \eta_0(x + c_s t)] \tag{11.12}$$

According to this expression, the waveform at any time $t > 0$ is obtained by shifting the initial shape (including its extension) by ct towards the right and by the same amount towards the left side and subsequently averaging both shifted functions, as shown in the lower part of Figure 11.6. The extension makes sure that both ends of the string remain at rest for any time, that is, $\eta(0, t) = \eta(L, t) = 0$.

Figure 11.7 represents waveforms constructed in this way for successive instants; the time interval between two figures is 1/10 of the vibrational period. Here it was assumed that the point where the string has been plucked is at a distance $0.3 \cdot L$ from the left end of the string.

In a similar way the waveform of a string struck by some narrow 'hammer' can be determined. We assume that the hammer hits the string at time $t = 0$, until this instant the string be at rest. From the condition $\eta_0(x) = 0$ we find $f(x) = -g(x)$ (see eq. (11.10)). By the hammer's action it assumes the initial speed $V_0(x)$. Again, this function is continued unevenly beyond the ends of the string at $x = 0$ and $x = L$, and the modified $V_0(x)$ becomes

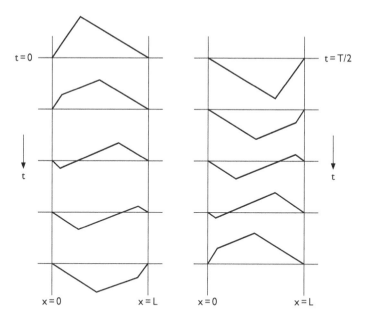

Figure 11.7 Successive shapes of a plucked string during one cycle, in time intervals of T/10.

a periodic function with the period 2L. Instead of eq. (11.11) we have now the condition:

$$\left(\frac{\partial \eta}{\partial t}\right)_{t=0} = -c_s f'(x) + c_s g'(x) = V_0(x)$$

or

$$-2f(x) = \frac{W(x)}{c_s} \quad \text{with} \quad W(x) = \int_0^x V_0(x')dx'$$

Thus, the final solution for the struck string reads

$$\eta(x,t) = \frac{1}{2c_s} \cdot [-W(x - c_s t) + W(x + c_s t)] \tag{11.13}$$

In Figure 11.8 its displacement is represented at various successive instants. Here it was assumed that the hammer with width h has its centre at $x_0 = 3L/10$ and that the velocity V_0 produced by it is constant within the range from $x_0 - h/2$ to $x_0 + h/2$ with $h = L/10$.

Again, it should be noted that no vibrational losses have been taken into account which are responsible for the gradual decay of the produced sounds.

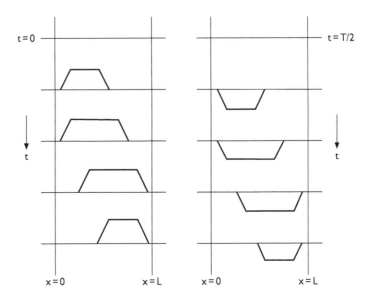

Figure 11.8 Successive shapes of a struck string during one cycle, in time intervals of T/10.

Accordingly, the sound signals are not exactly periodic and the spectra of the tones do not consist of sharp lines but of more or less sharp peaks with finite widths.

Maybe the best-known instruments with plucked strings is the guitar along with its numerous relatives, for instance, the lute or the balalaika. As mentioned in Section 11.3 the player varies the pitch of a tone by pressing the string against a so-called fret. The function of the bridge and the corpus of the guitar is similar to that of bowed instruments. Sometimes the strings of bowed instruments are also excited by plucking ('pizzicato').

Further instruments with plucked strings are the harp and the cembalo (also know as harpsichord). In these instruments it is always the full string which is set into vibrations; this means that the player cannot change the effective length of the string and that for each tone a separate string is needed. The strings of a harp are plucked by the fingers, and they transfer their vibrations to a long sound box to which they are attached at an angle of 30–40°. The strings of a cembalo are plucked with a metal pin called a 'plectrum' which is operated by a lever connected to the corresponding key. Here the sound is radiated by a large sound board which is excited by the vibrating strings via a bridge.

The most important musical instrument with struck strings is the piano. For most of the tonal range two or three strings (unisons) tuned to the same pitch are provided for each tone. They are excited by a small hammer covered with felt which is thrown towards the strings by a complicated mechanism when the corresponding key is pressed. Due to the soft cover, the hammer remains in contact with the string(s) for a short time. The decay time of the freely vibrating string is in the order of seconds at low and at medium frequencies; normally, the decay process is stopped by a damper when the key is released. However, all dampers can be prevented from functioning by pressing a pedal.

11.5 Wind instruments

A wind instrument consists basically of a pipe resonator in which standing waves are generated by a periodically interrupted air stream originating from the player's lungs. The interruptions are brought about either by the instability of an air jet directed towards an edge or by the vibrations of a thin piece of elastic material called a reed which acts as an oscillating valve. Accordingly, we distinguish between instruments with air jets and reed instruments. Brass instruments can be regarded as reed instruments with the player's lips playing the role of the valve. As with string instruments the kind of sound generation is of great influence on the tonal character of the instrument.

11.5.1 Wind instruments driven by air jets

An essential part of such instruments is a sharp edge towards which a narrow air stream is directed as shown in Figure 11.9. This edge does not, as one would expect, neatly split the air stream into two separate streams; instead, the stream will oscillate back and forth across the edge and will produce a succession of eddies. In this way a so-called edge tone is generated. Its frequency increases with the flow velocity; furthermore, it depends on the geometry of the arrangement. Edge tones produced by the wind are often observed in and around buildings and near other obstacles.

In Figure 11.9 the pipe resonator has to be imagined at the right side of the edge. On the one hand, the eddies moving downstream excite the longitudinal normal modes of the pipe, and on the other hand, the pressure

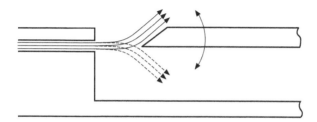

Figure 11.9 Generation of oscillations at an edge.

variations associated with these modes synchronise the oscillations of the air flow. This holds in particular for the lowest mode which usually determines the frequency of the tone and hence its pitch. However, the pitch depends also to some minor extent on the flow velocity of the air. In general, it grows with increasing velocity although not as strongly as with a pure edge tone which is not synchronised by a resonator. If the flow velocity exceeds a certain value the vibration frequency jumps to that of a higher mode. In any case there is a close interaction between the primary oscillator and the resonator coupled to it.

Probably the most common woodwind of this kind is the recorder. It is made of wood or plastic; the player must open and close the side holes with his fingers. In contrast, modern flutes are usually made of metal, for instance, of silver or even gold. The airflow is formed by the player's lips; it is blown perpendicular to the length of the instruments and hits the edge of a hole. The side holes are supplied with metal covers which are pressed down with the fingers or by means of a mechanical lever. Organ pipes are of both types, so-called flue pipes driven by an air jet as well as reed pipes. The latter are mostly of metal and have circular cross section. However, some of them are made of wood; in this case they have rectangular, mostly, quadratic cross section. The influence of the wall material on the tone quality is a disputed matter; traditionally, a lead-tin alloy is used for organ pipes.

The spectrum of wind instruments – no matter of which type they are – and hence their tone quality depends critically on the shape of their bore, which is, apart from slight deviations, either cylindrical or conical. In the first case, an important factor is the ratio of its diameter to its length. The larger this ratio, the softer is the timbre of the produced tone.

11.5.2 Wind instruments with reeds

In these instruments the primary oscillator is a blade of metal or some cane, called a reed, which is set into vibration by the air flow the player produces. Again, the oscillator interacts with a resonator – usually again a pipe – although to a smaller extent than in instruments driven directly by the air flow.

The function of a reed pipe is explained in Figure 11.10. Here the reed is replaced with a solid plate forming a valve together with its seat. In the illustrated situation the plate is close to the seat, and the valve is slightly opened. We assume that the air pressure at the left side of the valve is higher than that to the right. Hence some air will pass through the narrow slit between the plate and the seat. Now Bernoulli's law of flow mechanics tells us that

$$p + \frac{\rho}{2}v_f^2 = \text{const.} \tag{11.14}$$

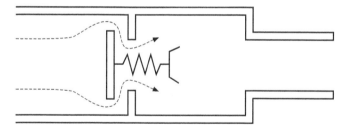

Figure 11.10 Function of a reed pipe (schematically).

with p denoting the local pressure of the air and v_s the local flow velocity; the density ρ is assumed as constant. (In reality, ρ does not remain constant, hence only qualitative conclusions should be drawn from eq. (11.14).) The flow speed is higher in the slit than outside; accordingly, there is an underpressure and the plate is moved towards the right. When it hits the seat, the air flow is interrupted and the underpressure caused by it vanishes. The plate springs back and the whole process is repeated, basically, with the resonance frequency of the mass-spring system. Additionally, the frequency of the plate motion can be increased by increasing the blowing pressure. On the other hand, the pressure fluctuations associated with the vibrational modes within the tube resonator imagined at the right of the valve tend to synchronise the motion of the plate. Consider, for example, the phase when the valve is opened: at this instant a pressure wave is emitted into the tube which will be reflected with reverse sign from its open end. If this underpressure wave arrives at the valve just when this is opened again it will accelerate the closing motion of the valve.

In any case the primary oscillator of a reed instrument forms the nearly closed end of the attached pipe resonator, in contrast to instruments described in Subsection 11.5.1. Hence the resonator is of the quarter-wavelength type with resonance frequencies given by eq. (9.3). Figure 11.11a and b shows the mouthpieces of two reed instruments. In the clarinet and also in the saxophone the vibrations are generated by a single reed made of cane which closes periodically an aperture in the mouthpiece. The oboe and the bassoon are equipped with two moveable reeds forming a small flat tube. In all these instruments the reed itself can be regarded as a resonator which is relatively strongly damped due to the material it is made from and the player's lips which are in close contact with it. This is different with reed organ pipes: here the reed is a small metal blade with a pronounced resonance the frequency of which is close to the fundamental frequency of the attached pipe.

It may be mentioned that the clarinet has a cylindrical bore while that of the oboe is conical. From Figure 9.2b the eigenfrequencies of a conical pipe are

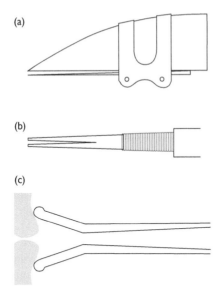

Figure 11.11 Mouthpiece of (a) the clarinet, (b) the oboe and (c) the trumpet.

not harmonic; in particular, the lowest one is significantly higher than that of a cylindrical pipe of equal length and termination. This has consequences not only for the timbre of the produced sounds but also for the fundamental tone. Thus the lowest tone produced by a modern oboe is higher by almost one octave than that of a clarinet although both instruments are nearly of equal length. Nevertheless, the overtones of an oboe are harmonic, of course, since the signal it produces is periodic.

11.5.3   Brass instruments

In brass instruments as, for instance, the trumpet, the trombone and the French horn it is the player's lips which play the role of a primary oscillator and excite the pipe resonator. As shown in Figure 11.11c, they are pressed against the cup of the mouthpiece and form the one termination end of a metal pipe of which the main part of the instrument consists. Suppose the lips are closed at first. If the blowing pressure is high enough they will be pressed apart, and an air stream begins to flow through the gap at relatively high speed. According to Bernoulli's theorem (see eq. (11.14)) this flow is associated with a reduction of the air pressure in the gap, in the same way as in the model shown in Figure 11.10. This reduction has the tendency to reverse the motion of the lips, and it is supported by the restoring force within the lips which will become closed again and the air flow ceases.

The resonator of a brass instrument consists usually of a cylindrical pipe connected to a conical section which is terminated by the bell. The latter is a short section of more rapid flare, its function is to increase the area and hence the radiation impedance of the aperture. This section can be approximately described as a Bessel horn. For circular cross section the shape of a Bessel horn is given by

$$r(x) = r_0 \left(1 - \frac{x}{a}\right)^{-\varepsilon} \quad \text{with} \quad a = \frac{L}{1 - q^{-1/\varepsilon}} \tag{11.15}$$

In this expression $r(x)$ denotes the varying radius of the horn and r_0 the radius at $x = 0$, q is the ratio of the final and the initial radius, and L is the length of the horn. Figure 11.12 depicts some horn shapes of this kind for $q = 10$; the parameter is ε. In the usual brass instruments ε is between 0.7 and 1. Instruments with short and widely opened bells as the trumpet and the trombone produce brighter tones than those of a more slender shape. The eigenfrequencies of the composed pipe are non-harmonic which makes a fine tuning of the shape necessary.

The total tube resonator of a brass instrument is of considerable length. Therefore the instrument is usually made more manageable by coiling the tube, either in a circular shape as in the French horn or in elongated loops as in the trumpet. Its effective length and hence the pitch of the tone produced can be altered either by valves which add extra lengths of air columns, or by a slide which allows continuous changes of pitch and also the production of vibrato effects.

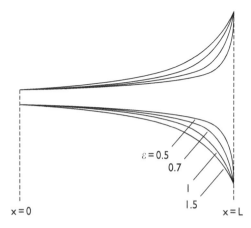

Figure 11.12 Various Bessel horns (see eq. (11.15)).

11.6   The human voice

With his voice man is capable of producing a large number of speech sounds of quite different and rapidly varying tonal character. As already mentioned in Section 11.1 we distinguish between different groups of speech sounds, called phonemes. The most essential ones are the vowels including semivowels, voiced and unvoiced fricatives and plosives which, as the name says, have impulsive character. The change between successive phonemes is effected by the articulation, that is, mainly by motions of the lips, the jaw and the tongue. The pitch is varied by higher or softer tension of muscles in the larynx.

This section explains how the different phonemes are produced. The functional scheme shown in Figure 11.3 applies also to the production of speech with the difference that – depending on the phoneme to be generated – different primary oscillators are employed.

Figure 11.13 represents a section through the human head as far as it is relevant for the generation of speech. In this process, the larynx including

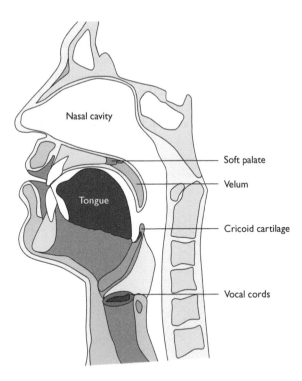

Figure 11.13 Section of the human head.

Source: E. Zwicker, *Psychoakustik*. Springer-Verlag, Berlin 1982. With kind premission of Springer Science and Business Media.

the voice cords and the vocal tract consisting of the pharynx and the mouth are involved, to some minor extent also the nasal tract, furthermore, the tongue, the palate, the teeth and lips and finally, the apertures of the mouth and nose. The energy required for speaking is supplied by exhaling the air stored in the lungs, and the radiation is effected by the opening of the mouth and by the nostrils.

11.6.1   The larynx

Figure 11.14 shows a cross section of the larynx. In this organ there are, expressed simply, two muscles of nearly triangular cross section opposite to each other. They are often referred to as vocal cords, or, with some more justification, as vocal lips. They are separated by a slit called the glottis. During breathing without speaking the glottis is widely opened.

When voiced phonemes, for instance, vowels, are produced the glottis is only slightly opened. Therefore the breathing air passes it with relatively high flow velocity. According to eq. (11.14) an underpressure will develop in the slit which draws the vocal cords together against their restoring force and hence arrests the air flow. At the same time, the underpressure due to Bernoulli's theorem vanishes and the elastic restoring force moves the vocal cords back into the direction of their initial position. Then the whole process starts again which is similar to the generation of vibrations in brass instruments as described in Subsection 11.5.3. However, the pitch of the produced oscillation is independent of any attached resonators. Instead, it is determined by the geometry of the larynx and the effective mass of the vocal cords; accordingly, the higher pitch of female and children's voice is due to the fact that their vocal cords are smaller than those of adult males. Furthermore, the frequency of the vibration depends on the muscle tension in

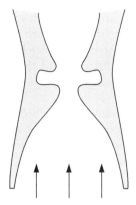

Figure 11.14 Section of the larynx.

the vocal cords which determines their elastic restoring force. By controlling this tension a speaker or singer can control the frequency of vocal cord oscillation within a wide range. Normally, the vocal range is about two octaves corresponding to a frequency ratio of $1:4$; for practiced singers it may be considerably larger.

In any case the flow of the breathing air is periodically interrupted by the larynx when the pitch is kept constant. Hence a regular succession of pressure impulses is generated, that is, a signal which contains many overtones due to its temporal structure but which is still quite unspecific. To become a speech sound its spectrum must be shaped in a characteristic manner.

11.6.2   The vocal tract, vowels

This change of signal character occurs when the primary signal produced by the larynx is transmitted through the cavity formed by the vocal tract and the nasal tract before being radiated from the mouth opening and the nostrils. As any cavity (see Chapter 9), it has several eigenfrequencies within the audible range – or, as we can say as well, resonances, which are not too pronounced because of the radiation losses. This is why they do not influence the fundamental frequency of the speech sound to any noticeable extent, hence there is no feedback to the primary source (see Fig. 11.3). The resonance frequencies depend on the size and shape of the cavity, in particular, on the position of the tongue and the opening of the mouth and are continuously changed during speaking. The same holds for the transfer function of the vocal tract which is formed by the eigenfrequencies; it enhances the spectral components of the primary signal in certain ranges while other components are weakened or suppressed.

Figure 11.15 represents schematically the spectra of some vowels. Each sequence of spectral lines shows three or four flat maxima. These ranges of increased spectral energy are characteristic for the particular speech sounds. They are called formants, and the frequencies where they occur are named formant frequencies. Of course, these frequencies are not completely fixed but show individual differences; they differ even for one and the same speaker. Usually, a particular vowel is sufficiently characterised by the first two formants.

Of course, the generation of vowels can be alternatively described in the time domain. Each impulse produced by the operation of the larynx excites a decaying pressure oscillation in the cavity; the particular speech sound can be imagined as the superposition of all these decays. They can be directly heard if the lips and the tongue are brought into the proper position for a particular vowel and the cavity is excited by snapping with the finger against one cheek. (This effect is most clearly heard for the vowels /0/ and /a/, and it is less pronounced for /u/ or /e/.)

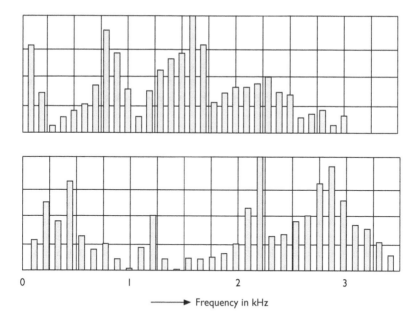

Figure 11.15 Spectra of two vowels.

Source: H. Fletcher, *Speech and Hearing in Communication*. Von Norstrand, New York 1958.

During speaking a speaker changes continuously the fundamental frequency of his vocal chords. This 'speech melody' does not only carry information, for example, on the state of the speaker's mind, or whether his utterance is a question or a statement, but is also important for the intelligibility and the naturalness of speech. This can be demonstrated with an electrical model consisting of a set of properly adjusted filters by which the spectrum of a particular vowel is simulated. If these filters are excited with a strictly periodical sequence of impulses the result reproduced by a loudspeaker is hardly recognised as a speech sound; it rather sounds like some 'technical' noise. This is different when the repetition rate of the impulse train is slightly varied, then the synthesised speech sound is immediately recognised.

Vowels can also be produced without engaging the vocal chords. This is the case in whispered speech. Here the vocal tract is excited by the flow noise occurring when the breathing air passes the glottis or other constrictions in the oral cavity. In this way the flow noise will be given the formant structure of the respective vowels. Of course, the spectra of these signals do not consist of discrete lines but are continuous.

The phonemes /m/, /n/, /ng/ and /l/ are called semivowels. Like the vowels they are voiced speech sounds. The first three of them are radiated from the nostrils. Thus /m/ is spoken with closed lips, for /n/ an occlusion is formed by the tongue pressed against the front part of the palate while for the phoneme /ng/ the air stream is interrupted by the tongue and the rear part of the palate. The distinction of these speech sounds is considerably facilitated by the transitions from the preceding or the following vowel.

11.6.3 The generation of consonants

Regarding the classification of sounds presented in Section 11.1, consonants are some kind of noise. Since they are represented by aperiodic signals they have broad and continuous spectra which may reach up to about 12 kHz. They are caused by turbulent air flows occurring when the breathing air passes certain constrictions formed by articulation within the channel consisting of the trachea, the larynx and the vocal tract. When the phoneme /h/ is spoken the opened glottis acts as a noise source. For articulating other stationary fricatives the speaker forms additional constrictions in his mouth, for instance, between the upper teeth and the lower lip for pronouncing the consonant /f/, or between the upper and the lower teeth for the phonemes /s/ or /sch/. In the latter case an additional resonator in front of the teeth is formed by pushing the opened lips forward, by which the bulk of the spectrum is shifted towards a lower frequency. These speech sounds can be voiced or unvoiced depending on whether the vocal chords are operated in addition to the noise source. In the first case the continuous spectrum is superimposed on the line spectrum produced by the larynx.

While the fricatives are stationary sounds, at least in principle, the stop consonants /p/, /k/, /t/, /b/, /g/ and /d/ are transient sounds. They are produced by forming at first a complete closure at some point in the vocal tract leading to increased air pressure. When the pressure is suddenly released a pressure impulse is formed followed by a rapidly decaying turbulent flow. In continuous speech a stop consonant is usually followed by a vowel or semivowel, that is, the voice cords start oscillating soon after the consonant. Again, this fact facilitates the correct recognition of the consonant. For producing the consonants /p/ and /b/, the closure is formed with the lips, for speaking the phonems /t/ and /d/ the air flow is interrupted with the tongue together with the front part of the palate. A stop in the rear part of the palate is employed to generate the phonemes /k/ and /g/. When the consonants /p/, /k/ and /t/ are spoken the full lung pressure acts onto the stop while for the 'soft' consonants /b/, /g/ and /d/ the overpressure is more local. Furthermore, the oscillation of the vocal cords starts immediately after the latter phonemes have been generated.

Chapter 12

Human hearing

Our interest in acoustics is mainly due to our ability to perceive sound directly with our hearing organ, that is, without any artificial aids. In this function our auditory system proves to be of amazing sensitivity – in fact, were it just slightly more sensitive we could hear the molecules of the air beating down on our eardrum, which would not be of any use. Another fascinating point is the wide range of pitch and loudness which can be processed by our hearing. Concerning the former it should be recalled that the frequency range of our eye is much more limited in that it spans just the ratio 1:2, that is, one octave. In contrast, the range of audible frequencies is from about 16 Hz to more than 20 000 Hz, which corresponds to more than three powers of ten or about ten octaves. Equally phenomenal is the wide range of sound intensities which can be processed by our hearing without being overloaded; it comprises more than 12 powers of ten – without any switching between different ranges as is usual with electrical measuring instruments. Furthermore, our hearing can detect slight differences in pitch and timbre. This admirable performance is made possible by the delicate anatomical – not to say, mechanical – construction of our hearing organ, by its non-linear properties and by the complicated way in which our brain processes the electrical nerve impulses into which the acoustical signal is converted by the inner ear.

In order to dwell a little more on the comparison of our auditory system with the visual sense it may be mentioned that the spatial range of auditory perception is extended over all directions in contrast to the field of vision which is limited to a relatively small solid angle. However, there is also a restriction of auditory performance: although our hearing can detect with relatively high accuracy the direction from which sound arrives if there is just one sound source, it is more difficult to distinguish between several different and spatially distributed sound sources. The reason for this limited resolution are the long acoustical wavelengths and the fact that only two acoustical 'sensors' are at our disposal each of them being small compared to the wavelength. In contrast, in our eye the extension of the sensitive organ – the pupil or the retina – covers many wavelengths. This is the reason why we cannot receive a true acoustical 'image' of our environment in the same sense as the image portrayed by our eyes.

The knowledge of the auditory performance and of the way it is brought about is interesting in its own right. Beyond this, however, studying the properties of our hearing has important practical aspects. Thus the quality of any natural or electroacoustic sound transmission, or the efficiency of some measure of noise reduction, cannot be assessed quantitatively if there is no criterion to do this. This holds not only for the judgement of actual situations but also for acoustical planning, for instance, of industrial sites or of buildings of any kind, for designing hearing aids or with respect to electroacoustical installations. Likewise, in modern techniques of sound recording certain properties of human hearing such as masking are deliberately exploited. For this reason we shall deal particularly with the performance of auditory perception which after all yields the yardstick for tasks of the sort mentioned earlier.

12.1   Anatomy and function of the ear

The human hearing organ as sketched schematically in Figure 12.1 can be divided into the outer ear consisting of the pinna, the ear canal and the eardrum, the inner ear being represented essentially by the cochlea with the vestibular apparatus, and the middle ear between which transfers the sound vibrations from the outer ear to the inner ear. Although the pinna is the antenna of the ear, so-to-speak, it is of minor relevance for the hearing process. In particular, it does not collect the sounds to any remarkable extent as a horn would do – in contrast to the pinnae of many animals such as, for instance, of sheep, dogs, horses and bats.

The sound arriving at the head enters first the opening of the ear canal which is a slightly curved tube about 2.7 cm long and with a diameter of 6–8 mm. After passing it the sound signal reaches the eardrum, a tightly stretched membrane with an area of about $1 \, cm^2$ which closes the ear canal at its inner end and is set into vibration by the sound wave. The eardrum partially reflects the arriving sounds, hence the ear canal acts as a damped $\lambda/4$-resonator (λ = acoustical wavelength in air) with a resonance frequency of about 3000 Hz according to eq. (9.3). This resonance is the reason for the increased sensitivity of the human ear in this frequency range.

On the inside of the eardrum there is the middle ear – an air-filled cavity, containing a chain of small bones called the *ossicles* – the hammer, the anvil and the stirrup – which are suspended by ligaments. To equalise the air pressures on both sides of the eardrum there are connections to the pharynx known as the Eustachian tubes. The three middle-ear bones are in contact with each other and transmit the vibrations of the eardrum to the entrance of the inner ear, the 'oval window'. They act as a lever system increasing the force from the eardrum to the oval window by a factor of about 2.

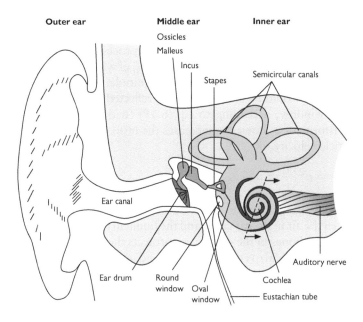

Figure 12.1 The human hearing organ.

Source: E. Zwicker, *Psychoakustik*. Springer-Verlag, Berlin 1982. With kind permission of Springer Science and Business Media.

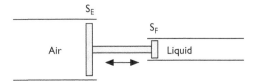

Figure 12.2 Pressure transformation in the middle ear.

An additional, even larger transformation ratio is effected by the different areas S_E und S_F of the eardrum and the oval window as sketched in Figure 12.2 in which the chain of bones is replaced with a rigid pole. Let p_C denote the pressure in the ear canal, then the force acting on the eardrum is $F_E = S_E \cdot p_C$. The force exerted on the oval window is $F_W = S_F \cdot p_W$. Since both forces must be equal we have

$$p_W = \frac{S_E}{S_F} \cdot p_C \tag{12.1}$$

In the human ear the ratio S_E/S_W is about 30; together with the lever effect of the ossicles this results in a pressure transformation of about $\mu = 60$. This pressure increase which goes along with a reduction of vibrational velocity by the same factor is very important for the hearing process since the cochlea is filled with a liquid the characteristic impedance Z_0' of which surpasses that of the air within the ear canal (Z_0) by orders of magnitude, hence sound arriving from the ear canal would be almost perfectly reflected if it fell immediately onto the oval window, according to eq. (6.19) (with $\vartheta = \vartheta' = 0$). The pressure transformation, however, increases the ratio of the sound pressure and the particle velocity to

$$Z_0' = \frac{p_W}{v_W} = \mu^2 \frac{p_C}{v_C} \approx \mu^2 \cdot Z_0 \tag{12.2}$$

Hence the middle ear improves – much like an electrical transformer – the match between the air in the ear canal and the liquid in the inner ear. Furthermore, the chain of bones in the middle ear protects the inner ear to a certain degree against mechanical overload at high sound intensities. This is achieved by some reflex by which two small muscles change the mutual position of the bones and hence reduce the efficiency of vibration transmission.

The inner ear consists essentially of the cochlea, which basically is a coiled tube and is, as already mentioned, filled with a viscous liquid, the perilymph. It has about $2\frac{3}{4}$ turns and is embedded in a very hard bony structure. The same holds also for the vestibular apparatus, semicircular channels which communicate with the cochlea and belong to our sense of equilibrium. Imagined as a straight channel as represented in Figure 12.3a the length of the cochlea is about 3 cm. Its cross section (see Fig. 12.3b) reveals that in reality it is divided into two parallel channels, the *scala vestibuli* and the *scala tympani*. The partition between them is composed of a bony projection, called *lamina spiralis*, and the basilar membrane attached to it. The very thin Reissner's membrane separates off a third channel, the *scala media* which, however, can be regarded as belonging to the *scala vestibuli* from the mechanical standpoint. At the left end of the cochlea in Figure 12.3a the basilar membrane is rather narrow, and it becomes wider towards the right end. Furthermore, it is relatively tightly stretched in a lateral direction but not very much in the longitudinal direction. There are openings for pressure balance: the round window covered by a thin membrane which is at the left end of the cochlea, and the *helicotrema* at its tip which connects the *scala tympani* and the *scala vestibuli*. The basilar membrane carries the organ of Corti which we shall describe below in some detail.

Thus, in mechanical terms the cochlea consists of two channels coupled laterally to each other and filled with a liquid which may be regarded as nearly incompressible. The cross section and the compliance of the coupling membrane vary along its length. When the oval window is set into vibrations by the stirrup, pressure differences between both channels will

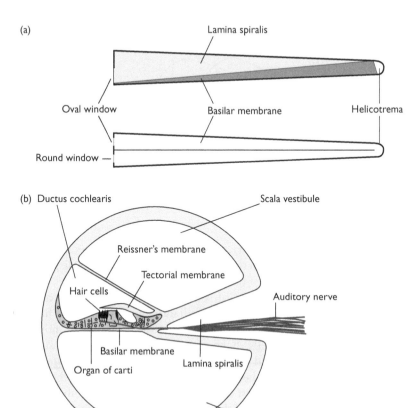

Figure 12.3 The cochlea, represented as an unwound channel. (a) Longitudinal section, (b) cross section.

Source: I. Veit, *Technische Akustik*, 4. Aufl. Vogel Buchverlag Würzburg 1988.

occur. They displace the basilar membrane from its resting position, and this displacement propagates from the oval window to the helicotrema in the form of a progressive wave. On account of this somewhat involved geometrical and mechanical situation the amplitude of this wave is not locally constant but exhibits a pronounced maximum the position of which depends on the frequency. With increasing frequency this maximum is shifted towards the left side; at very high frequencies only the region next to the oval window will be in motion. Hence the inner ear performs some kind of spatial frequency analysis: Each place on the basilar membrane is associated with a certain characteristic frequency at which maximum deflection occurs. Figure 12.4 represents the envelope of membrane vibration for three

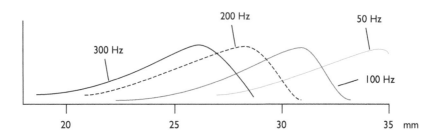

Figure 12.4 Displacement amplitude of the basilar membrane when excited with pure tones of various frequencies. Abscissa is the distance from the oval window.

Source: E. Meyer und E. G. Neumann, *Physikalische und Technische Akustik*, 2. Aufl. F. Vieweg, Braunschweig 1974.

frequencies. For the sake of clarity, the deflections have been exaggerated; at a sound pressure level of 60 dB the maximum amplitude of the displacement is in the order of magnitude of only $10^{-4}\,\mu$m.

The organ of Corti is the seat of some 20 000 sensory cells, called the hair cells, which convert mechanical stimuli into nerve signals. The hair cells are arranged in four rows, one row of inner hair cells and three more rows formed by the outer hair cells. Each of these cells carries on its surface a bundle of thin interconnected filaments of different lengths, the *stereocilia*. They project into the space between the organ of Corti and the tectoral membrane, the tips of the longest ones contact the tectoral membrane. When the basilar membrane is deflected the stereocilia undergo some deformation leading to the release of electrochemical impulses, so-called 'spikes' which are conveyed by the nerve fibres to the brain. The pattern of these impulses which are also known as action potentials is by no means a replica of the sound signal, in reality, the latter will be coded in a complicated way. Each hair cell is connected with several nerve fibres, and each fibre has a characteristic frequency at which it responds most easily. This frequency is related to the characteristic frequency due to the spatial frequency analysis performed in the cochlea. Finally, it may be mentioned that not all of the approximate 30 000 nerve fibres conduct signals from the inner ear to the brain but that many of them convey information in the reverse direction thus providing for some sort of feedback.

The transmission of sound through the middle ear is by no means linear, hence eqs. (12.1) and (12.2) are only an approximate description of the real process. Likewise, the interaction of the stereocilia with the tectoral membrane is non-linear. Therefore distortion products are generated in the ear, that is, additional spectral components (see Section 2.11) which are not present in the original sound signal. A particular obvious phenomenon of this kind is the difference tone when the sound signal arriving at the ear consists of two tones of different frequencies. It can often be heard with two flutes, or during tuning of a string instrument.

Finally, we want to mention that there is still another way of exciting the inner ear other than via the middle ear, namely, directly through the bone in which it is embedded. In this case the external sound receiver is the whole skull. Since there is no impedance matching by the middle ear the sensitivity of hearing by bone conduction is much lower than that through the regular sound path. Nevertheless, bone conduction is relevant for perceiving one's own voice and hence for self-control during speaking.

Next we shall deal with the perceptional performance of human hearing. This is in the realm of psychoacoustics which tries to provide insight into our subjective loudness and pitch scale as well as on many other important questions of auditive perception. The only way to do this is by performing numerous time-consuming and delicate experiments with test persons who are supposed to be cooperative and unbiased. These persons are exposed to different acoustical stimuli which they are asked to assess according to certain prescribed criteria. It is self-evident that results of such tests are afflicted with strong fluctuations, and that only averaging over many single assessments leads to reliable results.

12.2   Psychoacoustic pitch

In Section 11.2 we dealt with musical tone scales which are basically derived from the principle of consonance, the frequency of 440 Hz is internationally accepted as a reference. Thus, for the musician the question of pitch or tone height is settled.

However, one arrives at different results if the problem of pitch is approached from the psychoacoustic point of view. Suppose an unbiased listener is asked to select or to adjust the pitch of a tone T_2 in such a way that he perceives it 'twice as high' as that of a given reference tone T_1 when both tones are alternately presented to him. Figure 12.5a is to illustrate this experiment and its result. The abscissa of this diagram is the frequency of the tones, the ordinate is the perceived pitch associated with them; its exact definition is left open for the time being. Both axis are logarithmically divided. Thus doubling the pitch is indicated by a vertical line of constant length. At relatively low frequencies the test person will decide in favour of a tone with twice the frequency of that of the reference. This is different when the frequency of the reference is 500 Hz. Now to arrive at a tone twice as high the frequency must be augmented not to 1000 Hz but to 1140 Hz, and one step more of pitch-doubling leads to a tone with the frequency with 5020 Hz! Analogous results are obtained when the listener is requested to indicate tones with half the pitch than that of a reference tone. Performing such investigations with numerous test persons and averaging their assessments leads to a relation between the subjective pitch and frequency as shown in

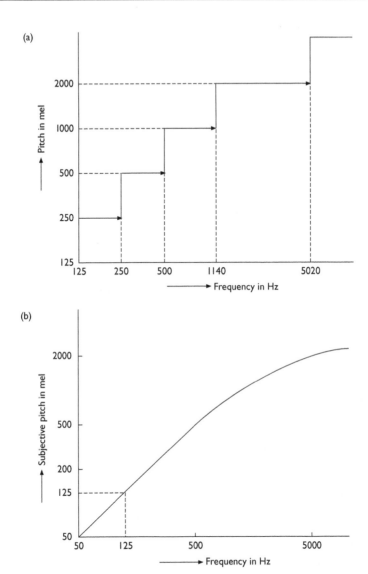

Figure 12.5 Subjective pitch: (a) pitch-doubling and (b) relation between subjective pitch and frequency.

Figure 12.5b. The unit of this kind of pitch is the 'mel' and is defined by the requirement that the frequency of 125 Hz corresponds to a pitch of 125 mel and that each doubling (halving) of pitch corresponds to doubling (halving) the number of mels.

We arrive at a similar result if we ask for the difference limen for tone height, that is, for the smallest frequency change which leads to a just notice-able change of pitch. It can be determined with sine tones the frequencies of which fluctuate between the values $f + \Delta f$ and $f - \Delta f$, or, in the lan-guage of communication engineering, with frequency modulated tones. The frequency of fluctuation is a few Hertz. If the range $2\Delta f$ of frequency fluc-tuation is very small a tone of constant pitch is heard; only if the fluctuation exceeds a certain threshold $2\Delta f_s$ a variation of pitch is perceived. Figure 12.6 plots this threshold as a function of the mean frequency f of the signal. Below 500 Hz it is nearly constant and has the value 3.6 Hz; for higher frequencies it increases steadily with an average slope of about $0.007 \cdot f$. By arranging such critical frequency steps and the associated pitch changes side by side a stair-like curve similar to that of Figure 12.5a can be constructed with steps of constant height but variable width $2\Delta f_s$. A smooth curve connecting the lower edges would have the same shape as the curve shown in Figure 12.5b. Moreover, it turns out that 640 steps are needed to cover the frequency range from 0 to 16 kHz. In other words: a normal hearing person can distinguish about 640 different pitch values.

In this context another remarkable observation is worth mentioning. As described in Section 12.1 there is an unambiguous assignment of sound frequency to the position at which the displacement amplitude of the basi-lar membrane reaches its maximum. If we replace frequency with pitch this assignment becomes very simple; it turns out that there is a linear

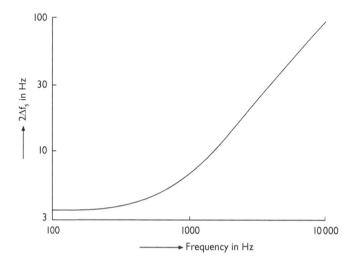

Figure 12.6 Difference limen of pitch discrimination.

relationship between the pitch and the coordinate of the envelope maximum. Or expressed in a different way: the mel scale is the natural length scale along the basilar membrane, or still more simply: 1 mm of the basilar membrane corresponds to pitch increase of 75 mels, or the difference limen of the pitch corresponds to a length of 50 μm on the basilar membrane.

Although the pitch of a tone depends primarily on its frequency its intensity is also of influence although to a minor extent. This fact can be easily demonstrated by placing a struck tuning fork immediately in front of the ear channel. We hear, of course, a loud tone. If the vibrating tuning fork is slowly removed from the ear, the tone will not only become weaker but its pitch will slightly rise. Generally, increasing the intensity reduces the pitch at frequencies below 2000 Hz; at higher frequencies the shift of pitch is in the reverse sense, that is, the pitch of the tone increases when its intensity is enhanced. An analogous phenomenon occurs also with visual perception, namely, a change of the apparent colour at high light intensities.

As mentioned in Section 11.2 the pitch of a complex tone with harmonic overtones is determined by the frequency of its fundamental. This is so even if, for some reason, the physical spectrum of the tone does not contain the fundamental, or the fundamental and the lowest overtones. Apparently, the ear is capable of reconstructing the missing fundamental from the mutual frequency distances of the overtones. This kind of pitch is called the 'residue' or 'virtual pitch'. Due to this remarkable property we can overcome certain deficiencies of sound transmission. For instance, some musical instruments produce only faint fundamentals at their lowest tones since their body is just too small, measured in wavelengths. This is the case for the viola or the double bass. Nevertheless, we perceive the correct pitch intended by the player (or the composer). Likewise, in a telephone conversation we have no difficulties in recognising the pitch of a male voice although the fundamentals of the vowels produced by the speaker are missing because of the limited frequency bandwidth of the transmission channel.

Thus, we arrived at the somewhat confusing result that at frequencies above 1000 Hz the musical pitch derived from the principle of consonance differs widely from the psychoacoustic pitch as determined from systematic listening tests. It is easy to demonstrate that this difference may be relevant in music: suppose you strike one octave on a piano, at first on the left half of the keyboard (say at 100 Hz) and then on its right half (6000 Hz), then we observe that the latter interval sounds 'narrower' in some way than the former one. The same can be heard when a simple melody is played on different parts of the keyboard. Fortunately, tones with high fundamentals are rather rare in music. For instance, the solo part of the quite virtuoso violin concerto by F. Mendelssohn-Bartholdy (first movement) contains less that 15% tones with fundamentals above 1000 Hz, and for the accompanying orchestra this percentage is still considerably lower.

12.3   Hearing threshold and auditory sensation area

As early as in Chapter 1 a preliminary statement on the frequency range was made in which sounds can be perceived by the human hearing. In this section this question will be discussed in more detail. Figure 12.7 (lower curve) presents the threshold of audibility of a normally hearing person for pure tones, that is, the sound pressure level of a just audible sine tone as a function of its frequency. The measurement of this threshold is of great diagnostic relevance in otology. Nowadays it is carried out more or less automatically with so-called audiometers. In the Bekesy audiometer a sinusoidal test tone with gradually increasing or decreasing sound pressure level and slowly increasing frequency is presented to the subject, usually by earphones. The subject is instructed to alternate the sense of level change by pressing a button whenever the tone becomes just audible or just vanishes. Recording the level of the test tone as it is presented to the subject yields a curve with very dense fluctuations the average of which is the threshold of hearing.

According to Figure 12.7 the hearing threshold, starting from low frequencies, falls very steeply at first, then with a decreasing rate. Between 3 and 4 kHz the sensitivity of the ear becomes a maximum as indicated by the dip in the curve. Then the threshold rises again with increasing steepness. This shows that the frequency range of human hearing is also a matter

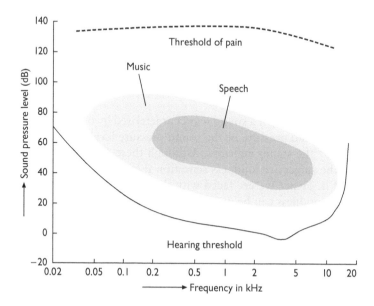

Figure 12.7 Threshold of audibility and threshold of pain; regions in the frequency-level plane occupied by speech and music.

of intensity, strictly speaking. In fact, we can also perceive sound with frequencies below 16 Hz, so-called infrasound, provided its intensity is high enough although in this case the sensation is an awareness of a pressure fluctuation rather than that of a tone. Likewise, intense sounds with frequencies of 30 or 50 kHz can be heard in a way, we can even assign some kind of pitch to them. Generally, the hearing threshold at high frequencies shows large individual differences; the range of steep increase, that is, the upper limit so-to-speak, gradually shifts towards lower frequencies with age.

Figure 12.7 shows as a dashed curve the so-called threshold of pain which is the sound pressure level of a sine tone causing an unpleasant or even painful sensation and thus marks the upper limit of the intensity of useful sound signals. This threshold is less clearly defined than the lower threshold of hearing since it is hard to draw a sharp limit between a strongly unpleasant sensation and slight pain.

The region between the threshold of hearing and the threshold of pain is sometimes called the 'auditory area'. It contains the frequencies and pressure levels of all pure tones which our hearing can receive and convert into auditory sensations without causing any harm to our hearing organ. Moreover, the figure indicates the areas in the frequency-level plane occurring in speech and music. (However, Hifi fans will probably not accept that the upper frequency limit is as low as 10–12 kHz.)

Another question of interest is how large a level change must be in order to be perceptible. This difference limen of sound pressure level is about 0.5 dB for white noise if the pressure level is above 40 dB; towards less intense sound the difference limen becomes noticeably larger. For pure tones the just perceptible level difference is a little smaller than for wideband noise.

12.4   Loudness level and loudness, critical frequency bands

Until now the strength of a sound was characterised by objective quantities – the effective sound pressure or the sound pressure level, or, which is equivalent to it, by the sound intensity. Although these quantities are doubtless related to the subjective loudness sensation they are by no means a general measure of it. Hence we are faced with the question of how the subjective sensation of loudness can be quantified.

To tackle this question one should keep in mind that the sensitivity of our hearing is frequency dependent, which means that two tones of different frequencies but with equal sound pressure level will not sound equally loud. This frequency dependence can be determined by asking subjects to adjust the level of a reference tone with 1000 Hz such that it appears equally loud as a given sound. Then the level of the reference tone can be regarded as a measure of the loudness of the given tone. It is called the loudness level, and its unit is called 'phon'. Again, this procedure requires averaging over many individual assessments and therefore is very long-winded and time-consuming, but in the end it leads to consistent results. Applied to pure tones it results in the famous 'contours of equal loudness level' which are presented in Figure 12.8 for the case of sound waves incident from ahead. Each of these curves connects those points of the frequency-level plane which cause the same loudness sensation; the figures indicate the loudness level in phon. According to its definition the loudness level at 1000 Hz is identical with the sound pressure level in decibels. This diagram expresses the frequency dependence of loudness sensation. It tells us, for instance, that the sound pressure level of a 100 Hz tone must be raised to 56 dB if it is to appear equally loud as a 1000 Hz tone of 40 dB. It should be noted that these contours are not exactly parallel to each other. Anyhow, they assign a certain loudness level to each sine tone of given frequency and pressure level; if necessary, the loudness level can be determined by interpolation.

However, the loudness level or the phon scale misses one important feature of any reasonable scale, namely, that doubling the quantity means doubling the number of units. Thus a distance 100 km is clearly twice as far as one of 50 km, but a sound signal with a loudness level of 100 phon is not twice as loud as a signal with 50 phon but much louder. This is the reason why we call the quantity measured in phon not just loudness but somewhat more cautiously 'loudness level'.

To arrive at a subjective quantity which fulfils that condition one proceeds in an analogous way as in defining a subjective scale of pitch (see Section 12.2): subjects are asked to compare sounds and to decide when

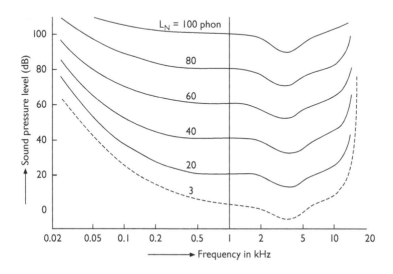

Figure 12.8 Equal-loudness contours for pure tones at frontal sound incidence.

one sound is twice as loud or half as loud as another. The subjective quantity defined by such comparisons is called the 'loudness', and its unit is the 'sone'. Every doubling (halving) of loudness is associated with twice (half) the number of sones. To convert this scale in an absolute one, 1 sone is defined arbitrarily as the loudness of a 1000 Hz tone at a sound pressure level of 40 dB (equivalent to 40 phon). Now both quantities, loudness and loudness level, are related to each other in an unambiguous manner. For pure tones or small band signals this relation is represented by the curve shown in Figure 12.9, both axes are divided logarithmically. Fortunately, for the range above 40 phon this curve can be approximated by a straight line which is mathematically expressed by

$$N = 2^{(L_N - 40)/10} \text{ sone,} \tag{12.3a}$$

or reversely

$$L_N = 33.2 \cdot \log_{10} N + 40 \text{ phon} \tag{12.3b}$$

Evidently, increasing the loudness level L_N by 10 phon is tantamount to doubling the loudness. Applied to the numerical example earlier this means that a sound with a loudness level of 100 phon is 32 times as loud as another one of 50 phon or still more precisely: the former tone has a loudness of 64 sones while the less intense sound signal has only 2 sones. Because of the simple relation between both quantities the curves shown in Figure 12.8 refer to both, loudness and loudness level; the difference is just in the numbering of the curves. However, the lowest curve is denoted by '0 sone' in the first case.

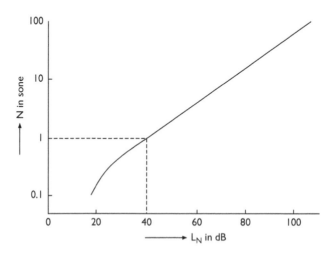

Figure 12.9 Relation between loudness N and loudness level L_N.

Equation (12.3) can be interpreted in yet another way: according to eq. (3.34) we express the loudness level in eq. (12.3a) by the root-mean-square sound pressure $\tilde{p}$:

$$L_N = 10 \cdot \log_{10}(\tilde{p}^2/p_b^2)$$

and obtain, after omitting all insignificant numerical factors, $N \propto 2^{\log_{10} \tilde{p}^2}$ or, since $2^{\log_{10} x} = x^{\log_{10} 2} \approx x^{0.3}$:

$$N \propto \tilde{p}^{0.6} \qquad\qquad (12.4)$$

Hence, the loudness is proportional to the 0.6th power of the root-mean-square sound pressure! This holds, as assumed earlier, for sound signals of small frequency bandwidth; for wideband signals the exponent reduces to about 0.5.

Investigating the loudness of sounds leads to further interesting results. Suppose, for instance, that the sound signal is random noise with an initially very small bandwidth Δf. Now Δf is gradually increased while the total intensity of the noise is kept constant. Then its loudness will remain unaltered up to a critical bandwidth Δf_k. After exceeding this value, however, the perceived loudness will continuously increase. This means that for bandwidths below Δf_k the loudness sensation is formed only from the total intensity contained in the signal while signals with a greater bandwidth are processed by our hearing in a more complicated way. It is obvious that the critical bands play an important role for the loudness sensation. This has to be taken into regard if we are looking for a correct method of measuring loudness which we will discuss in Section 12.6. Accordingly, the whole frequency range of hearing can be thought of as being subdivided into bands with width Δf_k. Table 12.1 lists the limiting frequencies of these bands. At low frequencies the critical bands have constant width of 100 Hz, and at higher frequencies the bandwidths rise until 3500 Hz. These values reflect in a way the complex mechanical properties of the inner ear and are related to the mel scale introduced in Section 12.2. In fact, the width of the critical bands can be expressed much simpler in terms of subjective pitch:

width of a critical band = 100 mel

= 1.3 mm distance on the basilar membrane

We conclude this section with a note on phase hearing: according to the acoustical Ohm's law, phase differences between the spectral components of a complex tone should be inaudible. However, this is not generally true; in fact, pairs of certain sound signals can be generated which sound quite different although they differ only in their phase spectra while their amplitude

Table 12.1 Critical frequency bands

Number of band	Lower frequency limit in Hz	Upper frequency limit in Hz	Band width in Hz
1	0	100	100
2	100	200	100
3	200	300	100
4	300	400	100
5	400	510	110
6	510	630	120
7	630	770	140
8	770	920	150
9	920	1080	160
10	1080	1270	190
11	1270	1480	210
12	1480	1720	240
13	1720	2000	280
14	2000	2320	320
15	2320	2700	380
16	2700	3150	450
17	3150	3700	550
18	3700	4400	700
19	4400	5300	900
20	5300	6400	1100
21	6400	7700	1300
22	7700	9500	1800
23	9500	12000	2500
24	12000	15500	3500

spectra are equal. For the loudness sensation, however, such phase relations are not relevant.

12.5 Auditory masking

Important insights into the kind of signal processing performed by our hearing are obtained by investigations in the so-called masking effects. It is a common experience that very loud sounds can conceal softer sound signals, that is, make them inaudible. Thus, for instance, next to a working pneumatic hammer we cannot understand conversational speech; we cannot even hear if somebody is speaking because the noise of the hammer masks the speech sounds. Everyday life offers many other examples of masking.

To study systematically the masking phenomena we again determine the threshold of hearing, this time, however, with a masking tone present. In such an experiment, the frequency variable tone plays the role of a test signal by which we probe out the excitation of the basilar membrane caused by the masker. To avoid beats it is practical not to use pure tones in such experiments but random noise of very small bandwidth thus averaging out any phase effects.

Figure 12.10a shows masking by noise with the bandwidth of a critical band and with the mid-frequency 1000 Hz. The sound pressure level of the masking noise is 80 dB and 100 dB. Far away from the frequency of the masking noise the masked threshold agrees with the unmasked one, shown as a dashed line, that is, there is no masking. When the frequency of the test tone approaches the frequency of the masker from the left, the threshold shows a steep ascent, then reaches a maximum at 1000 Hz and drops towards higher frequencies. With increasing level of the masking sound the high-frequency slope becomes flatter which indicates that some non-linear process is involved. The steepness of the left flank which reaches 100 dB per octave is an indication of the excellent frequency discrimination by our hearing. A steeply rising flank on the low-frequency side and a flat trail at the high-frequency side of the masking band is also observed if low-pass or high-pass filtered noise is used as a masker. Generally, it may be concluded from this finding that tones with low pitch are masked to a lesser degree by high-frequency sounds than vice versa.

What has been described earlier is masking by stationary sounds in the frequency domain. Besides, masking occurs also in the time domain with non-stationary sound signals. This is relevant for the perception of rapidly varying sound signals as in certain musical passages. Figure 12.10b presents a typical example. It shows a masking signal consisting of a rectangular tone impulse of 200 ms duration (shaded area). The probing signal is a short impulse the strength of which can be varied. It is remarkable that the masked threshold of hearing (upper line) already rises even before the ear has received the masking impulse. This phenomenon is called backward masking. The horizontal part is the range of simultaneous masking, while the trail at the right side is due to forward masking. It indicates that the hearing needs a certain time of recovery before it is ready to perceive subsequent sound signals. On the other hand, backward masking is attributed to the fact that the hearing organ needs some more time to process and hence to perceive a weak test signal than that required for the much stronger masking signal.

12.6 Measurement of loudness

For noise control and, in particular, for the assessment and comparison of noisy environments as well as for many other questions of practical acoustics, methods are needed by which the loudness of sounds can be reliably measured. A measuring procedure which is beyond any doubts from its very principle would be the direct comparison of the sound to be measured with a 1000 Hz tone the level of which can be adjusted; after

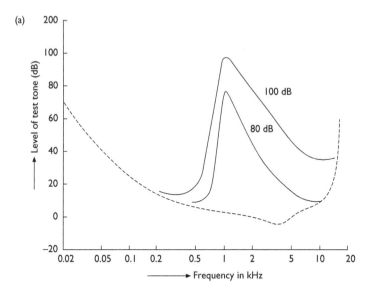

(a)

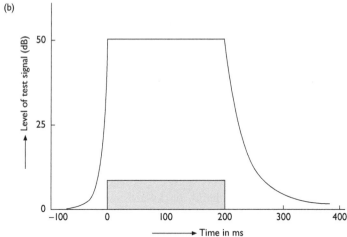

(b)

Figure 12.10 Masking: (a) in the frequency domain (broken line: normal threshold of audibility), (b) in the time domain (shaded area: masking signal).

Source: E. Zwicker, *Psychoakustik*. Springer-Verlag, Berlin 1982.

repeating this comparison many times and averaging the results one would
eventually arrive at the correct loudness level in phon. It is clear that such
an awkward and time-consuming procedure is not well-suited to practical
purposes.

If the signals the loudness of which is to be determined are pure tones or narrow band noise it is sufficient to measure their frequencies and sound pressure levels; from these data the loudness level can be determined by using the curves shown in Figure 12.8. More convenient are sound level meters containing a filter which imitates more or less the shape of these curves. They consist (see Fig. 12.11a) of a calibrated microphone, which picks up the sound signal and converts it into an electrical signal, of the 'ear filter', a quadratic rectifier and a meter the scale of which indicates the level in decibels. In more modern instrument the meter is replaced with a digital display.

One principal problem of this method is that the characteristics of the correct weighting filter depends on the quantity to be determined, namely, the loudness. In fact, several filter curves (A, B, C, etc.) for different loudness

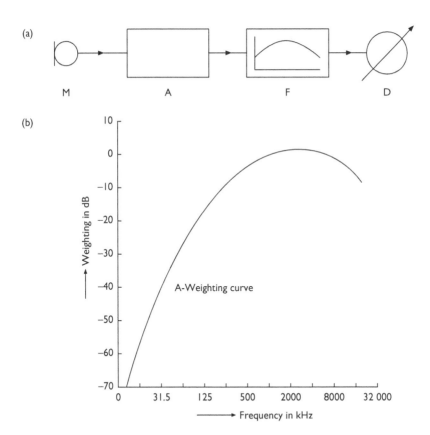

Figure 12.11 Measurement of the weighted sound pressure level: (a) block diagram of a weighting sound level meter (M: microphone, A: amplifier, F: weighting filter, D: meter and display), (b) A-weighting curve.

ranges have been developed and internationally standardised. The only one which is still in common use today is the so-called A-weighting curve which is shown in Figure 12.11b. The quantity measured with it is called the 'A-weighted sound pressure level', and its unit is named dB(A).

Even more severe is the fact that the spectrum of most everyday noises covers a wide frequency band while the contours of equal loudness refer to pure tones or narrowband signals. If we try to measure the loudness level of a wideband sound with a weighting sound level meter of the kind described the values obtained are systematically too low, and the error can be as high as 15 dB. The reason is that the contours of equal loudness do not account in any way for the mutual masking of different spectral components as described in the preceding section. Nevertheless, the measurement of weighted sound levels as described has found general acceptance since the instruments are handy and easy to manage. Furthermore, all technical guidelines and legal regulations are based upon the A-weighted sound pressure level.

For a correct measurement of loudness the sound signal must be fed to a set of bandpass filters by which it is split into partial signals with the critical bandwidths mentioned in Section 12.4. Then the levels of the different frequency bands are converted into loudness figures. The result may be represented in the form of a stair-like curve as shown in Figure 12.12. According to Zwicker the mutual masking is accounted for by attaching a trail to the right side of each step which essentially corresponds to the right hand flank of the masked hearing threshold represented in Figure 12.10a. If curves intersect the higher curve portion is regarded as the valid one. Then the total loudness is obtained by evaluating the area below the modified loudness curve. The reader who is interested in a more detailed description of this procedure is referred to the literature.[1]

This method which has only been briefly outlined here appears relatively involved which is not surprising since it attempts nothing less than to imitate the complex signal processing as is carried out in our hearing. For its practical implementation graphical aids as well as computer programs have been developed. Moreover, compact instruments to measure the loudness directly are available nowadays. A widespread application of it, however, is hampered by the fact that, as mentioned, the existing guidelines and regulations refer to the weighted sound pressure level despite the proven shortcomings of that quantity.

12.7 Spatial hearing

Our ability to localise a lateral sound source is due to the fact that we usually hear with two ears and unconsciously compare the acoustic signals received

---

1 E. Zwicker and H. Fastl, *Psychoacoustics*. Springer-Verlag, Berlin 1999.

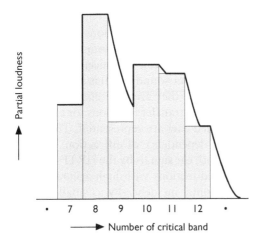

Figure 12.12 Measurement of loudness (schematically).
Source: E. Zwicker, *Psychoakustik*. Springer-Verlag, Berlin 1982.

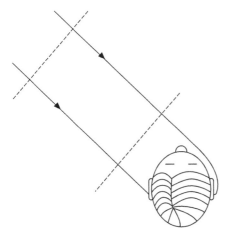

Figure 12.13 Directional hearing.

by them. If a sound wave is incident from a direction which is not within the vertical symmetry plane of the head (the median plane) one of both ears is more or less shaded by the head while the other one is fully exposed to the arriving sound. Additionally to the amplitude difference caused by this effect, differences in the transit times of both ear signals occur. The situation is explained in Figure 12.13. In reality, matters are somewhat more complicated than indicated by this figure since the incident wave is diffracted by the head and the pinnae, and the result of this process is different for both ears if the sound arrives from a lateral direction. This process is quantitatively described by the 'head-related transfer functions (HRTF)', that is, transfer functions referring to the transmission of a sound signal from far away to the entrance of the ear canal. They can be conceived as filter functions by which the spectrum of an incident sound signal is altered; of course, for a given direction of sound incidence they are different for both ears. In Figure 12.14 the absolute values of both transfer functions for exactly lateral sound incidence (exposed and shaded ear) are represented. Their irregular shapes reflect the strong frequency dependence of diffraction. The spectral differences which are imposed on both ear signals by the HRTFs are detected by our brain and are attributed to directions with high accuracy. Thus an angular shift of the source from frontal incidence by as small as about 2^0 can be detected. At lateral sound incidence the uncertainty of localisation increases to about $\pm 10^0$; if the sound wave arrives from the rear the error is within about $\pm 5^0$.

This explanation fails if the sound wave hits the head from the median plane, that is, from ahead, from the rear or from above, etc. Then the signals arriving at both ears are identical because they are filtered by the same or almost the same HRTFs; therefore, no interaural differences can be evaluated. But still the HRTFs depend on the direction of incidence and thus influence the timbre of both ear signals because of the asymmetries of the human head (see Fig. 12.14b). Moreover, it appears that the pinnae play a certain role, at least at higher frequencies. In any case, for localising sound sources in the median plane we have to rely on these changes in timbre which are common to both ears; obviously, they are sufficient for a certain localisation within the symmetry plane. *Blauert* was even able to show that we associate certain frequency bands with certain directions in the median plane ('direction-determining bands').

The localising properties of our hearing as described earlier refer to a listener in a reflection free environment who is exposed to just one plane (or spherical) wave. In contrast, a listener in a closed room will receive many sound waves all of them originating from the same source, namely, those waves which have been reflected once or repeatedly from the walls of the room (see Chapter 13). They transmit the same or at least a similar signal as the wave reaching the ear by the shortest possible path, the so-called direct sound. However, because of the longer paths they arrive at somewhat later moments. Now our hearing has the remarkable ability of localising the sound source in the direction from which the direct sound arrives; the reflected sounds do not influence our judgements of direction although the energy carried by them may considerably surpass that of the direct sound. This fact is known as 'precedence effect' or the 'law of the first wavefront'.

Still further we have the phenomenon known as the 'Haas effect'. According to it a sound source can be correctly localised even if one particular

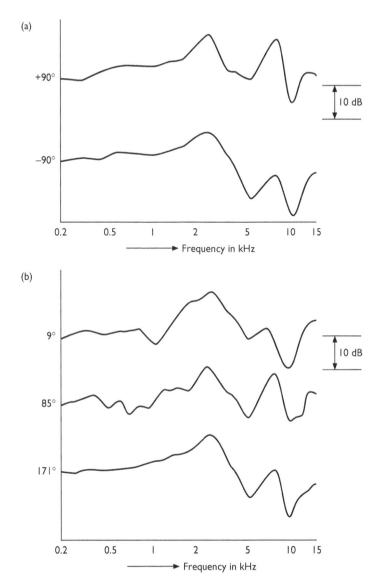

Figure 12.14 Some head-related transfer functions (HRTF) (magnitude). The curves within each diagram are vertically offset by 20 dB: (a) horizontal plane, sound incidence at ±90°, (b) vertical plane (median plane), sound incidence from various elevation angles.

Source: S. Mehrgardt and V. J. Mellert, Transformation characteristics of the external human ear. *Journal of the Acoustical Society of America* **61** (1977), 1567.

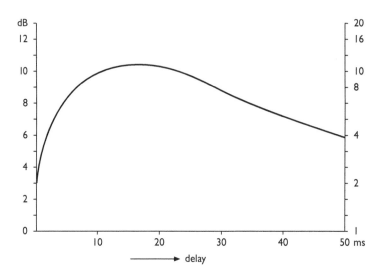

Figure 12.15 Haas effect: just tolerable level increase of the secondary sound.

reflection (or delayed repetition of the source signal) is more intense than the primary sound and arrives from a different direction. This applies to a threshold represented in Figure 12.15. In this diagram the abscissa is the delay of the secondary sound signal with respect to the arrival of the direct sound. The ordinate indicates how much the level of the secondary sound may be raised above that of the direct sound without destroying the listener's illusion that all the sound he hears is supplied from the primary source. Hence, in favourable cases this level difference may be as high as 10 dB which means that the localisation is only impaired if the intensity of the secondary sound is more than ten times that of the direct sound. This effect is particularly important for the design of electroacoustic sound reinforcement systems in which most of the sound energy delivered to the audience is produced by loudspeakers (see Subsection 20.3.2).

Chapter 13

Room acoustics

Most people of our cultural environment spend the main part of their lives in closed rooms, in offices, workshops or factories, or in their homes, in hotels and restaurants. Therefore the question of which way the various sorts of rooms influence the sounds produced in them has relevance in our everyday life, in particular, speech and music and also noise. This holds even more for auditoria and other spaces where people attend a lecture, enjoy an opera or concert, watch a sports events, etc. Every layman knows that a theatre or a concert hall may have good or less good 'acoustics'. This is true even for such rooms in which – as in churches – the transmission of acoustical information is not of primary interest. In working rooms such as in factories or open-plan offices the acoustical room properties determine the level of working noise and the intelligibility of conversations or telephone calls of neighbours. All these things are important for our well-being, for the acoustical comfort of the environment and sometimes for work efficiency.

The physical foundations of room acoustics have been dealt with already in Chapter 9. As we have seen any sound field in an enclosure can be thought of as being composed of characteristic, three-dimensional standing wave patterns called normal modes. These modes depend on the shape of the room and on the physical properties of its boundary. However, the calculation of just one single normal mode of a realistic room with all its details turns out to be quite difficult. In view of the enormous number of normal modes occurring within the frequency range of audible sounds it becomes evident that it is hopeless, from a practical point of view, to compute sound fields in rooms by computing its normal modes. Moreover, such a computation would be useless since it would not yield any useful information on the 'acoustics' of a room, on measures to improve it if necessary, etc. Hence, the concepts described in Chapter 9 are correct and mandatory for a real understanding of sound propagation in rooms; for practical purposes, however, other ways of sound field description are more profitable. In using them one has to sacrifice certain details; in exchange they show more clearly how the sound field is related to the data of the room on the one hand and to what a listener perceives in a room on the other.

In the following discussions we restrict ourselves to rooms or to frequency ranges for which the 'large-room condition' (9.38) is fulfilled which may be shown here once more:

$$f > 2000\sqrt{\frac{T}{V}} \tag{13.1}$$

Here the frequency f is in Hertz, V denotes the room volume in m³ and T the reverberation time in seconds. This condition ensures that the resonance curves associated with the normal modes strongly overlap and hence cannot be separately observed. It is not very restrictive: for a room with a volume as small as 400 m³ and a reverberation time of 1 s the frequency range indicated by eq. (13.1) covers most of the audible frequencies, namely, those above 100 Hz.

13.1 Geometric room acoustics

An illustrative way of describing the propagation of sound in a room is offered by geometric acoustics which is – in analogy to geometric optics – limited to the range of very high frequencies where diffraction and interferences can be neglected. Then it is convenient to think of sound rays as the carriers of sound energy instead of extended sound waves. As in optics a ray may be imagined (see Fig. 13.1) as an infinitely narrow section of a spherical wave emerging from a small sound source. From this concept it can be immediately concluded that the total energy of a sound ray remains constant during its propagation – provided we neglect attenuation in the air; however, its energy density is inversely proportional to the distance from its origin, as in any spherical wave. In what follows a sound ray will be represented by a straight line; nevertheless, we should not forget its physical meaning.

The most important law of geometric room acoustics is the law of 'specular' reflection which reduces to the simple rule

angle of incidence = angle of reflection

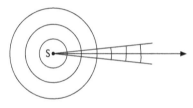

Figure 13.1 Definition of a sound ray.

Strictly speaking, this law holds only for the reflection of a plane wave from an infinitely extended surface. However, for practical purposes it may also be applied to surfaces of finite extension provided their dimensions are large compared to the wavelength. About the same holds for the reflection of sound rays from curved walls or ceilings; the reflection law may be applied to such surfaces if their radius of curvature is large, again in comparison with the wavelength. Concerning sound reflection from 'rough' surfaces which scatter the incident sound we refer to Section 7.5.

Figure 13.2 represents schematically the longitudinal section of a room. In it some ray paths are marked along which the sound travels from a sound source A to some observation point P. The sound portion which reaches P on the shortest path is called the direct sound. All other contributions are due to reflections from the boundary. Apart from those rays which have undergone just one reflection, there are other ones which suffer multiple reflections before reaching the observation point. Thus one of these rays is reflected first from the rear wall of the stage and then from the ceiling, another one hits first the floor and the rear wall of the stage before it arrives

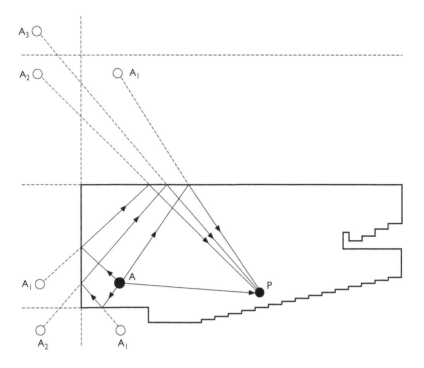

Figure 13.2 Longitudinal section of an auditorium with a few ray paths and image sources. A: sound source, A₁: first order image sources, A₂: second order image sources, etc.

at the ceiling which directs it eventually to the observation point P. The construction of such rays is a simple way to examine the contributions of wall and ceiling portions to the supply of an audience with sound. However, if many multiple reflections are to be taken into account, this picture will become too confusing and looses its clarity.

Viewed from the observation point P the reflected rays in Figure 13.2 seem to originate from virtual sound sources which are mirror images of the original sound source. Thus the sources denoted by A_1 are mirror images of the source with respect to the floor, the rear wall of the stage and the ceiling. The ray which has undergone two reflections can be attributed to a second order image source (A_2, above), which itself is an image of the source A_1 (behind the stage) with respect to the plane containing the ceiling. And the ray including three reflections seems to arrive from an image source of third order A_3 which is the image of the lower virtual source A_2 with respect to the ceiling plane. This process of successive mirroring can be extended to all walls and may be continued ad lib. Since the number of image sources increases rapidly with increasing order it is useful to employ a digital computer for the practical implementation of this procedure. One has to assume, of course, that the original source and all its images produce the same signal. The energy loss of a ray due to imperfect wall reflections is approximately accounted for by attributing a reduced power output to the image source. Let $\alpha_1, \alpha_2, \alpha_3, \cdots, \alpha_n$ denote the absorption coefficients of the walls involved in the construction of a particular image source of nth order, then the power reduction factor of this image source is

$$(1 - \alpha_1)(1 - \alpha_2)(1 - \alpha_3) \cdots (1 - \alpha_n)$$

The construction of image sources is particularly simple for a rectangular room. Because of the symmetry of this enclosure many of its image sources coincide. In their totality they form the regular pattern shown in Figure 13.3, which must be extended into the third dimension, of course.

Once we have determined a sufficient number of image sources which become weaker and weaker with increasing order, the boundary of the room is no longer needed; we can obtain the total sound signal in the observation point by adding the contributions of all image sources plus that of the direct sound, taking into account the energy loss and the time delay of each component, both grow with increasing distance.

Strictly speaking, for calculating the resulting sound signal in the receiving point P we ought to add the sound pressures of all contributions. If the room is excited by a sine tone the total intensity in P is obtained as

$$I = \frac{1}{2Z_0} \left| \sum_n p_n \right|^2 = \frac{1}{2Z_0} \sum_n \sum_m p_n p_m^*$$

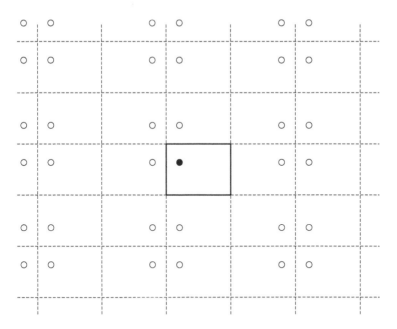

Figure 13.3 Image sources of a rectangular room.

the asterisk denotes the complex conjugate quantity. The contributions p_n and p_m have equal frequencies but their phase angles are quite different because of the different distances they have travelled. Therefore the terms with $n \neq m$ cancel if the number of components is very large, and

$$I \approx \frac{1}{2Z_0} \sum_n |p_n|^2 = \sum_n I_n \qquad (13.2)$$

This means, it is permissible to add just the intensities instead of sound pressures, which is much simpler.

The principle of image sources is limited to plane walls. If a reflecting surface is curved one must construct the wall normal for each point which is hit by a sound ray. This normal is the reference for the angles of incidence and reflection.

13.2   Impulse response of a room

With the aid of geometric acoustics we can examine not only the spatial distribution of stationary sound energy within a room but also the temporal succession in which the reflected sounds arrive at a given point in the room. Suppose that the sound source emits any signal represented by the sound pressure s(t). A listener will receive the direct sound first since this component travels the shortest distance. The contributions of the image sources, that is, the wall reflections, are weaker, and they are delayed by t_n with respect to the direct sound, according to the longer paths lengths. Then the signal received by the listener can be represented as

$$s'(t) = \sum_n a_n s(t - t_n) \tag{13.3}$$

If the signal emitted by the sound source is a very short impulse idealised as a Dirac delta function eq. (13.3) is the impulse response of the room:

$$g(t) = \sum_n a_n \delta(t - t_n) \tag{13.4}$$

In Figure 13.4a, which shows such an impulse response, the first vertical line indicates the direct sound, and every additional line represents a reflection. The temporal density of reflections increases with the square of time. Thanks to the limited temporal resolution of our hearing we do not perceive such an impulse response as a rattling noise but as a decaying, more or less uniform kind of noise. We encountered this gradual decay of sound energy already in Section 9.6 where it was named reverberation. It should be noted, how-ever, that the impulse response, due to its fine structure, yields information on the quality of sound transmission in a room which reaches far beyond the assessment of reverberation; in fact, it can be regarded as the 'acoustical fingerprint' of a room. This holds even more for impulse responses as mea-sured in real rooms. One example is represented in Figure 13.4b. Its more complicated structure compared to that in Figure 13.4a has at least two reasons. One of them is the frequency dependence of wall reflection factors which modifies the spectrum of the signal and hence the signal itself. Another one is that real walls do not exclusively reflect the incident sounds specularly but scatter part of it. This holds particularly for traditional concert halls and theatres with their columns, niches, coffered ceilings, etc., or for baroque churches with their rich decorations.

Generally, our hearing does not perceive reflections with delays of less than about 50 ms as separate acoustical events. Instead, such reflections enhance the apparent loudness of the direct sound, therefore they are often referred to as 'useful reflections'. The remaining reflections with longer delays are responsible for what is perceived as the reverberation of the room. The relative contribution of useful reflections may be characterised by several parameters derived from the impulse response. One of them is the 'definition' or 'Deutlichkeit' defined by

$$D = \frac{\int_0^{50\text{ms}} [g(t)]^2 \, dt}{\int_0^\infty [g(t)]^2 \, dt} \cdot 100\% \tag{13.5}$$

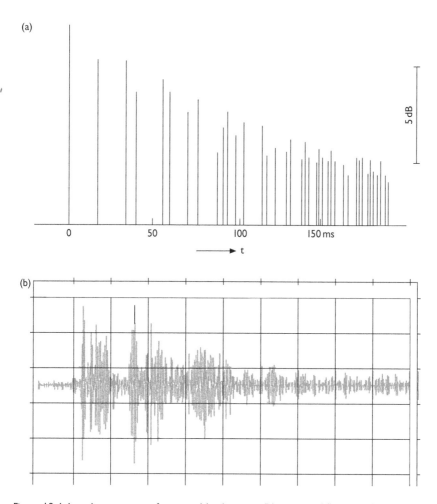

Figure 13.4 Impulse response of a room: (a) schematic, (b) measured (linear ordinate scale).

It can serve as an objective measure for speech intelligibility. Furthermore, we mention the 'clarity'

$$C = 10 \cdot \log_{10} \left[\frac{\int_0^{80\text{ms}} [g(t)]^2 \, dt}{\int_{80\text{ms}}^{\infty} [g(t)]^2 \, dt} \right] \text{ dB} \tag{13.6}$$

which is used to characterise the transparency of musical presentations.

Sometimes it happens that the impulse response of a room shows a pronounced peak occurring at a delay time exceeding 50 ms. Such a peak is perceived as an echo. It can be caused by sound reflection from a concavely curved wall or by accidental accumulation of many weaker reflections with almost equal delays.

Another feature of a reflection is the direction from which it arrives at the listener's position. Usually, sound reflections do not impair a listener's ability to localise a sound source in a room although most of the sound energy he receives arrives from other directions than from that of the direct sound. This interesting effect is due to the 'law of the first wavefront' mentioned by the end of the preceding chapter. It states that a listener localises the sound source from the direction of the direct sound. Nevertheless, the listener has a certain perception of the great variety of directions involved in the transmission of reflected sounds; it creates what may be called the subjective sensation of space, or the impression of being enveloped by sound.

13.3   Diffuse sound field

In a closed room the sound waves (or sound rays) are repeatedly reflected from its boundary, and with each reflection they change their direction. Therefore, at any given point sound waves arrive from quite different directions. Let us denote with $I'(\phi, \theta)d\Omega$ the intensity of all waves or rays which cross this position within a solid angle $d\Omega$ specified by the angles ϕ and θ. Then the energy density associated with this intensity is $dw = I'd\Omega/c$. The total energy density is obtained by integrating this quantity over all directions, that is, over the full solid angle 4π:

$$w = \frac{1}{c} \iint_{4\pi} I'(\phi, \theta)d\Omega \tag{13.7}$$

Furthermore, we calculate the energy per second incident on a wall element with the area dS (see Fig. 13.5). The portion of energy arriving from the direction ϕ, θ is

$$dE(\phi, \theta) = I'(\phi, \theta) \cdot \cos\theta dS d\Omega \tag{13.8}$$

the factor $\cos\theta$ accounts for the projection of dS perpendicular to the considered direction. Since all contributions stem from half the space this expression must be integrated over the solid angle 2π. The result, divided by the area dS, reads:

$$B = \iint_{2\pi} I'(\phi, \theta) \cos\theta d\Omega \tag{13.9}$$

The quantity B is the so-called irradiation density of the wall as already mentioned in Section 7.5.

The calculation of the integrals in eqs. (13.7) and (13.9) becomes trivial if the quantity I' is independent of the angles ϕ and θ. Then we obtain with

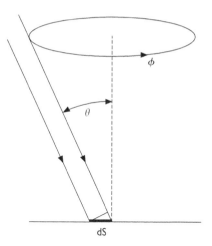

Figure 13.5 Derivation of eq. (13.8).

$\mathrm{d}\Omega = 2\pi \cdot \sin\theta \mathrm{d}\theta$:

$$w = \frac{4\pi I'}{c} \quad \text{and} \quad B = 2\pi I' \int_0^{\pi/2} \cos\theta \sin\theta \mathrm{d}\theta = \pi I'$$

Eliminating I' from both formulae yields the important relation:

$$B = \frac{c}{4}w \qquad\qquad (13.10)$$

The independence of the 'differential intensity' I' on the angles ϕ and θ as was assumed here means that all directions participate equally in the sound propagation. Such a sound field can be called isotropic; in room acoustics the term 'diffuse' is more common. Of course, it cannot be exactly realised, otherwise there would be no net energy flow within the room. This, however, is impossible, since the inevitable wall losses 'attract' a continuous energy flow originating from the sound source. Nevertheless, the structure of a sound field in a real room is usually closer to the ideal case of a diffuse sound field than to that of a plane wave. This holds, in particular, if the enclosure is irregularly shaped and is bounded by diffusely reflecting surfaces.

In what follows we shall derive two further properties of diffuse sound fields. Usually, the absorption coefficient α of a wall element dS depends on the angle of incidence θ. To find the way it can be averaged we consider once more the energy $\mathrm{d}E(\phi, \theta)$ falling per second onto dS from the direction ϕ, θ. The wall absorbs the fraction $\alpha(\theta)$ of this energy. To obtain the total energy absorbed by dS per second we integrate $\alpha(\theta) \cdot dE(\phi, \theta)$ over all directions using eq. (13.8) (with constant I'):

$$E_a = I'dS \iint_{2\pi} \alpha(\theta) \cos\theta \, d\Omega = 2\pi I'dS \int_0^{\pi/2} \alpha(\theta) \cos\theta \sin\theta \, d\theta$$

Now we introduce an average absorption coefficient by the relation $E_a = \alpha_m BdS$. Equating this with the earlier expression leads to

$$\alpha_m = 2 \int_0^{\pi/2} \alpha(\theta) \cos\theta \sin\theta \, d\theta \Omega \tag{13.11}$$

since $B = \pi I'$. This expression is known as Paris' formula.

If the wall impedance is known the absorption coefficient $\alpha(\theta)$ can be obtained from eq. (6.23). In this case the integration leads to a closed expression provided the considered wall portion is of the locally reacting type (see Section 6.4) which means that the wall impedance does not depend on the angle of incidence. The result is shown in Figure 13.6 in the form of contours of constant absorption coefficient. This diagram is in a way the counterpart of Figure 6.5 valid for normal sound incidence. It tells us that at random sound incidence the absorption coefficient cannot exceed the value 0.951 which occurs for $\xi = 1.567$ and $\eta = 0$.

Next we imagine the sound field as consisting of small energy portions called sound particles: they are thought to travel along straight paths through the room until they hit a wall and are reflected from it. Of course, these particles do not have any physical reality; rather, they offer a convenient possibility of considering the time history of sound propagation. Let us denote the energy of such a particle with ε_0, then its contribution to the mean energy density is $w = \varepsilon_0/V$ with V indicating the room volume. Suppose a particle hits a wall $\bar{n}$ times per second on average, then it transports the power $\bar{n}\varepsilon_0$ towards the boundary. Dividing this by the area S of the boundary yields the particle's contribution $B = \bar{n}\varepsilon_0/S$ to the irradiation density of the boundary. Now we insert these expressions for w and B into eq. (13.10) and obtain immediately the average rate of wall collisions of a sound particle (or of a sound ray):

$$\bar{n} = \frac{cS}{4V} \tag{13.12}$$

The reciprocal of this expression is the average time between two collisions or reflections, or, multiplied with the sound velocity c, the average distance which a particle travels between two successive wall reflections:

$$\bar{l} = \frac{4V}{S} \tag{13.13}$$

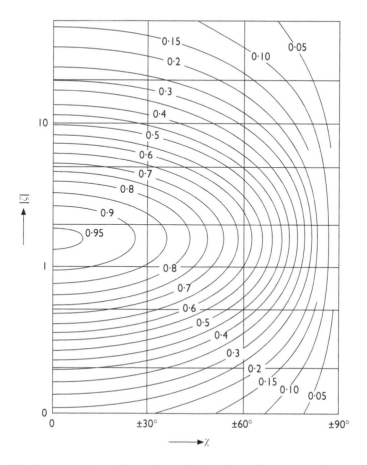

Figure 13.6 Contours of equal absorption coefficient of a locally reacting surface at random sound incidence. The abscissa is the magnitude, the ordinate the phase angle of the specific impedance of the surface.

Usually, this quantity is called the 'mean free path length' of a sound particle or a sound ray.

Basically, the quantities $\bar{n}$ and $\bar{l}$ are defined as temporal averages, valid for one particular particle. However, if the sound field is diffuse it is safe to assume that the 'fates' of all particles, although being different in detail, are similar as far as their statistical behaviour is concerned. Then the difference between time averages and averages over many particles becomes meaningless, so that the quantities in eqs. (13.12) and (13.13) can be regarded as representative for all particles.

13.4 Steady-state energy density and reverberation

In the preceding section there was no mention of any sound source; we regarded the sound field as a given fact. Now our main concern is the energy density which will be established when a certain acoustical power is supplied to a room. In this discussion assume from the outset that the sound field in the room is diffuse. Moreover, we restrict our discussion to statistical averages for which some simple relations can be derived.

It is intuitively clear that the energy density in a room will be the higher, the more acoustical power is produced by a sound source operated in it, and the lesser energy per second will be lost by dissipative processes, in particular, the lower the absorption coefficient of the boundary. This leads us immediately to the energy balance:

Temporal change of energy content = energy supplied by the source

− absorbed energy

All energies on the right hand side are per second. The energy content is the energy density w multiplied with the room volume V, hence the left side of this equation is $V \cdot dw/dt$. The energy supplied to the room per second is the power output P of the source. To calculate the absorbed energy we divide the boundary into partial surfaces S_i along which the absorption coefficient α_i can be assumed as uniform. Each of this partial surfaces absorbs the energy $\alpha_i B S_i$ per second, or, after expressing the irradiation density B with eq. (13.10) by the energy density, $\alpha_i S_i c w / 4$. Finally, we introduce the 'equivalent absorption area'

$$A = \sum_i \alpha_i S_i \tag{13.14}$$

Hence the mathematical expression for the energy balance reads

$$V\frac{dw}{dt} = P(t) - \frac{c}{4}Aw \tag{13.15}$$

This is a differential equation of first order for the energy density. It can be solved in a closed form for any time-dependent source power P(t). Here we shall consider two special cases: first the source power P is assumed as constant, then the same holds for the energy density. This leads to

$$w = \frac{4P}{cA} \tag{13.16}$$

This formula agrees with what we would have expected. However, it does not yet reflect the full truth: next to the sound source, the prevailing part of the energy density is doubtless that which is produced directly by it, that is, without any influences of the room. Suppose the source emits a spherical wave with the intensity $I_d = P/4\pi r^2$ where r is the distance from the sound source. This corresponds to the energy density $w_d = I_d/c$ or

$$w_d = \frac{P}{4\pi c r^2} \tag{13.17}$$

The distance r_c where both energy densities, w and w_d, are equal is called the 'critical distance' or 'diffuse-field distance' (see Fig. 13.7). By equating eqs. (13.16) und (13.17) we obtain:

$$r_c = \sqrt{\frac{A}{16\pi}} \tag{13.18}$$

With this quantity the total energy density $w_{tot} = w + w_d$ can be expressed by

$$w_{tot} = \frac{P}{4\pi c}\left(\frac{1}{r^2} + \frac{1}{r_c^2}\right) \tag{13.19}$$

The first term describes the direct field, the second one represents what is called the reverberant field.

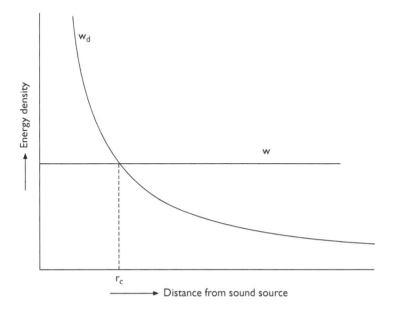

Figure 13.7 Definition of the critical distance r_c (w, w_d: energy density in the reverberant and in the direct field, respectively).

On the other hand, if the source radiates the sound non-uniformly because of its directivity then the first term in eq. (13.19) must be multiplied with the gain γ introduced in eq. (5.16) by which the intensity in the direction of main radiation is enhanced:

$$w_{tot} = \frac{P}{4\pi c}\left(\frac{\gamma}{r^2} + \frac{1}{r_c^2}\right) \tag{13.19a}$$

Hence, the critical distance where the direct and the reverberant energy density are equal becomes $r'_c = r_c\sqrt{\gamma}$.

The above formulae show in which way the energy densities and hence the noise level, for instance, in a work room can be controlled, namely, by increasing the absorption area A which can be achieved by a sound-absorbing lining particularly on the ceiling. However, only the energy density in the reverberant field, that is, the second term in eq. (13.19) or (13.19a), can be influenced in this way. Therefore, this method has its limitations when the critical distance r_c becomes comparable to the room dimensions. Furthermore, the sound field in such a room is usually so far from the diffuse state that the earlier relations yield at best a clue for the attainable level reduction.

In the second case we suppose that the sound source was in operation until the time $t = 0$ and has created a certain energy density w_0 until that instant. Then the source is switched off ($P = 0$). Now the differential equation (13.15) is homogeneous, and its solution is

$$w(t) = w_0 e^{-cAt/4V} \quad \text{for } t > 0 \tag{13.20}$$

It describes the decay of sound energy in a room and agrees with eq. (9.36) with $\langle\delta\rangle = cA/8V$, which was obtained in a different way. From eq. (9.37) we obtain the reverberation time:

$$T = \frac{24 \cdot \ln 10}{c} \cdot \frac{V}{A} \tag{13.21}$$

Inserting the sound velocity of air leads to:

$$T = 0.163\frac{V}{A} \tag{13.22}$$

Here all lengths are in metres.

Along with the reflection law, this equation is the most important relation in room acoustics. It goes back to the pioneer of room acoustics, W. C. Sabine, and is named after him.

However, it can claim validity only as long as the absorption area is small compared to the area S of the boundary walls. To see this let us assume that the whole boundary is totally absorbing, that is, by setting all absorption coefficients in eq. (13.14) equal to unity. Then eq. (13.22) predicts a finite reverberation time although there are no reflecting walls at all.

To arrive at a more exact reverberation formula we realise that during sound decay the energy is not continuously diminished as in eq.(13.15) but in finite steps. As in the preceding section we consider the fate of a hypothetical sound particle. Each reflection reduces its energy by a factor $1 - \alpha$. Since it undergoes $\bar{n}$ reflections per second the particle energy left after t seconds is the fraction $(1 - \alpha)^{\bar{n}t} = \exp\left[\bar{n}t \ln(1 - \alpha)\right]$ of its initial energy. What is true for one single particle holds for the total energy density too which now decays according to

$$w(t) = w_0 e^{\bar{n}t \ln(1-\alpha)} \tag{13.23}$$

This relation must be completed by taking into account two factors: since the absorption coefficient is generally not constant along the entire boundary, α must be replaced with the arithmetic average

$$\bar{\alpha} = \frac{A}{S} = \frac{1}{S} \sum_i \alpha_i S_i \tag{13.24}$$

Furthermore, the sound energy is dissipated not only on the room walls but also during its propagation in air. This is accounted for by an additional factor $\exp(-mct)$ in eq. (13.23). Here m denotes the attenuation constant defined in eq. (4.19). Finally, we set $\bar{n} = cS/4V$ and obtain the more correct formula

$$w(t) = w_0 \exp\left[\frac{cSt}{4V} \ln(1 - \bar{\alpha}) - mct\right] \tag{13.23a}$$

$$T = 0.163 \frac{V}{4mV - S\ln(1 - \bar{\alpha})} \tag{13.25}$$

This relation is usually referred to as 'Eyring's formula'. For $\bar{\alpha} \ll 1$ we can use the approximation $\ln(1 - \bar{\alpha}) \approx -\bar{\alpha}$; then eq. (13.25), the Eyring's formula, becomes identical with the Sabine decay formula, eq. (13.22), however, with the somewhat extended version of the equivalent absorption area which includes the attenuation in air:

$$A = \sum_i \alpha_i S_i + 4mV \tag{13.26}$$

In most cases, however, the term 4mV may be neglected unless the decay formula is applied to large rooms and for elevated frequencies.

13.5 Sound absorption

Whichever description we prefer, the geometrical one as in Section 13.1 or the statistical one as in Section 13.4, in any case the sound field in a room and hence what we hear in it is largely determined by the absorption of its boundary. Since the basic mechanisms of sound absorption have already been described in Section 6.6 we can restrict this discussion to some amendments.

At first we must remember that walls without any sound absorption do not exist. Even a completely rigid wall with a smooth surface has a finite absorption coefficient. Similar to the inner surface of a pipe (see Section 8.1), a boundary layer is formed on any smooth surface which is exposed to a sound wave, and in this layer viscosity and heat conductivity of the air cause vibrational losses. The wall can be thought of being covered with a skin in which sound is absorbed (see Fig. 13.8a). For this reason, a wall or ceiling will hardly have an absorption coefficient below 0.01. Roughness of the wall increases the thickness of this layer (see Fig. 13.8b). A further, quite dramatic increase of absorption occurs if the wall has pores. When these are sufficiently narrow they are completely filled with the boundary layer as shown in Figure 13.8c. In this way a relatively high amount of energy is extracted from the intruding sound wave.

In Subsection 6.6.2 the absorption of a porous layer arranged immediately in front of a rigid wall was dealt with. The result is presented in Figure 6.9. Although this diagram is based on the highly idealised Rayleigh model of a porous material and is valid for normal sound incidence only it shows the typical properties of porous absorption. In particular, it demonstrates that high absorption is only achieved with layers the thickness of which is a significant fraction of a wavelength which may be difficult to reach, especially at low frequencies.

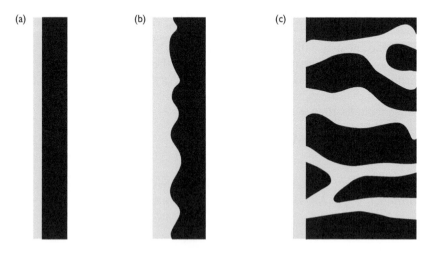

Figure 13.8 Lossy boundary layer: (a) in front of a smooth surface, (b) in front of a rough surface and (c) in front of and in a porous material.

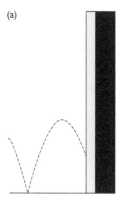
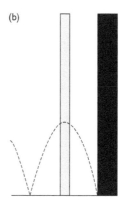

Figure 13.9 Porous layer in front of a rigid wall (dotted line: particle velocity when the layer is absent): (a) layer immediately on the wall, (b) layer mounted with air backing.

The low-frequency absorption of a porous layer of given thickness can be considerably enhanced by mounting it not directly onto a rigid wall or ceiling but with an air space in between (see Fig. 13.9). Then the absorber is no longer situated close to a zero of the particle velocity but at a position where air can be forced through the pores. In the limit of a very thin sheet layer we ultimately arrive at the stretched fabric already dealt with in Subsection 6.6.3 (second part).

Commercial absorption materials are often made of incombustible granules or fibres (glass wool or rock wool) by pressing them together, using a binding agent. Alternatively, foamed polymers with open pores can be used as absorbents. Since the appearance of such materials is usually not very pleasant they are often covered with a thin perforated sheet. This consists typically of highly perforated panels of metal, wood or gypsum which at the same time protect the soft surfaces against damage and prevent particles from polluting the air.

In any case, the typical range of application of a porous absorber is that of medium and high frequencies. The low-frequency range can be covered, if desirable, by resonance absorbers as described in Subsection 6.6.4 which find extended use in room acoustics. Practically, such an absorber consists of a panel made of wood, gypsum, metal, etc. mounted in such a way that it is free to perform flexural vibrations when exposed to a sound field (see Fig. 13.10a). For this purpose it is fixed by a supporting construction on the wall to be lined. Provided the panel is not too thick and the mutual distance L of supports is not too small the influence of the bending stiffness on the resonance frequency can be neglected against that of the air cushion

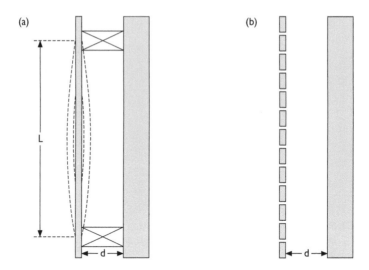

Figure 13.10 Resonance absorber: (a) with vibrating panel, (b) with perforated panel.

behind. Then the resonance frequency is given by eq. (6.52) which may also be written in the form

$$f_0 = \frac{600}{\sqrt{m'd}} \text{ Hz} \tag{13.27}$$

where m' is the specific mass in kg/m² and d the thickness of the air cushion in centimetres. In this form it is valid for normal sound incidence, for random incidence, that is, in a diffuse sound field the figure 600 has to be replaced with 850.

According to Subsection 8.3.3 a rigidly mounted, perforated panel in front of a rigid wall acts also as a resonance absorber (see Fig. 13.10b). Again, the resonance frequency is given by eq. (13.27), and the specific mass associated with the panel is

$$m' = \frac{S_1}{S_2} \rho_0 l \tag{13.28}$$

Here S_2/S_1 is the ratio of openings to the total area of a specimen. This ratio has the same meaning and effect as the porosity σ as introduced earlier. The variable l is the geometric thickness of the panel plus twice the end correction

$$l = l_{geo} + 2\Delta l \tag{13.29}$$

For circular holes with the radius a the latter term is $\pi a / 2$ (see eq. (7.11)).

It is self-evident that in both cases absorption occurs only if the vibration of the resonator is associated with losses. One loss process is radiation, and other losses are caused by elastic dissipation in a panel performing flexural vibrations, or, in case of a perforated panel, by the viscosity of the air flowing through its holes. For both types they can be enhanced by filling the back space behind the panel partially or completely with porous material. The frequency-dependent absorption coefficient of resonance absorbers with different loss resistance is presented in Figure 6.12.

The resonance frequency of a wall lined with unperforated wood panels is typically in the range of 80–100 Hz; using perforated panels the resonance frequency can be varied within wide limits. For instance, it is about 250 Hz for a panel of 10 mm thickness which is perforated at 3% with holes of 8 mm diameter and is mounted at 8 cm distance from the wall.

In most auditoria the principal sound absorption is caused by the audience. Therefore, the knowledge of audience absorption is of crucial importance for any reliable prediction of the reverberation time. Unfortunately, this quantity depends in a complicated manner on several circumstances, for instance, the kind of seats, the seating density of the audience and on the division of the whole audience area into seating blocks, etc., to a certain extent also on the kind of clothing worn since it is the latter which is ultimately responsible for the absorption.

For calculating the reverberation time the absorption of audience can be accounted for in two different ways. First, a certain absorption area δA can be attributed to each person present – including the seat it occupies. Then the total absorption area of the room is

$$A = \sum_i \alpha_i S_i + N \cdot \delta A \tag{13.30}$$

the sum represents the contribution of the boundary after eq. (13.14) and N is the number of listeners. The more common is the second way, namely, to attribute an absorption coefficient to a closed area of listeners. Then eq. (13.14) can be applied without any modification. However, the audience area S_a is not just the geometrical floor area the audience occupies, but it has to be augmented by $L_a / 2$ where L_a is the total length of the circumferences of the seating blocks. This additional area is to account for the effect of sound scattering which occurs at the edges of a block and which increases the effective absorption of the audience.

Finally, for somewhat larger rooms the attenuation of the air must be taken into account with a term 4mV as in eqs. (13.25) and (13.26). Table 13.1 lists

Table 13.1 Intensity-related attenuation constant m (in 10^{-3} m^{-1}) of air at normal conditions

Relative humidity (%)	Frequency (Hz)					
	500	1000	2000	4000	6000	8000
40	0.60	1.07	2.58	8.40	17.71	30.00
50	0.63	1.08	2.28	6.84	14.26	24.29
60	0.64	1.11	2.14	5.91	12.08	20.52
70	0.64	1.15	2.08	5.32	10.62	17.91

Table 13.2 Typical absorption coefficients (random incidence)

Material	Octave band centre frequency (Hz)					
	125	250	500	1000	2000	4000
Hard surfaces (concrete, brick walls, plaster, hard floors, etc.)	0.02	0.02	0.03	0.04	0.05	0.05
Linoleum on felt layer	0.02	0.05	0.1	0.15	0.07	0.05
Carpet, 5 mm thick, on solid floor	0.02	0.03	0.05	0.10	0.30	0.50
Slightly vibrating surfaces (suspended ceilings, etc.)	0.10	0.07	0.05	0.04	0.04	0.05
Acoustic plaster, 10 mm thick, sprayed on solid wall	0.08	0.15	0.30	0.50	0.60	0.70
Polyurethane foam, 27 kg/m^3, 15 mm thick on solid wall	0.08	0.22	0.55	0.70	0.85	0.75
Rockwool, 46.5 kg/m^3, 30 mm thick, on concrete	0.08	0.42	0.82	0.85	0.90	0.88
Same as above, but with 50 mm air space, laterally partitioned	0.24	0.78	0.98	0.98	0.84	0.86
Metal panels, 0.5 mm thick with 15% perforation, backed by 30 mm rockwool and 30 mm additional air space, no partitions	0.45	0.70	0.75	0.85	0.80	0.60
Plush curtain, flow resistance 450 Ns/m, deeply folded, distance from solid wall ca. 5 cm	0.15	0.45	0.90	0.92	0.92	0.95
Fully occupied audience, upholstered seats	0.50	0.70	0.85	0.95	0.95	0.90

some values of the attenuation constant m. Table 13.2 shows the absorption coefficients of various materials and linings; it includes also those of audience. However, these values should be understood rather as typical examples than as generally valid absorption data.

13.6 On the 'acoustics' of auditoria

Which objective sound field properties are responsible for what is called good or less good 'acoustics' of a lecture room, of a theatre or a concert hall? Are there quantitative criteria for assessing the quality of acoustics, and is it possible to design a hall in such a way that it will exhibit good or even excellent listening conditions?

At first there are some simple and obvious requirements which must be met in any hall. The first one concerns the noise level caused by technical installations such as, for instance, air conditioning, but also the noise intruding from outside. It must remain below certain limits prescribed by general standards. Furthermore, a hall must be free of audible echoes. This requirement is related to another one, namely, that the sound energy be uniformly distributed over the whole audience thus ensuring sufficient and as near as possible equal loudness at all places. This is mainly a matter of the shape of a room. Thus, curved wall or ceiling portions concentrate the reflected sounds predominantly into particular regions and prevent a uniform distribution of sound energy. In combination with long delays of the reflected sounds, this effect may lead to audible echoes. Particular risks in this respect must be expected in rooms with circular or elliptical ground plan, but other regular room shapes may also turn out as problematic.

In any case, however, it is important to deal with the limited sound energy in an economic way, that is, to direct it by suitable design of the walls and the ceiling to those places where it is needed, namely, at the listeners. Thus all surfaces which produce 'useful' reflections with small delays must reflect efficiently, in particular, they must not be lined or covered with any kind of sound absorbent. Hence, the decorative curtain which is seen so often at the rear walls of stages is completely out of place.

Another important condition is that the room has a reverberation time which favours the kind of presentation. If the room is to be used mainly for teaching, for lectures, for discussions or for drama performances its reverberation time should be relatively low, since long reverberation mixes the various speech sounds and syllables and hence reduces the intelligibility of speech. In principle, such a room would not need any reverberation. This could be achieved by a heavily absorbing treatment of the walls and the ceiling which, however, would also prevent useful reflections mentioned earlier which are so important for the sound supply. A useful compromise is a reverberation time in the range of about 0.5 to 1.3 seconds with the longer reverberation time for larger halls. Especially at low frequencies the reverberation must be not too long; otherwise, it would mask the medium and high-frequency spectral components which are particularly important for speech intelligibility.

Matters are different when it comes to concert halls. Music is not supposed to be 'understood' in the same sense as speech. Thus the bowing noises of the string instruments or the air noise of the woodwind should not be perceived, and the same holds for the inevitable imperfections of synchronism and intonation. For creating the typical orchestral sound some blending of the different musical sounds is indispensable, and also subsequent tones of a passage should merge to some extent. This spatial and temporal smoothing effect is achieved by reverberation; according to experience the duration of it should be somewhere in the order of 2 seconds in a large concert hall. Often an increase of reverberation time towards low frequencies is found desirable because this is said to add 'warmth' to the sound of an orchestra.

Besides adequate reverberation it is important that the projection of sound towards the audience by suitably oriented wall and ceiling portions is not exaggerated. If all the sound produced by the musicians would be directed exclusively onto the highly absorptive audience it would not have the chance of exciting the room's reverberation to any sufficient degree, and not much of it would be perceived even if the calculated reverberation time were in the correct range (see Fig. 13.11). Furthermore, the stage enclosure must be designed in such a way that it reflects some sound back to the performing musicians in order to establish the mutual auditory contact they need.

Adequate reverberation and good mixing of sound is not the only signature of an acoustically good concert hall. The concert-goer unconsciously expects that he is enveloped by the music, so-to-speak, or that the sound field creates an acoustically spatial impression. Nowadays, there is general agreement that this impression is brought about by reflections arriving from lateral directions at the listener. Since the listener's head diffracts the arriving sound waves these reflections give rise to different sound signals at both ears. As explained in Section 12.7 these interaural differences enable us to localise the direction of sound incidence in the field of a single sound source. In the more complicated sound field in a concert hall they contribute to the subjective impression of space.

Concerning the opera theatre we should expect that a compromise between the relatively long reverberation times of a concert hall and the

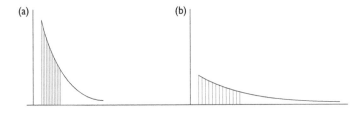

Figure 13.11 Reasons for insufficient reverberance of a concert hall: (a) reverberation time too short, (b) insufficient excitation of reverberation. (The vertical line represents the direct sound.)

shorter values favouring good speech intelligibility would create optimum listening conditions. Now older opera houses have rather short reverberation times in the range of about 1 second while the decay times of more modern opera theatres come pretty close to those which are regarded optimum for concert halls. Probably, the modern opera goer gives more preference to beautiful sounding arias and a full and smooth orchestra sound than to easy understanding of the sung text which the true connoisseur knows anyway by heart.

Now we shall go into the third of the questions asked at the beginning of this subsection. The reverberation time of a room is easily predicted by one of the formulae presented earlier; usually eq. (13.22) is used for this purpose. How realistic the results of such a calculation are depends on the inserted absorption coefficients, of course. Furthermore, the significant features of the impulse response essential can be calculated in advance from the architect's drawings. For this purpose the method of image sources as mentioned in Section 13.1 may be employed. In another very powerful method known as 'ray tracing' the individual paths of numerous sound particles are computed which are emitted in all directions at a given instant. It has the advantage of being applicable not only to plane and smooth surfaces but to curved or scattering ones as well. Furthermore, procedures involving combinations of both methods have been developed. From the impulse responses virtually all relevant data can be evaluated, that is, not only the reverberation time but also criteria which may differ from one seating location to another, as is the case with definition or clarity index, or criteria indicating the spatial impression.

A very interesting and promising procedure called 'auralisation' enables us to process and to present music samples in such a way as though the listener were attending a performance at a particular seat in the considered hall, which, maybe, exists only on paper. This is carried out with the aid of a digital frequency filter which simulates the binaural impulse response, calculated separately for each of the listener's ears. The result is presented to the listener by headphones or preferably with loudspeakers using a procedure to be described in Subsection 20.1.3. In this way listening conditions, for instance, at different places of a concert hall or in different concert halls can be directly assessed and compared with each other.

Finally, we want to refute the opinion that acoustical faults of a room can be cured by a carefully designed electroacoustical installation. In concert halls such systems are usually refused both by musicians and by music lovers. The same holds for the theatre, although to a somewhat minor extent; for the presentation of musicals electroacoustic sound systems are generally accepted. But even if the use of electroacoustic amplification is indispensable as is certainly true for large lecture halls, parliaments, sports arenas, etc., the 'natural' sound transmission should be as favourable as possible for the particular kind of presentation. In Section 20.3 the close relation between the acoustical properties of a room and the function of the sound system installed in it will become evident.

13.7   Special rooms for acoustic measurements

For many acoustical measurements environments with well-defined properties are needed. This is achieved with specially designed rooms, in particular, the anechoic room and its counterpart, the reverberation room or reverberation chamber.

The purpose of an anechoic room is to create free field conditions, that is, the absence of any reflections. This is needed, for instance, for the free field calibration of microphones, for measuring the transfer function and the directional characteristics of loudspeakers, for psychoacoustic investigations and for many other purposes. Since it is impossible to suppress all reflections completely the practical requirement is that the reflection factor of the floor, the side walls and the ceiling be less than 0.1 corresponding to an absorption coefficient exceeding 0.99. This means that the sound pressure level of a plane wave is reduced by 20 dB when it is reflected. This is easily achieved at medium and high audio frequencies but not in the low-frequency range. In most anechoic rooms the boundary is lined with pyramids or wedges of porous material forming a gradual transition from the characteristic impedance of air to that of the solid wall behind. The range of high sound absorption extends down to about that frequency for which the length of the wedges is one-third of the wavelength. In the lining shown in Figure 13.12 each group of three wedges is combined into one block with parallel edges, and adjacent blocks are mounted with their edges running perpendicular to each other. The lower limit of the useful frequency range can be further reduced by providing for an air space behind the wedges which, together with gaps between the wedges, acts as a resonance absorber. Since

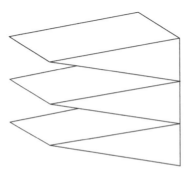

Figure 13.12 Wall element of an anechoic room.

the floor of the anechoic room is also lined with wedges, access to the room is given by a grille or a net which can be regarded as sound transparent.

In contrast to the anechoic room, the sound field in a reverberation room should be as diffuse as possible. This requires a highly reflective boundary which means that all walls, the floor and the ceiling of the room must be as smooth and as heavy as possible and completely free of pores. An irregular room shape is quite favourable in that it forces the sound waves to change their direction frequently and hence improves diffusion. Further improvement is achieved by regular or irregular surface structures such as cylindrical or spherical segments. Very effective are freely suspended scatterers such as bent panels, for instance. They improve mixing of sound field components by scattering them again and again. A reverberation room should have a volume in the range of 200 m^3 and a reverberation time of at least 5 seconds.

One important application of a reverberation room is in measuring the total output power of sound sources. This method is based upon eq. (13.16) according to which

$$P = \frac{cA}{4} \cdot w = \frac{cA}{4Z_0} \cdot \tilde{p}^2 \tag{13.31}$$

The latter expression follows from eq. (5.8); the effective sound pressure $\tilde{p}$ is determined with a calibrated sound level meter. The absorption area A is obtained from the reverberation time of the room.

Furthermore, the reverberation room is employed for the measurement of the sound absorption of walls, wall linings of any kind but also of persons, chairs and other single absorbing objects. While absorption measurements with the impedance tube (see Section 6.5) are restricted to small samples of locally reacting surfaces and to normal sound incidence, the reverberation room yields data determined with a diffuse sound field which are usually more relevant in room acoustics.

To determine the absorption coefficient in a reverberation room, a sample of the material under test is placed on the floor or mounted on a wall of the room. From the reverberation time the mean absorption coefficient $\overline{\alpha} = A/S$ (S = area of the boundary) is determined by using eq. (13.22) or, preferably, eq. (13.25). Because of the small volume of the room the term 4 mV is usually omitted. On the other hand, according to eq. (13.24) the mean absorption coefficient is

$$\overline{\alpha} = \frac{1}{S} [S_s \alpha_s + (S - S_s)\alpha_0] \tag{13.32}$$

In this expression S_s is the area of the sample and α_0 denotes the absorption coefficient of the bare boundary. The latter must be determined from the reverberation time of the empty chamber. Then the absorption coefficient α_s

of the sample is:

$$\overline{\alpha}_s = \frac{1}{S_s} \left[S(\overline{\alpha} - \alpha_0) + S_s \alpha_0 \right] \tag{13.33}$$

However, it should be mentioned that there is often disagreement between results obtained with the same sample, but in different laboratories. This shows that achieving diffuse-field conditions is neither a trivial matter nor easy.

Chapter 14

Building acoustics

In a certain sense building acoustics is the counterpart of room acoustics because both refer to sound propagation in buildings. However, the objectives of both areas of acoustics are quite different. While the goal of room acoustics is to optimise sound transmission and listening conditions in a room, in building acoustics we are endeavouring to impede sound transmission between adjacent rooms of a building or to prevent external noise from entering the building. Thus, building acoustics has to do with noise control in buildings.

Viewed from the acoustical standpoint a building consists essentially of wall, floors and ceilings which separate different rooms from each other or from the exterior. Hence, a necessary prerequisite of good noise protection in a building is sufficiently high sound insulation of such elements. The same holds for doors and windows. It is the goal of this section to describe the factors on which sound insulation depends.

In building acoustics it is customary to distinguish between airborne and structure-borne sound excitation. In the former case the vibrations of a partition are produced by sound waves in air originating from speakers, from musical instruments, or more typically, from the television loudspeakers or from external sound sources. In contrast, structure-borne sound is generated by sources which are in direct mechanical contact with a wall or floor and exert alternating forces on it. Typical sources of structure-borne sound are the shoes of walking persons, elevators, water installations or rotating technical devices. In any case the vibrations of the partition are converted by radiation into audible sound. Furthermore, they can travel within the structure of the building in the form of structure-borne sound waves and can be converted or reconverted into airborne sound at some more distant place. Of course, both forms of transmission lead to undesired effects.

At first a preliminary comment on the kind of waves we expect as carriers of structure-borne sound. The velocity of longitudinal waves in common building materials is of the order of some 4500 m/s. On the other hand the frequency range which is of primary interest in building acoustics reaches up to slightly over 3 kHz. At this frequency the wavelength of longitudinal waves is about 1.5 m, that is, it is large compared to common wall thicknesses. From this it follows that walls and ceilings of a building can be regarded as plates in the meaning of Section 10.3. Accordingly, the propagation of sound in a building takes place in the form of extensional or quasi-longitudinal waves, and bending waves.

14.1 Characterisation and measurement of airborne sound insulation

The transmission of airborne sound through a partition between adjacent rooms – for instance, a wall or a ceiling – is characterised by comparing the sound intensities of an impinging and transmitted wave. Let us denote these intensities by I_0 and I_t, respectively, then the sound reduction index or sound transmission loss of the considered element is defined by

$$R_A = 10 \log_{10} \left(\frac{I_0}{I_t} \right) \tag{14.1}$$

assuming plane waves.

Measurements according to this definition will be performed only exceptionally because the direct measurement of intensities requires special equipment and may turn out to be relatively time-consuming. Moreover, one is mostly interested in sound insulation with regard to random sound incidence. The typical arrangement for measuring the sound insulation consists of two adjacent rooms separated by the wall to be examined as shown in Figure 14.1. Instead of intensities, sound powers are compared with each other: let P_0 denote the total power incident on the partition wall while P_t is the power the partition emits on its far side. Then we arrive at the following definition of the sound reduction index:

$$R_A = 10 \log_{10} \left(\frac{P_0}{P_t} \right) \tag{14.2}$$

which is equivalent to eq. (14.1) if we set $I_0 = P_0/S$ and $I_t = P_t/S$ ($S =$ area of the partition).

Under the assumption that the sound fields in both rooms are diffuse we can express powers by sound pressure levels. The incident power is $P_0 = BS$ with B denoting the 'irradiation density' (see Section 13.3). The latter is related by eq. (13.10) to the energy density w_1 in the 'sending room', hence we obtain

$$P_0 = \frac{c}{4} S \cdot w_1 \tag{14.3}$$

The power P_t is easily obtained from eq. (13.16)

$$P_t = \frac{c}{4} A \cdot w_2 \tag{14.4}$$

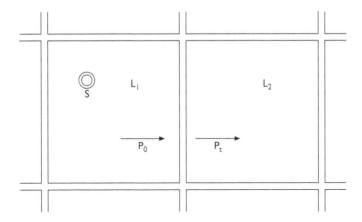

Figure 14.1 Sound transmission through a partition. P_0, P_t: incident and transmitted sound power; L_1, L_2: sound pressure level in the sending and the receiving room.

A is the equivalent absorption area of the receiving room. Inserting these expressions into eq. (14.2) yields:

$$R_A = 10 \log_{10} \left(\frac{w_1}{w_2} \right) + 10 \log_{10} \left(\frac{S}{A} \right)$$

or, since the first term is the difference of sound pressure levels in both rooms:

$$R_A = L_1 - L_2 + 10 \log_{10} \left(\frac{S}{A} \right) \tag{14.5}$$

Hence, with the assumptions of diffuse sound fields the determination of the transmission loss reduces to the measurement of the difference of two sound pressure levels; the absorption area A of the receiving room is obtained from the measured (or estimated) reverberation time by using the Sabine decay formula (13.22). Since the transmission loss depends markedly on frequency this measurement is usually carried out in frequency bands mostly of third-octave bandwidth covering the range from 100 Hz to 3.15 kHz.

Measurement of the transmission loss can be performed in special testing facilities as well as in completed buildings. Particularly, in the latter case the result is usually influenced by the fact that the sound can reach the receiving room not only by traversing the partition under test but also by transmission through flanking building elements. For instance, the primary sound field may excite bending waves in the adjacent elements which lead to radiation into the receiving room thus circumventing the test specimen. This and other flanking paths are shown in Figure 14.2. Errors due to flanking transmission

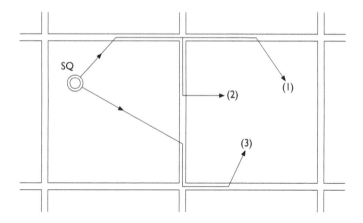

Figure 14.2 Flanking transmission.

can be avoided by using test facilities in which transmission along other paths is suppressed.

The quality of sound insulation of partitions is rated by means of an internationally standardised reference curve. This contour, valid for third-octave bands, is represented in Figure 14.3, along with a typical measurement result. On the one hand, it reflects what is technically feasible and reasonable since achieving high sound insulation at low frequencies is much more difficult and costly than in the high-frequency range. On the other hand, it takes into account that low-frequency spectral components are not as loud and annoying as those at higher frequencies, as may be seen from the curves of equal loudness (see Fig. 12.8).

The frequency dependence of the sound reduction index is of high interest because it may give clues to the reasons for unsatisfactory sound insulation. Nevertheless, it is often useful to characterise the transmission loss of a partitioning element by a single number. This can be obtained by shifting the reference curve upward or downward until it exceeds the measured results by just 2 dB, averaged over the frequency range from 0.1 to 3.15 kHz. ('Negative' excesses are not included in the averaging process.) Then the value of this shifted reference contour at 500 Hz is the single-number rating we are looking for; it is called 'weighted sound reduction index', abbreviated to R_w. According to international standards this quantity must be at least 53 dB for partition walls and 54 dB for floors if these elements are to separate different apartments.

The measured result shown in Figure 4.3 is the transmission loss of a partition of 24 cm thickness consisting of brick with plaster on both sides. Its sound reduction index is slightly higher than the reference curve. In fact,

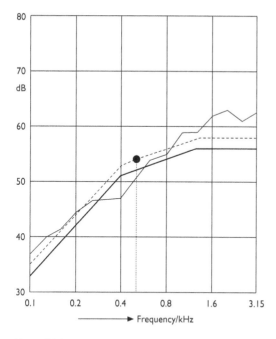

Figure 14.3 Airborne sound insulation of partitions. Solid curve: reference curve for the transmission loss; broken curve: shifted reference curve, thin curve; sound reduction index of a brick wall 24 cm thick with plaster on both sides.

the latter can be shifted upward by 2 dB until the condition mentioned is met. Its weighted sound reduction index R_w is 54 dB.

14.2   Airborne sound insulation of compound partitions

Often a partition wall is composed of two or more elements with different transmission losses. A common example is a wall with a window or a door in it. The goal of this section is to find the sound reduction index of such a multi-element partition.

Suppose the wall consists of two components with sound reduction indices R_{A0} and R_{A1} with $R_{A0} > R_{A1}$. The total wall area is S, its components have areas S_1 and $S - S_1$ (see Fig. 14.4a). The energies penetrating both of them per second are, after eq.(14.2):

$$P_{t1} = P_0 \cdot 10^{-0,1R_{A1}} \cdot \frac{S_1}{S} \tag{14.6a}$$

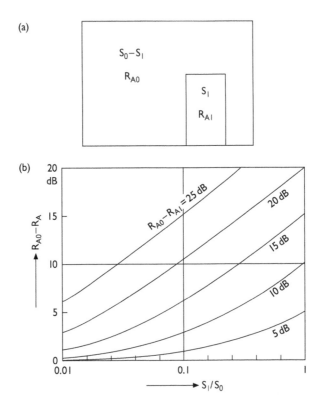

Figure 14.4 Airborne sound insulation of compound partitions: (a) representation, (b) reduction of the transmission loss. S_0, R_{A0}: area and transmission loss of the main wall; S_1, R_{A1}: area and transmission loss of the inserted wall element. Parameter: the difference $R_{A0} - R_{A1}$ of the transmission losses of both wall components alone.

and

$$P_{t0} = P_0 \cdot 10^{-0,1R_{A0}} \cdot \frac{(S - S_1)}{S} \tag{14.6b}$$

The total sound power P_t transmitted is the sum of both powers, hence the sound reduction index of the composed partition is, according to eq. (14.2):

$$R_A = -10 \log_{10} \left(\frac{P_t}{P_0} \right) = -10 \log_{10} \left[\left(1 - \frac{S_1}{S} \right) \cdot 10^{-0,1R_{A0}} \right.$$

$$\left. + \frac{S_1}{S} \cdot 10^{-0,1R_{A1}} \right] \tag{14.7}$$

The content of this somewhat awkward formula is represented in Figure 14.4b. It shows, as a function of the ratio S_1/S, the deterioration $R_{A0} - R_A$ of the transmission loss caused by inserting the element S_1 into the wall. The parameter of these curves is the difference $R_{A0} - R_{A1}$ of the transmission losses of both wall components. This diagram tells us, for instance, that an inserted element with a sound reduction index 20 dB below that of the main wall ($R_{A0} - R_{A1} = 20$ dB) and with $S_1 = S/10$ reduces the transmission loss of the total wall by slightly more than 10 dB.

Very often the sound reduction index R_{A0} of one part is much larger than that of the other which may be an inserted element with the sound reduction component R_{A1}. If, at the same time, the ratio S_1/S is not too small, the first term in the bracket of eq. (14.7) can be neglected which leads us to the approximate formula

$$R_A \approx R_{A1} - 10 \log_{10} \left(\frac{S_1}{S} \right) \qquad (14.7a)$$

This case is represented in the upper part of Figure 14.4b where the curves are almost straight lines.

However, the formulae presented earlier are only valid as long as the dimensions of the inserted element are significantly larger than the acoustical wavelength. If this condition is not fulfilled which may well be the case for a small window and for frequencies of up to about 500 Hz, the situation is complicated by sound being diffracted at the edges of S_1.

14.3 Airborne sound insulation of single-leaf partitions

In what follows we shall examine more closely the sound transmission through a homogeneous partition. We imagine the partition is of infinite extension, and the primary sound is assumed to be a plane wave. Furthermore, it is supposed that the wall itself is free of losses, hence its 'absorption' is caused in reality by transmission of sound to the far side. According to the definition of the absorption coefficient we can write for the transmission loss, using either eq. (14.1) or eq. (14.2):

$$R_A = 10 \log_{10}(1/\alpha) \qquad (14.8)$$

For normal sound incidence we immediately find the sound reduction index by applying eq. (6.48):

$$R_A = 10 \log_{10} \left[1 + \left(\frac{\omega m'}{2Z_0} \right)^2 \right] \qquad (14.9)$$

As earlier, m' is the specific mass (mass per m^2) of the partition.

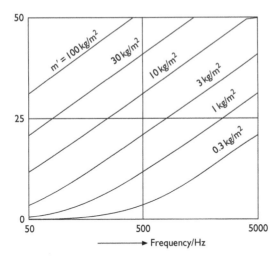

Frequency/Hz

Figure 14.5 Sound transmission loss of a single-leaf partition at perpendicular sound incidence.

This is the famous mass law which represents the upper limit of the transmission loss which can be achieved with a single-leaf partition. It is illustrated in Figure 14.5; the abscissa is the frequency $f = \omega/2\pi$. The first term in the square bracket of eq. (14.9) can be neglected unless the partitions are very light, then

$$R_A \approx 20 \log_{10} \left(\frac{\omega m'}{2Z_0} \right) \tag{14.9a}$$

The curves in Figure 14.5 become straight lines with a slope of 6 dB/octave equivalent to 20 dB/decade. Equations (14.9) and (14.9a) show that a single-leaf wall between two walls has the effect of an acoustical low-pass filter. This agrees with our everyday experience: when we hear in a hotel or at home the television from our neighbour's room we can easily decide whether a speaker is male or female but we cannot understand – apart from extreme cases – what he or she is saying since the medium and high-frequency spectral components which are particularly important for speech intelligibility are strongly muted by their transmission through the wall. Likewise, the enjoyment of music is reduced to the perception of the bass.

At oblique sound incidence matters become much more complicated since the primary wave excites bending waves on the partition while at normal incidence all its surface elements vibrate with the same amplitude and phase. These bending waves reduce the sound insulation in a characteristic way as will be shown in the following discussion.

Suppose the primary sound wave arrives at the wall at an angle ϑ. The alternating pressure it exerts on the front of the wall is, according to eq. (6.8) (with x = 0):

$$p_1(y) = \hat{p}(1 + R)e^{-jky \sin \vartheta} \cdot e^{j\omega t} \tag{14.10}$$

as earlier R denotes the reflection factor. The pressure acting on the rear side of the partition is the sound pressure of the transmitted wave. According to eq. (6.15), again with x = 0, this is:

$$p_2(y) = \hat{p}Te^{-jky \sin \vartheta} \cdot e^{j\omega t} \tag{14.11}$$

(T = transmission factor). The difference of both pressures enforces a wave-like deformation of the wall with the same periodicity as the y-periodicity of the incident, the reflected and the transmitted wave. This is shown in Figure 14.6 which presents another example of trace fitting as already mentioned in Section 6.1. The flexural deformation travels in the y-direction with the speed $c/\sin \vartheta$; the displacement of the wall is given by:

$$\xi(y) = \hat{\xi}e^{-jky \sin \vartheta} \cdot e^{j\omega t} \tag{14.12}$$

At first glance one might expect that this makes the wall more resistant to sound transmission than merely because of its mass inertia. However, this is not so, at least not at low frequencies, since the elastic restoring force counteracts the inertial force as in a simple resonance system.

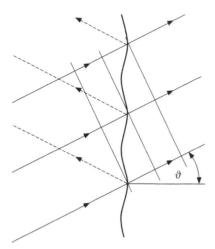

Figure 14.6 Oblique sound incidence on a wall.

To set the displacement ξ in relation to the pressure difference $p_1 - p_2$ we go back to Subsections 10.3.1 and 10.3.3. Of course, to adapt the formulae of those sections to our present coordinates we have to replace x with y and η with ξ. Then the force balance in eq. (10.19), completed by a term $p_1 - p_2$, reads:

$$m'\frac{\partial^2 \xi}{\partial t^2} + \frac{\partial F_x}{\partial y} = p_1 - p_2$$

or, since $F_x = B\dfrac{\partial^3 \xi}{\partial y^3}$:

$$m'\frac{\partial^2 \xi}{\partial t^2} + B\frac{\partial^4 \xi}{\partial y^4} = p_1 - p_2$$

According to eq. (14.12) differentiation with respect to t is tantamount to a factor $j\omega$ while differentiation with respect to y corresponds to a factor $-jk \cdot \sin \vartheta$. Hence we obtain:

$$p_1 - p_2 = (-\omega^2 m' + Bk^4 \sin^4 \vartheta) \cdot \xi$$

or, after expressing the displacement ξ of the wall by its velocity $v = j\omega\xi$ and replacing k with ω/c:

$$p_1 - p_2 = j\omega m' \left(1 - \frac{B\omega^2 \sin^4 \vartheta}{m'c^4}\right) \cdot v = j\omega m'_{\text{eff}} v \qquad (14.13)$$

In the latter expression the 'effective specific mass' m'_{eff} of the wall was introduced. It can be expressed in a very concise form by introducing the characteristic frequency of the wall (see Subsection 10.3.4):

$$\omega_c = c^2 \sqrt{\frac{m'}{B}}$$

This leads to:

$$m'_{\text{eff}} = m' \left(1 - \frac{\omega^2}{\omega_c^2} \sin^4 \vartheta\right) \qquad (14.14)$$

Increasing frequencies and increasing angles of incidence reduce the effective mass, and the wall appears to become lighter. In other words, with decreasing wavelength of the forced bending wave, or with increasing curvature of the wall its elastic reaction becomes more noticeable. At the frequency

$$\omega = \omega_\vartheta = \frac{\omega_c}{\sin^2 \vartheta} \qquad (14.15)$$

the effective specific mass even becomes zero; the wall has disappeared from the acoustical viewpoint. This phenomenon is called the 'coincidence effect'. At still higher frequencies m'_{eff} will become negative, then the sound insulation of the wall is predominantly controlled by its bending stiffness.

Replacing ω in eq. (10.29) with ω_ϑ yields $c_B = c/\sin\vartheta$ as the phase velocity of a free bending wave. This, however, agrees with the speed with which the deformation travels in y-direction (see eq. (14.12)): at this frequency the deformation of the wall imposed by the sound field is identical with the free bending wave and hence can be maintained with very low expenditure. Incidentally, it is not surprising that the characteristic frequency ω_c turns up in our discussion since the excitation of bending waves by a sound field is the reverse process as the sound radiation from a vibrating plate.

To calculate the sound reduction index we observe that the sound pressure p_1 at $x = 0$ at the left side of the partition (see Fig. 14.6) is the sum of the sound pressures p_i and p_r of the incident and the reflected wave; therefore, we can rewrite eq. (14.13) in the following way:

$$p_i + p_r - p_2 = j\omega m'_{eff}v \tag{14.13a}$$

On the other hand, at $x = 0$ the normal component of the particle velocities at both sides of the wall must be equal to each other and to the velocity v of the wall:

$$v_{xi} + v_{xr} = v_{x2} = v$$

or, since $v_{xi} = (p_i/Z_0)\cos\vartheta$, $v_{xr} = -(p_r/Z_0)\cos\vartheta$, $v = v_{x2} = (p_2/Z_0)\cos\vartheta$:

$$p_i - p_r - p_2 = 0$$

Adding this relation to eq. (14.13a) yields

$$2p_i - 2p_2 = j\omega m'_{eff}v = \frac{j\omega m'_{eff}}{Z_0}p_2\cos\vartheta$$

or

$$R_A = 10\log_{10}\left|\frac{p_i}{p_2}\right|^2 = 10\log_{10}\left[1 + \left(\frac{\omega m'_{eff}}{2Z_0}\cos\vartheta\right)^2\right] \tag{14.16}$$

which formally agrees with eq. (14.9). For partitions with negligible bending stiffness the characteristic frequency ω_c tends to infinity and $m'_{eff} \to m'$.

Figure 14.7 plots the sound reduction index as given by eq. (14.16) for a few angles of incidence as a function of frequency. Apart from the curve for normal sound incidence ($\vartheta = 0$), all curves show a very sharp dip reaching 0 dB at the frequency ω_ϑ after eq. (14.14); in the frequency range above they rise very steeply, namely, at 18 dB per octave. With increasing angle of

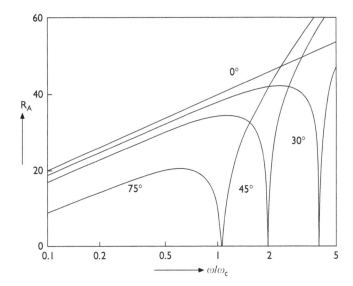

Figure 14.7 Sound transmission loss of a single-leaf partition at various angles of incidence.

incidence the frequency where the zero occurs approaches the characteristic frequency ω_c.

In real situations the primary sound will not arrive from just one direction but its incidence will be more or less random. Then the sharp dip in Figure 14.7 will be distributed continuously over the range above the characteristic frequency where it causes a significant deterioration of sound insulation compared to that predicted by the mass law eq. (14.9a). Figure 14.8 shows schematically the sound reduction index as a function of frequency at random sound incidence. Far below the characteristic frequency it follows the simple mass law; however, because of random sound incidence it is 3 dB lower than that predicted by eq. (14.9a).

For practical building acoustics this result is of considerable relevance. This may be seen from Table 14.1 which lists the characteristic frequencies of some partitions. Obviously, we can expect only thin leaves such as glass panes etc. to obey the mass law in the whole frequency range of interest. For thicker partitions the characteristic frequency is so low that the coincidence effect influences the sound insulation at virtually all frequencies. This may also be seen from Figure 14.9 which plots the weighted sound reduction index R_w as a function of the specific mass m', assuming common building materials. After an initial rise R_w approaches a constant value since with increasing thickness of a wall the critical frequency becomes smaller above which the sound insulation is impaired by the coincidence effect. Thus, the

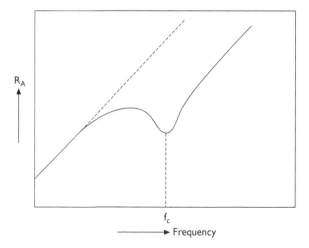

Figure 14.8 Sound transmission loss of a single-leaf partition at random sound incidence, schematically, f_c = critical frequency.

Table 14.1 Critical frequency of a few partitions

Material	Thickness in cm	Critical frequency in Hz
Glass	0.4	2450
Chipboard panel	0.9	4000
Gypsum	8	370
Brick	12	180
Dense concrete	24	65

sound insulation cannot be improved in this range just by increasing the mass of the wall. Only with very thick and heavy partition a further rise of the transmission loss can be achieved, which, however, falls far behind the prediction of the mass law. Paradoxically, the sound insulation of a thin wall is relatively higher than that of a thick one. Hence, the statement that the sound insulation of a partition depends in the first place on its weight is at the very least questionable.

The earlier discussions are meant to serve as a basic understanding of the processes which are relevant for the transmission of airborne sound through a wall. They are by no means a substitute for the examination of a wall by measurement. Moreover, one should note that the dimensions of real walls are finite; consequently, the bending waves excited in them by oblique sound waves are reflected by their boundary. This modifies the

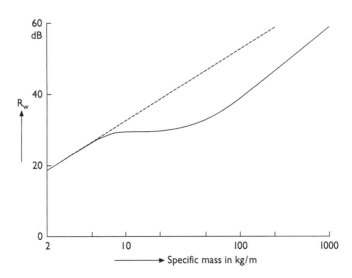

Figure 14.9 Weighted sound reduction index R_w of single-leaf walls of usual construction as a function of their specific mass.

picture we have portrayed, particularly if the walls are small and of a material with a low loss factor (see Subsection 10.3.5). This holds for the following section too.

14.4   Airborne sound insulation of double-leaf partitions

As is evident from Figure 14.9 the possibilities to reach greater and greater sound insulation with single-leaf partitions are limited for reasons of practical feasibility. However, much better results are obtained with partitions consisting of several solid sheets or 'leaves' separated by air layers; generally, they provide sound insulation surpassing that of a single layer with the same specific mass. The subsequent treatment is restricted to the sound transmission through double-leaf elements of infinite extension at normal sound incidence.

We regard the double-leaf partition sketched in Figure 14.10a; its equivalent electrical circuit is shown in Figure 14.10b, and Z_0 represents the characteristic impedance of the air adjacent at the rear of the partition. It is easily seen that the double-leaf partition is an acoustical low-pass filter. Both layers have the specific masses m_1' and m_2', respectively. The thickness d of the air layer between them is supposed to be thin compared to the wavelength; accordingly, it can be modelled as a spring with the pressure-related

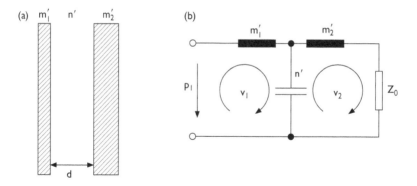

Figure 14.10 Double-leaf partition: (a) section, (b) equivalent electrical circuit.

compliance (see Subsection 6.6.1):

$$n' = \frac{d}{cZ_0} \tag{14.17}$$

By applying Kirchhoff's rule it follows from Figure 14.10b:

$$p_1 = j\omega m_1' v_1 + \frac{1}{j\omega n'}(v_1 - v_2) \tag{14.18}$$

and

$$0 = \frac{1}{j\omega n'}(v_2 - v_1) + (j\omega m_2' + Z_0)v_2$$

or

$$v_1 = (1 + j\omega n' Z_0 - \omega^2 n' m_2')v_2 \tag{14.19}$$

In these equations p_1 and v_1 are the sound pressure and the particle velocity at the front surface of partition; p_2 and v_2 refer to its rear surface. As in the preceding section, p_1 and v_1 represent sums of an incident and a reflected component:

$$p_1 = p_i + p_r \quad \text{and} \quad v_1 = v_i + v_r$$

or, with $v_i = p_i/Z_0$ and $v_r = -p_r/Z_0$:

$$Z_0 v_1 = p_i - p_r$$

By eliminating p_r which is of no interest in the present discussion we obtain, after invoking eq. (14.18):

$$2p_i = p_1 + Z_0 v_1 = \left(Z_0 + j\omega m_1' + \frac{1}{j\omega n'} \right) v_1 - \frac{1}{j\omega n'} v_2$$

Next we express v_1 by v_2 using eq. (14.19) and introduce the sound pressure $p_2 = Z_0 v_2$ of the transmitted wave:

$$2p_i = \frac{1}{j\omega n' Z_0} \left[\left(1 + j\omega n' Z_0 - \omega^2 n' m_1' \right) \left(1 + j\omega n' Z_0 - \omega^2 n' m_2' \right) - 1 \right] p_2$$

or, after separating the real and the imaginary part and dividing by $2p_2$:

$$\frac{p_i}{p_2} = 1 - \omega^2 n' \frac{m_1' + m_2'}{2} + j \left(\frac{\omega n' Z_0}{2} + \omega \frac{m_1' + m_2'}{2Z_0} - \omega^3 n' \frac{m_1' m_2'}{2Z_0} \right)$$

$$(14.20)$$

With this expression the sound reduction index can be computed using the general definition

$$R_A = 10 \log_{10} \left(\frac{I_i}{I_t} \right) = 10 \log_{10} \left| \frac{p_i}{p_2} \right|^2$$

Figure 14.11 shows the sound reduction index of a typical double wall as a function of frequency.

For the discussion of eq. (14.20) we disregard the first term in the bracket which is justified unless the leaves are very light. At very low frequencies all terms containing higher powers of ω than the first can also be neglected. Then we arrive at

$$\frac{p_i}{p_2} \approx 1 + j \, \omega \frac{m_1' + m_2'}{2Z_0} \qquad (14.20a)$$

This leads to the simple mass law eq. (14.9) with $m' = m_1' + m_2'$; the sound reduction index rises at a rate of 6 dB per octave. Furthermore, eq. (14.19) tells us that both layers vibrate with almost equal phases and amplitudes. Hence, the double-leaf construction is of no use in this frequency range. However, with increasing frequency the last term in eq. (14.20) becomes

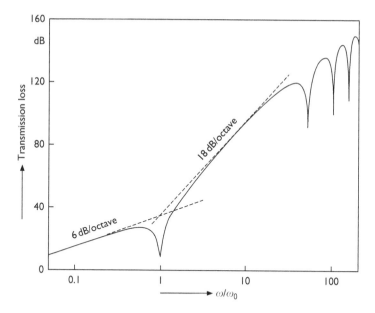

Figure 14.11 Sound transmission loss of a double-leaf partition at perpendicular sound incidence.

more and more noticeable. Finally, at the frequency

$$\omega_0 = \sqrt{\frac{1}{n'}\left(\frac{1}{m_1'} + \frac{1}{m_2'}\right)} \qquad (14.21)$$

the bracket disappears; the system is at resonance. Now both layers vibrate in equal phases with

$$\left(\frac{v_1}{v_2}\right)_{\omega_0} \approx -\frac{m_2'}{m_1'} \qquad (14.22)$$

and the ratio of sound pressures reduces to

$$\left(\frac{p_i}{p_2}\right)_{\omega_0} \approx -\frac{1}{2}\left(\frac{m_1'}{m_2'} + \frac{m_2'}{m_1'}\right) \qquad (14.23)$$

Beyond the resonance frequency the low-pass filter in Figure 14.1 is operated in its stop band; in eq. (14.20) the last term becomes the only significant one and

$$R_A \approx 20 \log_{10} \left(\frac{\omega^3 n' m_1' m_2'}{2Z_0} \right) = R_{A0} + 40 \log_{10} \left(\frac{\omega}{\omega_0'} \right) \qquad (14.24)$$

In this formula R_{A0} is the sound reduction index of the first leaf alone; the second term in which

$$\omega_0' = \frac{1}{\sqrt{n' m_2'}} \qquad (14.24a)$$

contains the improvement achieved by the second layer. In this frequency range the sound reduction shows a very steep rise, namely, 18 dB per octave, and lies far above the value which can be achieved with a single-leaf partition with specific mass $m_1' + m_2'$.

At very high frequencies the increase of the sound reduction index flattens to 12 dB per octave; moreover, the curve shows sharp dips. Both are due to the fact that in this range the thickness of the air cushion becomes comparable with the acoustical wavelength. Hence it can no longer be regarded as a lumped element; instead, it must be treated as an acoustical transmission line (see Section 8.2). In any case standing waves will be established in the space between both layers; at angular frequencies

$$\omega_n = n \cdot \frac{\pi c}{d} \quad \text{with} \quad n = 1, 2, \dots.$$

resonances will occur (see Section 9.1), which are responsible for the sharp dips.

Best overall sound insulation is achieved if the fundamental resonance after eq. (14.21) is very low while the first of the higher resonances (ω_1 in the earlier equation) should be high as possible. This requires relatively heavy leaves at a small distance d. Furthermore, it is advantageous if specific masses of both leaves are different; otherwise, the pressure ratio in eq. (14.23) will be unity which indicates vanishing sound insulation.

At random sound incidence, however, as will be common in practical situations, matters will be modified by coincidence as discussed in the preceding section. This is another reason to provide for layers with different thickness and hence with different characteristic frequencies. Then their coincidence dips shown in Figure 14.7 will not occur at the same frequency.

Double-leaf constructions are often applied in practical building acoustics since they have not only high sound transmission losses but also good heat insulation. For their function it is important that any rigid connections between the layers are avoided since each of them would act as a 'sound bridge' which reduces or even destroys the desired effect. Common examples

Figure 14.12 Improving the sound insulation by adding a panel.

Table 14.2 Weighted sound reduction index of some partitions

Type	Specific mass in kg/m²	Weighted sound reduction index R_w in dB
Brick 115 mm	260	49
Brick 240 mm	460	55
Concrete 120 mm	280	49
Gypsum board 60 mm	83	35
Chipboard panel 2 cm	11	22
140 mm concrete, 60 mm air gap with rock wool inside, 120 mm concrete	630	76
60 mm gypsum, 20 mm rock wool, 25 mm gypsum	89	48
Single-pane window	10	15
Single-leaf door	15	22

are partitions in terraced houses, highly insulating lightweight walls in apartments, bureaus, etc., furthermore, sound-proof doors. The air space between both layers is often filled with sound-absorbing material such as glass fibre, either partially or completely. On the one hand, this measure leads to additional damping of the resonances and thus improves the sound insulation, particularly, in critical frequency ranges; on the other hand, it helps to avoid sound bridges. With double-glazed windows additional damping can be achieved by a sound-absorbing treatment in the reveals.

The double-leaf principle is also applied to improve the acoustic performance of an existing wall or ceiling the sound insulation of which has proved insufficient. For this purpose a relatively thin panel, which may be of gypsum board, is mounted on the existing element (see Fig. 14.12) with an air space of a few centimetres and some porous material behind it. To avoid direct transmission between the wall and the panel, resilient mounts should be used. Table 14.2 lists the weighted sound reduction index R_w of some typical wall constructions.

14.5 Structure-borne sound insulation

Strictly speaking, any kind of vibrations of building elements can be regarded as structure-borne sound propagated in the form of extensional waves or bending waves, as explained in the introduction to this chapter. In a more restricted sense, however, we speak of solid-borne sound when these vibrations are generated by direct mechanical excitation of the building structure. If the source is relatively small it can be modelled as a point source from which the structure-borne sound propagates in the form of circular waves, similar to the water waves we observe after throwing a stone into a lake. Since the wave energy emerging from the source is distributed over larger and larger circles with circumferences proportional to the distance r from the source, the intensity decreases according to:

$$I = \frac{P}{2\pi r} \tag{14.25}$$

with P denoting the power output of the source.

The most important source of structure-borne sound in apartment buildings are the shoes of persons walking on a floor which, of course, is the ceiling of the room below. Since these sounds have impulsive character we speak of 'impact sound' and 'impact sound insulation' in this case. Structure-borne sound can also be generated by playing children or by certain musical instruments such as the piano or the violoncello. Furthermore, it arises from technical equipment such as pumps with rotating, vibrating or otherwise moving components. Other typical sources of structure-borne sound are water valves and pipes rigidly connected to walls or floors. Even a wrongly constructed light switch can be an annoying source of structure-borne sound.

The strength of solid-borne sound can be characterised by the normal component v_n of the vibrational velocity of some building element although this quantity does not include extensional waves. The velocity level based on the root-mean-square of v_n is:

$$L_v = 10\log_{10}\left(\frac{\tilde{v}_n}{v_0}\right) \tag{14.26}$$

The reference velocity v_0 is usually $5 \cdot 10^{-8}$ m/s.

14.5.1   Impact sound level and impact sound insulation

In principle, the structure-borne or impact sound insulation of a floor could be characterised by the pressure level of the sound it emits in relation to the exciting force. However, for practical tests, a procedure of this kind would be too complicated, particularly, for field measurements. A more practical method uses an electrically driven, standardised tapping machine which models several persons walking on the floor. It consists of five hammers each with a mass of 500 g arranged in a line, and the distance between the most separate hammers is 40 cm. Each of them is a cylinder with 3 cm diameter the lower end of which is slightly curved. The hammers fall freely from

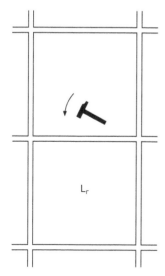

Figure 14.13 Measurement of the impact level L_r.

a height of 4 cm onto the floor thus producing an impact frequency of 10/s. At the same time the sound pressure level L_r is measured in the receiving room below (see Fig. 14.13). Under the condition of a diffuse sound field the energy density in this room is inversely proportional to its equivalent absorption area A (see eq. (13.16)). Therefore the result is normalised by using a reference of $A_0 = 10\,m^2$. The 'normalised impact sound pressure level' is defined by:

$$L_n = L_r + 10\log_{10}\left(\frac{A}{10\ m^2}\right) \qquad (14.27)$$

The measurement is carried out in frequency bands of third-octave bandwidth and with mid-frequencies reaching from 100 Hz to 3150 Hz. As in airborne sound insulation, its result is judged by comparing it with a standardised reference curve as shown in Figure 14.14. Now, however, the sense of the ordinate is the reverse of that in Figure 14.3, hence values above the reference curve indicate less satisfactory sound insulation. Again, the reference curve accounts for the fact that it is technically easier and less expensive to avoid high impact sound levels at high frequencies than at low frequencies. Fortunately, this tendency fits well into the frequency dependence of our loudness sensation.

As in airborne sound insulation it is often desirable to characterise the impact sound transmission through a ceiling by a single-number rating. This is achieved by fitting the reference contour to the measured curve in such

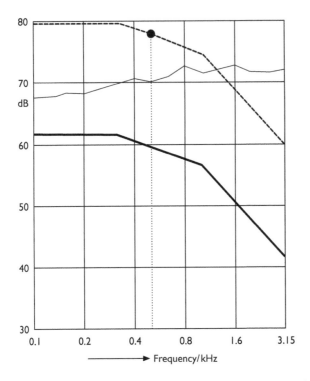

Figure 14.14 Impact sound insulation of ceilings. Solid curve: reference contour for the normalised impact level, broken line: reference curve shifted upwards, thin curve: normalised impact level for a ceiling of 12 cm concrete.

a way that the frequency average of unfavourable deviations of the measured data is as close as possible to 2 dB. Then the 'weighted impact sound level' $L_{n,w}$ is the ordinate of the shifted reference curve at 500 Hz. According to the relevant ISO standard, $L_{n,w}$ must not exceed 53 dB for a floor which separates different apartments.

The measured data shown in Figure 14.4 as an example represent the normalised impact sound pressure level of a ceiling made of 12 cm reinforced concrete. Obviously, this floor is acoustically unsatisfactory since the data points are far above the reference contour. In fact, the reference curve must be shifted upwards by as much as 18 dB to meet the fitting criterion mentioned earlier. Accordingly, the weighted impact sound level $L_{n,w}$ of this ceiling is as high as 78 dB. Table 14.3 shows the weighted impact sound levels of some typical floors.

Table 14.3 Weighted impact sound level of ceilings

Type	Specific mass in kg/m²	Weighted impact sound level $L_{n,w}$ in dB
Wooden joist ceiling	160	65
140 mm reinforced concrete	350	75
140 mm reinforced concrete, with floating floor on resilient material	430	59
200 mm reinforced concrete, hollow plates	160	87
200 mm reinforced concrete, hollow plates, with suspended panels underneath	175	73

14.5.2 Improvement of impact sound insulation

It is common experience that noise from footsteps can be reduced in the same room as well as in the room below with the aid of soft floor coverings. Suitable materials are rubber, soft plastics and, of course, carpets of any kind. Such layers alter the spectrum of the exciting force by suppressing the high frequency components which otherwise would strongly contribute to the weighted impact noise level. Further improvement is due to the mechanical losses inherent to these materials.

A common way of improving the impact sound insulation particularly in apartments and other dwellings is by using floating-floor constructions. A floating floor consists of a slab of concrete, gypsum, asphalt, etc. about 3–6 cm thick built above the structural floor with a resilient layer in between. The latter is usually made of sheets or mats of glass wool or mineral wool and has a thickness of about 1 cm when compressed by the floor slab. Somewhat less favourable are sheets of foamed plastics. The main task of the resilient layer is to store the air enclosed in it and hence to act as a spring. Thus the floating floor is a double-leaf construction which also improves the airborne sound insulation. It is important that sound bridges in the form of solid connections are avoided. Therefore the soft layer should be pulled up over the lateral end of the solid slab. The reduction of the impact level is given by the last term of eq. (14.24):

$$\Delta L = 40 \log_{10} \left(\frac{\omega}{\omega_0'} \right) \tag{14.28}$$

together with eq. (14.24a) with m_2' denoting the specific mass of the solid slab and n' being the compliance of the resilient layer after eq. (14.17).

14.5.3   Propagation of structure-borne sound in buildings

As already mentioned vibrations of building elements as excited by airborne or structure-borne sound can be transferred to adjacent elements. This transfer is not only the reason for flanking sound transmission mentioned in Section 14.1 but the sound waves within the building structure may travel to more distant parts of a building. In the course of this process the amplitudes of the waves will be diminished, on the one hand, according to the geometrical spreading law (see eq. (14.25)). On the other hand, the structure-borne sound will be attenuated by losses occurring within the materials. The relevant attenuation constants are related to the loss factor η of the material by eqs. (10.34) and (10.36). The corresponding level reduction per metre is:

$$D_E = 27.3 \frac{\eta}{\lambda_E} \text{ dB/m} \quad \text{(extensional waves)} \tag{14.29}$$

$$D_B = 13.6 \frac{\eta}{\lambda_B} \text{ dB/m} \quad \text{(bending waves)} \tag{14.30}$$

since $D = 10 \log_{10}(e^m) \text{ dB/m} = 4.343 \cdot m \text{ dB/m}$. As a rule the attenuation due to interior losses is quite small. For common building materials such as concrete or brick they are below 0.1 dB/m in the mid-frequency range.

To a much larger extent the propagation of structure-borne sound is influenced by discontinuities, for instance, by changes of cross section, by edges and junctions. At each of these discontinuities an incident wave is split up into a reflected and a transmitted wave. A further complication is wave type conversion which means that at each discontinuity both wave types – extensional and bending waves – are partially converted into each other. If, for instance, the primary wave is a bending wave, the reflected sound wave contains both a bending and an extensional component. The same holds for the transmitted sound.

A particularly efficient suppression of structure-borne sound propagation is achieved with a soft layer separating two structural elements. In Figure 14.15 the simplest case is sketched, namely, an extensional wave travelling on a plate or a beam which is interrupted by a thin layer with the stress-related compliance $n' = (\xi_1 - \xi_2)/\sigma$ where ξ_1 and ξ_2 are the longitudinal displacements at both sides; v_1 and v_2 are the corresponding particle velocities. Finally, $\sigma = \sigma_1 = \sigma_2$ is the tensile stress which is the same in all three components. Hence we have

$$v_1 - v_2 = j\omega n' \sigma$$

This expression agrees formally with eq. (14.13). A derivation similar to that following eq. (14.13) with $\vartheta = 0$ leads to

$$R_s = 20 \log_{10}\left(\frac{v_i}{v_2}\right) = 10 \log_{10}\left[1 + \left(\frac{1}{2}\omega n' Z_{0E}\right)^2\right] \tag{14.31}$$

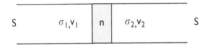

Figure 14.15 *Figure 14.15* Impact sound insulation by a soft layer with compliance n. σ_1, σ_2: tensile stress, v_1, v_2: particle velocity on both sides.

with v_i denoting the particle velocity in the incident wave. This formula corresponds to the mass law of airborne sound reduction as given in eq. (14.9). Hence we can use Figure 14.5 to find the structure-borne transmission loss R_s of the resilient layer; in this case, the parameter of the curves is now $n' Z_{0E} Z_0$ which also has the dimension kg/m^2.

Suitable materials for such resilient layers of the kind described are cork, rubber or soft plastics. Of course, the intermediate layer does not need to be homogeneous, therefore, perforated rubber plates or suitably formed steel springs can be used as well.

Now we return to the distance law of structure-borne sound addressed at the beginning of this subsection. Equation (14.25) holds only for a homogeneous plate of infinite extension. In a building, however, the free propagation of extensional and bending waves is impeded by a great number of discontinuities. Nevertheless, it should be intuitively clear that the energy of solid-borne sound is increasingly 'diluted' the farther it departs from the point of excitation, even if there are no or only very small losses in the material. The situation is illustrated in Figure 14.16. It shows schematically a section through a very large building which is regularly subdivided by walls and ceilings (just called 'walls' hereafter) into many rooms or 'cells' of equal dimensions. Of course, we must imagine this picture to be extended in the third dimension. We consider a sphere with arbitrary radius r. The circle shown is its projection into the x-y-plane of a coordinate system. The sound source is situated in its centre; it injects the power P into one particular building element. Let us denote with N the number of elements which are intersected by the indicated sphere, then sound power carried by each of them is $P' = P/N$ on average.

To find the number N we note that the circle in Figure 14.16 contains roughly $\pi r^2 / L_x L_y$ rectangles with dimensions L_x and L_y. Each of them contributes two walls to N which are perpendicular to the x-y-plane. Hence the whole sphere (front and rear side) is intersected by

$$N_z = 2 \cdot 2 \cdot \frac{\pi r^2}{L_x L_y}$$

of those walls. Similar expressions hold for N_x and N_y the number of intersected walls perpendicular to the x-z-plane and the y-z-plane. It must be

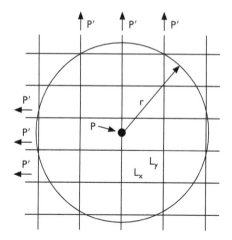

Figure 14.16 Propagation of impact sound in a large building.

noted, however, that each wall is counted twice if we just add these three expressions since each wall which is perpendicular to one coordinate plane is at right angles to the other one as well. Hence the total number N of walls intersected by the sphere is:

$$N = \frac{1}{2}(N_x + N_y + N_z) = 2\pi r^2 \left(\frac{1}{L_xL_y} + \frac{1}{L_xL_z} + \frac{1}{L_zL_x}\right) = \frac{\pi r^2}{2} \cdot \frac{L}{V}$$

(14.32)

Here V is the volume of a room and L the length of all edges in it. The final result reads:

$$P' = \frac{2PV}{\pi L r^2}$$

(14.33)

This formula shows that the propagation of structure-borne sound in a regularly structured building follows the same spreading law as in a homogeneous three-dimensional medium: its energy (or intensity) diminishes in inverse proportion with the square of the distance; if the distance is doubled the structure-borne level is reduced by 6 dB on the average.

Chapter 15

Fundamentals of noise control

According to the common definition, noise is any kind of undesired sound. It is clear that we cannot derive some physical yardstick from it which would tell us whether a particular sound will appear as noise or not, rather it is the specific circumstances and also the attitude of those exposed to the sound which are relevant for the distinction between noise and other sounds. In particular, it follows that one and the same acoustical event or signal may be felt as a pleasant sound in a certain situation while it is annoying under different circumstances. Maybe the most common example is a passenger car passing by with an open window and with the stereo on; the driver doubt-less enjoys what he is hearing, but the pedestrians nearby, to say the least, do not.

Even the loudness cannot be regarded as an unambiguous criterion whether a sound is to be rated as noise or not. Thus an enthusiastic discotheque freak would certainly protest against the assertion that what he hears is noise. On the other hand even low noises or sound signals can severely interfere with the concentration needed for mental work. This holds in particular if the sounds are intermittent or contain any information.

This, however, does not mean that the loudness is of secondary or no relevance for annoyance by noise. On the contrary, at sound pressure levels exceeding 85 dB(A) one must expect that the hearing organ exposed to it will suffer temporary or permanent damages. Such levels are easily reached in a discotheque, but also in the orchestra pit of an opera house. And, of course, many employees in factories and workshops are exposed to such high noise levels. Sounds of high intensity reduce the sensitivity of the sensory cells which manifests itself in an upward shift of the hearing threshold to a degree which depends on the duration and the strength of the sound. This threshold shift may be temporary, that is, the hearing can recover from noise exposure. With longer and repeated exposures to sounds of moderate to high intensities this recovery may be incomplete resulting in a permanent and ever increasing hearing loss caused by degeneration of the hair cells.

Even sounds of relatively low intensities are not necessarily harmless when it comes to health risks. At levels exceeding 60 dB(A) they can cause vegetative disorder which mainly concerns blood circulation and metabolism. At sound levels slightly above 30 dB(A) sensitive persons may suffer disturbance of sleep or mental concentration. This is, by the way, the range in which the irritation by noise depends to a particularly high degree on the mental state of the person exposed to it. This subjective component makes it so difficult to find a general standard for the annoyance by noise. On the other hand, such a standard would not change the fact that different persons are differently sensitive to noise; at best it would yield an average rule without much meaning in individual cases. At least it can be stated that sounds of low frequency or with predominant low-frequency spectral components are less annoying at equal intensity than those which contain strong high-frequency components in their spectrum.

This chapter deals mainly with noise originating from technical equipment and installations, that is, from machinery in the widest sense. This includes, of course, all motor-driven vehicles. Noise control in buildings, in particular, the reduction of living noises, has already been treated in the preceding chapter and can be disregarded here.

In the discussion of technical solutions for noise control we follow the usual classification in primary and secondary methods. The former ones concern modifications and alterations to be carried out at the noise source itself with the aim of reducing or suppressing the generation of noise. Secondary noise control concerns measures which impede as far as possible the propagation of noise from its origin to man. (It should be noted that this classification is not completely unambiguous.) Consequently, tertiary methods of noise control would be those by which exposed persons are directly protected by ear plugs, earmuffs, etc.

15.1   Noise criteria

The usual basis of any quantitative assessment of noise exposure regarding health risks or tolerance is the A-weighted sound pressure level as described in Section 12.6. Here one has to account for the fact that this level is rarely constant but often shows more or less pronounced temporal fluctuations. For instance, fluctuations of highway noise result from variations in the spatial and temporal traffic density. It may also happen that the total acoustical energy density occurring at some immission point is composed of the contributions of several or many noise sources which are only temporarily in operation. This latter situation is typical for many factories and workshops.

To characterise a noise situation where the level fluctuates with time the so-called 'energy equivalent sound level' is widely used which is based on averaging the energy density over a certain period T_e. According to eq. (3.32) the energy density is proportional to the square of the effective sound pressure. Hence this average can be written as

$$\overline{\tilde{p}^2} = \frac{1}{T_e} \int_0^{T_c} [\tilde{p}(t)]^2 \, dt \tag{15.1}$$

It should be noted that the mean-root-square pressure itself is an average value – we could call it a short time average – while the average according to eq. (15.1) is aimed at slow changes. Thus the equivalent sound level derived from $\overline{\tilde{p}^2}$ reads

$$L_{eq} = 10 \log_{10} \left(\frac{\overline{\tilde{p}^2}}{p_b^2} \right) = 10 \log_{10} \left(\frac{1}{T_e} \cdot \int_0^{T_e} 10^{0,1 L(t)} \, dt \right) \tag{15.2}$$

Often L is replaced with the A-weighted noise level L_A (see Section 12.6). For special cases of noise immission, particularly, for aircraft noise, different averaging procedures are also in use. To be meaningful, the energy equivalent sound should not be evaluated for long periods T_e in which only a few noise events occur. Thus, for instance, it would be unreasonable to extend the average over a full day if the site to be characterised is a quiet residential area which is passed over just once a day by a helicopter. In such cases it would be preferable to characterise the noise situation by a level statistics or by indicating percentile noise levels, that is, by the level which is exceeded 10% or 1% of the time.

To keep the noise exposure within tolerable limits quite a number of ordinances and guidelines have been developed. We refrain from presenting some of them because they vary considerably from one country to the other.

15.2   Basic mechanisms of noise generation

In view of the large variety of technical noise sources it is impossible to give even a moderately complete overview on the various mechanisms of noise formation within the given frame. Instead, just a few typical processes of noise formation will be briefly described in this section while other sorts of noise such as, for instance, noise arising from electrical machines or the rolling noises of vehicles must be disregarded, in spite of their importance.

15.2.1   Impact noise

Impact noises are produced when two solid bodies, for instance, two parts of a machine, are coming into sudden contact with each other, whether to transfer forces as by gears or in transport processes in production installations, or to effect permanent changes in one of the partners. In the latter case mechanical energy is accumulated in some tool which is suddenly released when the tool hits the workpiece. An everyday example is driving a nail into a board with a hammer. This process – a moving tool hitting a workpiece

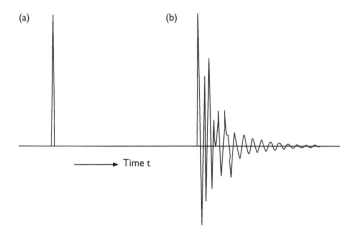

Figure 15.1 Noise of an impact: (a) on a massive body, (b) on a plate.

which is at rest – is the core of many technical or industrial procedures such as pressing, punching, forging, riveting, even sawing, etc.

Whenever a solid body hits onto another one with a certain speed both of them will be elastically deformed. These sudden changes of shape or volume and the reactions of both partners to them are one cause of sound generation. Another one is due to the fact that the bodies concerned are accelerated or decelerated. The same holds for the air dragged along by the moving body; when it is decelerated it will be first compressed, and afterwards it will expand thus producing sound. Furthermore, air between both partners may be suddenly expelled leading to a pressure disturbance. In any case, the primary sound is a short impulse (see Fig. 15.1a), which itself is often not very intense. The high loudness of many impact noises is due to ringing, that is, to the excitation of vibrations in one or both parts which gradually die out as shown in Figure 15.1b. If, for instance, one of both impact partners is a plate or is in rigid contact with a plate bending waves will be excited on the latter which are accompanied by intense sound radiation into the surrounding air as has been described in Subsection 10.3.4. This sound 'amplifying' effect can be easily demonstrated by dropping a stone either just to the ground or on a metal plate.

15.2.2   Flow noise

A very common source of noise is the flow of gases or liquids. Each of us knows the noise produced by water installation, or the noise originating from jet aircraft. Other examples of such noise sources are welding torches, fans or aircraft propellers. Likewise, with rapidly driven cars flow noise is a noticeable component of the total noise produced.

The cause of noise production is the instability which arises when the flow velocity exceeds a particular limit depending on the geometry of the flow and the viscosity of the fluid. If an air or liquid flow of relatively low speed strikes a resting body, for instance, a cylinder, eddies rotating in opposite sense are alternately shed from both sides, they form the well-known Kármán vortex street. These eddies exert transverse oscillatory forces on the body. We can convince ourselves easily of these forces by rapidly drawing a stick through water.

The noise these forces produce with propellers and similar bodies has a wide spectrum, however, with a peak occurring at a frequency determined by the frequency of vortex detachment. For a cylinder this frequency

$$f_v \approx 0.2 \frac{v_r}{d} \tag{15.3}$$

(v_r = relative speed between the fluid and the cylinder, d = diameter of the cylinder). Because of the symmetry of the vortex street the arising sound field has dipole character, and the radiated sound power increases as the sixth power of the flow velocity. In a similar way edge tones are produced which are responsible for the sound generation in certain woodwind instruments (see Subsection 11.5.1).

In any case, the edge tone is one component of the noise produced by propellers and fans. Furthermore, the displacement of air or liquid by the blades generates a tonal noise with many overtones. Its fundamental frequency, the blade frequency, is determined by the number of revolutions per second of the propeller and by the number of its blades. Superimposed on it are often non-periodic sound components which are caused by irregularities in formation vortex.

At still higher flow velocities the Kármán vortex street disintegrates into many small eddies randomly distributed in space and size. Now the local flow velocity has become a random function of space and time. This type of flow is called turbulent. The same condition prevails in a free jet as is used in jet-driven aircrafts (see Fig. 15.2). It consists in its simplest form of a gas stream issuing from a nozzle with high velocity. We assume the latter as significantly below the sound velocity. Then we can distinguish three regions: the so-called potential core which gradually disappears, the mixing zone surrounding the core, and the fully developed turbulent wake of the jet. Noise is mainly produced in the zone in which the exhausted gas becomes mixed with the stationary air. This zone is characterised by strong turbulence. The local fluctuations δv of flow velocity are connected by Bernoulli's law (see eq. (11.14)) to fluctuations of the pressure

$$\delta p \propto \rho (\delta v)^2 \tag{15.4}$$

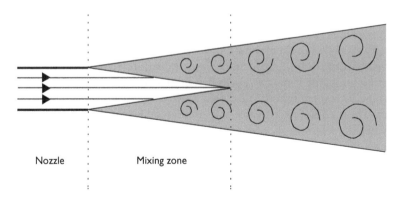

Nozzle

Mixing zone

Figure 15.2 Turbulent flow produced by a free jet.

(This relation is valid for incompressible fluids, strictly speaking, but it may be applied here as long as δv is significantly smaller than the sound velocity as assumed.) These pressure fluctuations cause in turn fluctuations of the local density.

Now a volume element which periodically or randomly expands and contracts may be regarded as a point source or as a small 'breathing sphere' (see Section 5.7). A turbulent flow contains many sources of this kind. They do not work quite independently but are mutually correlated within a 'coherence region' where they show similar temporal behaviour. Because of random phase differences the sound waves produced by them cancel completely, and any remainder would require some mass exchange across the boundary of a coherence region which does not occur. We conclude that a coherence region cannot emit a spherical wave.

A similar argument holds if we imagine pairs of such point sources with opposite phases combined as dipoles, each of them radiating according to Section 5.5. Like the 'monopole sources' discussed earlier, these dipole sources are also mutually correlated within the coherence region. They cannot produce any net radiation unless an external force acts on that region. (To understand this we can think of a dipole as a small reciprocating body moving to and fro as a whole which certainly requires a certain force.) This argument is not valid, however, for quadrupoles, each of them consisting of two dipoles of opposite polarity (see Section 5.5), since their operation is accompanied neither by mass exchange nor by the exchange of forces. Hence we conclude that the sound radiation issuing from a coherence region is mainly quadrupole in character. A typical feature of this kind of radiation is that the acoustical power produced increases as the eighth power of δv and hence of the flow velocity.

So far we have considered mainly flows of gases. In streaming liquids, the mechanisms of noise generation are essentially the same. However, there is one phenomenon which is specific to liquids: cavitation. By this term we understand the formation and activity of cavities in liquids with locally varying flow velocity or more precisely, in regions of underpressure on account of Bernoulli's law. This occurs, for instance, in pumps, turbines, ship propellers, constrictions or bents in pipes, valves and many more devices. When the underpressure disappears or the cavities are carried downstream the cavitation voids collapse rapidly emitting sharp pressure impulses. These impulses combine to form a hissing noise with a broad frequency spectrum. In the next chapter we shall encounter a somewhat different kind of cavitation.

15.2.3   Shock waves

The term shock wave or shock front denotes a surface, for instance, a plane, across which the state of a fluid, that is, its pressure, its density, its temperature, etc., undergoes a sudden change. Its existence and formation is due to the non-linearity of the basic hydrodynamic equations (3.6), (3.10) and (3.11) or their three-dimensional extensions. A shock wave propagating into a stationary medium travels with a velocity exceeding the sound velocity. When it arrives at our ear it is perceived as a sharp bang.

Shock waves may be formed in different ways. In Section 4.5 their generation by steepening the positive-going flanks of a plane and originally harmonic wave was described. In principle, this effect occurs in any plane wave which travels over a sufficiently long distance. In reality, this is not so since the high-frequency spectral components are continuously reduced by loss processes occurring in the medium. In weak waves the attenuation outweighs the non-linear generation of those components while in intense waves the latter process is the dominant one. Therefore the mentioned steepening process is observed with sufficiently strong waves only.

Practically, shock wave formation by steepening is observed whenever a gas volume is suddenly released into a tube or pipe. A common example is a combustion engine in combination with its exhaust pipe in which shock wave formation must be expected unless special means are taken to avoid it.

Shock fronts are also generated when a body moves through a medium faster than sound velocity, or when a supersonic flow hits a resting body. The former case occurs, for instance, with a projectile or an aircraft flying at supersonic speed. In this case the shock front which is dragged along with the moving body (see Fig. 15.3) has nearly the shape of a cone with the aperture angle $2\alpha = 2 \cdot \arcsin(c/v)$, the so-called *Mach* cone. In the second case the shock front is at rest. It can easily be observed with a stick held in a rapidly flowing brook. If the flow velocity of the water exceeds the velocity of surface waves – usually capillary waves – a wedge-like wave field emerges from the stick, which is it at rest relative to the stick (and the observer) and is bounded by a Mach cone. Likewise, the envelope of spherical waves issuing from a sound source moving with supersonic velocity forms a shock wave (see Fig. 5.3b).

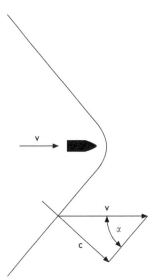

Figure 15.3 Shock wave in front of a body moving with supersonic speed v relative to the surrounding medium.

Perhaps the best-known example of acoustical shock waves is the 'double bang' resulting from an air craft travelling at supersonic speed. It consists essentially of two shock waves closely following each other (N-wave) which are formed by the main discontinuities of the aircraft, namely, its front and its end. Every further discontinuity causes an additional shock albeit of minor strength. This 'sonic boom' does not arise just once, namely, when the aircraft is 'breaking through the sound barrier', but is dragged along with it all the time.

The tips of the rotor blades of a helicopter move also with supersonic speed relative to the stationary air; they are responsible for the cracking character of the generated noise heard in the vicinity of a helicopter. The crack of a whip, by the way, is also due to parts of the lash accelerated beyond sound velocity.

15.3   Primary noise control

The possibilities of primary noise control are as diverse as the noise sources themselves. Therefore in this section only a few general aspects can be described which form the basis of primary noise control. It should be noted that their application in actual situations is often restricted by practical or economic limitations.

If the noise is caused by impacting rigid bodies, for instance, machine elements, the strength of the primary impact noise and hence the excitation

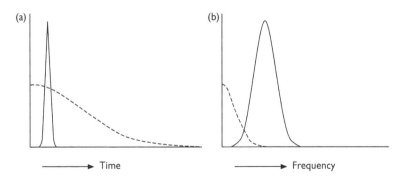

Figure 15.4 Time signal and spectrum of an impact: (a) between hard bodies, (b) between relatively soft bodies.

of other structures connected with them can often be reduced by slowing down the speed of force transfer. This can be achieved by choosing a softer material for the impact partners. Then the apparent loudness reduction is due to shifting the centre of the force spectrum towards lower frequencies as shown in Figure 15.4. Even if the total sound energy produced by the impact remains unaltered, the noise will be perceived as less loud and annoying. Correspondingly, the A-weighted sound pressure level will also be lower because it gives less emphasis to low frequencies than to higher ones. An additional level reduction is achieved if the chosen materials show high internal losses as, for instance, hard rubber or certain types of plastic. Of course, this method has its limits, after all a nail cannot be driven into a board with a rubber hammer. Nevertheless, there are many cases where the noise level can be diminished in this way. Thus the noise from gears can be significantly reduced by employing gearwheels made of plastics. The shape of the teeth is also of significant influence on the intensity and the spectrum of noise.

As already noted in Subsection 15.2.1 the level of impact noise attains especially high values if impacting machine parts are rigidly connected with sound radiating surfaces or plates, especially if the latter show pronounced bending resonances. If such a connection is inevitable for some technical reason, the radiating parts should be kept as small as possible and should have low bending stiffness and hence high characteristic frequencies. Furthermore, the loss factors of the materials they are made from should be high. The efficiency of the latter measure is easily demonstrated by comparing the noise emerging from a freely hanging metal plate with that of a plate of rubber or soft plastic when struck with a hammer.

The mechanical strength of metals can be combined with the elastic losses of damping materials by putting both together in a layered structure. If a metal sheet covered with a viscoelastic layer is excited into bending vibrations the lossy layer is forced to participate in the elastic deformations of the supporting plate and thus will withdraw motional energy from the plate and convert it into heat. The damping is more effective the higher the loss factor of the damping layer. In order to allow a significant fraction of the elastic energy to enter the damping material the Young's modulus of the latter must not be too low. In the simplest case, the damping layer can be glued on one side of the supporting plate, or applied in liquid form by painting or spraying. Of particular efficiency in this respect are sandwich plates consisting of two metal sheets or layers with the damping layer between.

Another way to reduce significantly the noise radiation from vibrating plates or sheets is by perforating them. Figure 15.5a shows a section of a perforated plate. The area of one aperture is denoted with S_a while the plate area per hole is called S.

If the plate vibrates as a whole with velocity v_0, only a certain fraction of the displaced air is used to built up the air pressures $\pm p$ on both sides which are the sound pressures of the transmitted and the reflected wave. The remaining part of the air flows through the opening with the speed v_a thus providing for some pressure equalisation. We have:

$$(S - S_a)v_0 = S_a v_a + S\frac{p}{Z_0}$$

The pressure difference 2p between both sides of the plate must overcome the mass reactance $j\omega m$ with $m = \rho_0 S_a d'$; d' is the thickness of the plate plus the end corrections (see Subsection 7.3.3). The effect of viscosity in the apertures is neglected. Hence we obtain:

$$2p = j\omega\rho_0 d' v_a$$

From these expressions v_a can be eliminated. After introducing as earlier the 'porosity' $\sigma = S_a/S$ and replacing ρ_0 with Z_0/c the sound pressure of the transmitted wave becomes:

$$p = \frac{1 - \sigma}{1 + 2\sigma c/j\omega d'} \cdot Z_0 v_0 \tag{15.5}$$

In Figure 15.5b the magnitude of $p/Z_0 v_0$ according to this expression is plotted as a function of the frequency for $d' = 1$ cm and for perforations $\sigma = 0.05, 0.2$ and 0.5. At low frequencies where the mass reactance of the air plugs in the apertures is very small the acoustic short-circuit impedes an efficient radiation even if the perforation is low. For $\sigma = 0$ the sound pressure becomes $Z_0 v_0$ as is to be expected.

Since the acoustic power output of a fluid flow grows with a high power of the flow velocity – of free jets with the eighth power, of other flow configurations with the sixth – the first and most important step of primary noise

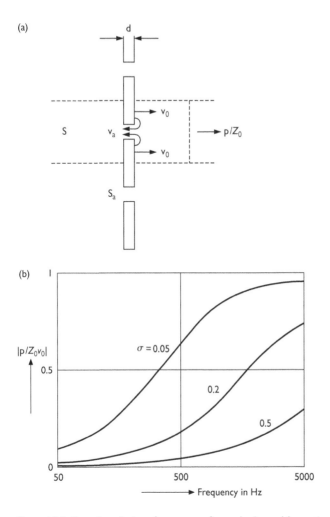

Figure 15.5 Sound radiation from a perforated plate: (a) section, (b) the sound pressure produced by the plate for various porosities σ ($d' = 1$ cm).

control must be the reduction of the flow velocity. With pipes carrying a flow this is principally achieved by increasing the cross-sectional area. Like-wise, in free jets such as of jet engines a somewhat lower speed leads to a significant reduction of the noise level. Modifying the shape of the nozzle may result in a further reduction of noise level without sacrificing too much thrust. Examples are annular nozzle shapes, or splitting up the nozzle into several smaller ones which are slightly inclined against the longitudinal axis

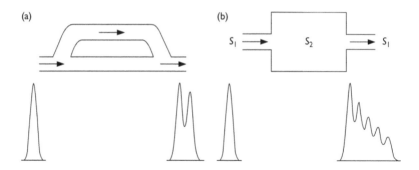

Figure 15.6 Splitting up a shock wave in a tube: (a) by a bypass, (b) by an enlarged section.

of the jet. All these measures have the effect of speeding up the mixing process and hence reducing the size of the mixing zone which is the main source
of jet noise.

Another way of reducing the sound level is by avoiding discontinuities
of adjacent solid bodies, for instance, in pipes, or on the outer surface of
vehicles. Each of them would produce vortices or turbulence, and both are
sources of flow noise as discussed in Subsection 15.2.2. Hence in pipes abrupt
changes of the cross section should be replaced with steady transitions, likewise, the direction of flow should not be changed by sharp bends but by
generously rounded sections. Since turbulence is accompanied by an increase
in the flow resistance the acoustical requirements agree in this case with the
need for economic mass transport.

Shock waves are inevitable because they are an inherent property of supersonic flow. Thus there is no way to avoid the N-wave, that is, the sonic boom
issuing from supersonic aircraft. However, the steepening of strong sound
impulses travelling within a tube can be counteracted up to some point by
splitting up the original impulse into smaller impulses as is explained in
Figure 15.6. The left part of this figure shows a simple bypass section which
delays the signal travelling in it. In the right part the disintegration of the
impulse is achieved by reflecting it to and fro within an enlarged section. Such
measures become useless, however, if a somewhat longer tube is attached
at the downstream side in which the single partial impulses would catch up
each other and again unify into one single shock.

15.4   Secondary noise control

The possibilities of reducing noise emission by appropriate construction of
machines and other technical equipment are often limited by non-acoustic
factors. Then the noise exposure of the environment must be further
diminished by secondary measures. This means one attempts to prevent

the propagation of sound. Regarded in this way, the measures described in Chapter 14 may be conceived as 'secondary noise control in buildings'. The contents of the next two subsections are also related to the matter in the preceding chapter.

15.4.1   Enclosure of noise sources

An obvious way of secondary noise control is by enclosing machines and other noise sources, that is, by surrounding them with a sound insulating box or cabinet. In most cases this is built of sheet metal; larger ones can be made of brick or concrete. In any case it is important that the enclosure is tight since even small gaps or slits may strongly reduce the sound insulation (see Subsection 7.3.3). This requirement competes with the need of access since all machines must be operated or at least maintained from time to time. Furthermore, the enclosure must be isolated against structure-borne sound arriving, for instance, via the floor.

The sound reduction attainable with an enclosure depends, of course, on the sound reduction index of its walls (see preceding chapter) and on its acoustic tightness. If carefully designed, the enclosure may reduce the sound level by 30 dB and more. Since the walls are usually not very large preventing flexural resonances is more important than in building acoustics. For thin-walled enclosures this can be achieved by applying damping layers to them as described in Section 15.3. Another fact to be regarded is that the sound produced by the source will undergo multiple reflections from its walls, thus increasing the sound level inside. This build-up of energy degrades the effect of the enclosure and can even make it useless. It can be reduced or avoided by placing sound-absorbing material inside the enclosure, usually by lining the walls with it as described in Section 13.5. Typical linings are layers of glass wool of foamed plastics, covered with perforated panels. If possible, the frequency dependence of the absorption should be adapted to the noise spectrum. A particular problem is the ventilation of the machinery inside the enclosure and the removal of heat. Both can be achieved by duct sections which are designed as dissipative silencer (see Subsection 15.4.7).

15.4.2   Vibration isolation

Many technical installations and devices operated in buildings contain moving, for instance, rotating, elements which produce vibrations. If they are in rigid contact with the floor or a wall of a building these vibrations are transferred to the building structure and are propagated through it in the form of structure-borne sound (see Section 14.5). Since vibrating building elements radiate airborne sound into the environment the noise of the original source may be heard not only behind or below the partition where they are mounted, but even in more remote parts of the building. Therefore the question arises how such vibration sources can be isolated from the building where they are installed.

As a typical example we consider rotating machines which produce vibrations by out-of-balance forces. The results can be applied to other sources of solid-borne sound such as a piano or a violoncello which produce not only the desired airborne sounds but also emit vibrations into the floor and hence into the ceiling of other residents living underneath.

The transfer of structure-borne sound from machines can be prevented or significantly reduced by mounting the vibration source on flexible supports, that is, on springs. For light equipment pads of cork, rubber, etc. may be employed; for the insulation of heavy machinery specially developed steel springs are in use for this purpose, both with and without additional damping. It may be advantageous to provide for a massive foundation between the source and the springs (see Fig. 15.7a). In any case, the machine and the foundation, having a mass m, and the spring with a compliance n form a resonance system, the electrical equivalent circuit of which is shown in Figure 15.7b. All losses, for instance, elastic losses of the springs, are represented by a resistor r. We assume that the floor is nearly rigid, so we can neglect its admittance. It is seen from the equivalent circuit that the vertical force transferred into the floor is given then by

$$F' = \frac{r + 1/j\omega n}{j\omega m + r + 1/j\omega n} \cdot F = \frac{1 + \frac{j}{Q}(\omega/\omega_0)}{1 + \frac{j}{Q}(\omega/\omega_0) - (\omega/\omega_0)^2} \cdot F \qquad (15.6)$$

Here F is the alternating force produced by the machine, $\omega_0 = 1/\sqrt{mn}$ is the resonance frequency and $Q = m\omega_0/r$ the Q-factor of the system. This equation agrees with eq. (2.32) which is not surprising since the example discussed at the end of Section 2.6 concerns the inverse problem, namely, the protection of some delicate equipment from vibrations of the ground. Hence this is another example of the rather general principle of reciprocity which was mentioned already in Section 5.2. The contents of eq. (15.6) are

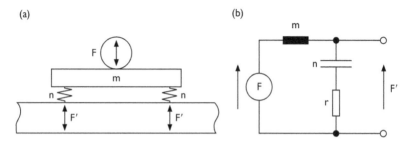

Figure 15.7 Reduction of solid-borne sound transfer by resilient supports: (a) arrangement (schematically), (b) equivalent electrical circuit.

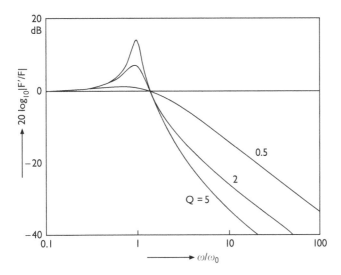

Figure 15.8 Structure-borne sound insulation by resilient supports. Parameter: Q-factor.

represented in Figure 15.8 which plots the quantity $20 \log_{10} |F'/F|$ as a function of the angular frequency with the Q-factor as a parameter. Insulation is only obtained in the range above the resonance frequency where, however, it may be quite considerable. For the resonance frequency of the system the following rule of thumb may be useful: if the load m causes a static compression of the spring by 1 mm, the resonance frequency of the system is about 16 Hz. A critical point is the resonance peak; if the number of rotations per second agrees with the resonance frequency the machine together with its foundation will attain excessive vibration amplitudes. Therefore it is important to pass this critical range as fast as possible when the machine is started or stopped. By providing for high damping the resonant peak can be flattened or removed. However, this must be paid for by a degradation of the achievable vibration insulation.

15.4.3 Noise barriers

It is a common experience that sound will be more or less weakened by an extended obstacle as, for instance, a wall or a building, and that the obstacle casts a 'shadow'. However, this shadow is not perfect since sound is diffracted around the boundaries of the obstacle (see Chapter 7). Therefore noise screens as they are often seen along highways with heavy traffic or along railway tracks have only a limited effect which is nevertheless quite useful.

In practical situations the approaching sound wave is not plane as has been assumed in Chapter 7 but is a spherical wave originating from a point or a limited region. The level reduction ΔL effected by a straight barrier of constant height and infinite length can be calculated with sufficient accuracy by using a semi-empirical formula of Kurze and Anderson:[1]

$$\Delta L = 20 \log_{10} \left[\frac{\sqrt{2\pi N}}{\tanh \sqrt{2\pi N}} \right] \, dB + 5 \, dB \tag{15.7}$$

The quantity N in this expression is a frequency parameter

$$N = \frac{2}{\lambda}(a_1 + a_2 - b) \tag{15.8}$$

the meaning of the lengths a_1, a_2 and b may be seen from Figure 15.9a. Evidently, the expression in the brackets is the detour which the barrier imposes on the sound path connecting the source with the observation point. The solid curve in Figure 15.9b shows the level reduction after eq. (15.7) in dependence of the parameter N.

Equation (15.7) can be used to determine the level reduction of a barrier with respect to the noise from a road with heavy traffic which can be regarded as a straight line source. Since the sounds issuing from the various length elements are incoherent the resulting noise immission can be obtained by adding the intensities. Of course, the different values of the parameter N must be taken into account; its maximum N_{max} occurs for the length element opposite to the observation point. The dashed curve in Figure 15.9b shows the result of this summation (or rather integration), again after Kurze and Anderson. Here the quantity of the abscissa is N_{max}. Evidently, the barrier is less efficient for a line source than for a point source.

These curves are only valid as long as the airborne sound insulation of the wall itself is significantly greater than the level differences effected by screening. This condition is easily fulfilled provided the wall is tightly closed, that is, that it is free of gaps, slits or other openings. Inversely, it follows that the interlaced fences which are so popular among garden owners may protect them against views from outside but are not well-suited as noise barriers.

At larger distances the level reductions shown in Figure 15.9b should be considered as a rough clue only since the curvature of sound rays by gradients of temperature and wind speed may noticeably modify the propagation (see Section 6.2). To avoid multiple reflections between the barrier and, say, a railway train, barriers are often lined with some sound-absorbing material, which, of course, must be weather-proof.

---

1 U. J. Kurze and G. S. Anderson, Sound attenuation by barriers. *Applied Acoustics* 4 (1971), 35.

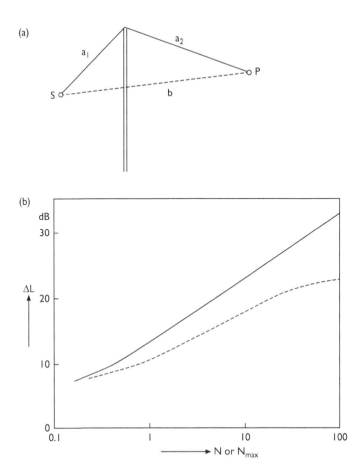

Figure 15.9 Effect of a noise barrier: (a) location of sound source S and observation point P, (b) level reduction for a point source (solid line) and for a line source (dotted line). $N = (2/\lambda) \cdot (a_1 + a_2 - b)$.

15.4.4 Noise protection by vegetation

Much less efficient than tight noise barriers are densely planted trees, shrubs, etc. the sound screening effect of which is often overrated by laymen. Thus a dense hedge may offer an excellent protection against views; the noise reduction effected by it, however, is rather of psychological than of physical nature. The reason for this should be clear: a row of plants is nearly transparent to sound because sound – unlike light – can pass leaves and branches due to diffraction. However, a small fraction of the sound energy may get lost on account of absorption and scattering.

The same holds for extended forests. Here the attenuation of sound waves depends strongly on the kind of trees; it is obvious that a coniferous forest influences the penetration by sound in a different way from a dense decid-uous wood in summertime which is interspersed with shrubs of different heights. Accordingly, the data published in literature is scattered over a wide range. At least, a rough average value may be distilled from the plentitude of measured data: for traffic noise over plane ground we can reckon on an additional attenuation of the order of 0.1 dB/m due to forest.

15.4.5 Noise control by absorption

In a closed room a significant reduction of noise level may often be achieved by sound-absorbing lining of walls and essentially the ceiling. In principle, it does not matter whether the noise intrudes from outside into the room – because of poor sound insulation of the walls, for instance – or whether it is produced within the room itself as in workshops, open-plan bureaus, airports or theatre foyers. We consider the noise issuing from a single point source with output power P. The basis of level reduction is eq. (13.19) which is repeated here in somewhat different form:

$$w_{tot} = \frac{P}{4\pi c}\left(\frac{1}{r^2} + \frac{16\pi}{A}\right) \tag{15.9}$$

Here w_{tot} is the total energy density at distance r from the noise source, and A denotes the equivalent absorption area of the room. Since the latter influences only the second term an increase of A is the more efficient the larger the distance of the observation point from the source. In the limiting case of very large distances doubling of the absorption area reduces the noise level by 3 dB. If there are many sound sources distributed throughout the room almost every point lies in the direct field of a source; therefore the level reduction according to eq. (15.9) is not too impressive. Nevertheless, a sound-absorbing treatment of the ceiling will make the room less noisier because at least the ceiling reflection is eliminated.

15.4.6 Reactive silencers

Silencers or mufflers are used to prevent the propagation of noise in ducts or pipes in which gases or liquids are moved. Main applications are exhaust pipes of combustion machines, and air conditioning systems in which fresh air is set into motion by a fan and is delivered into the rooms to be served. Furthermore, quite large silencers are employed in industrial plants.

Depending on the principle they are based on one has to distinguish between reactive silencers and dissipative ones. The former are partially transparent barriers, so-to-speak, which reflect a part of the incident sound energy. The latter silencers consist of ducts lined with sound-absorbing material which converts sound energy into heat.

The simplest reactive silencer is an abrupt change of cross-sectional area of a tube from S_1 to S_2 (see Fig. 8.3a). Its reflection factor is after eq. (8.15):

$$R = \frac{S_1 - S_2}{S_1 + S_2} \tag{15.10}$$

while its transmission factor is $T = 1 + R$. However, the sound insulation achieved by a single expansion or constriction is very modest. A much more efficient silencer is the combination of an expansion with a subsequent constriction as shown in Figure 8.3b or 15.6b. However, at present we are not discussing the prevention of shock formation but linear propagation of sound through such an expansion chamber. Accordingly, we consider sine waves with an angular frequency $\omega = ck$. In contrast to the discussion in Subsection 8.3.2 the length l of the chamber is not assumed to be small with compared to the acoustical wavelength λ. However, the lateral dimensions, for instance, the diameter of the channel, are – as earlier – supposed to be smaller than λ.

At first, the sound pressure p_i of a sound wave entering the chamber from the left is reduced by the transmission factor T. Then the wave is repeatedly reflected between both ends of the chamber. During each roundtrip its amplitude is altered by a factor R^2 and by a phase factor $\exp(-j2kl)$ since it has passed twice the length l of the chamber. Finally, the wave portion leaving the chamber is multiplied by the transmission factor $T' = 1 - R$ of the right end. Hence the sound pressure behind the chamber is

$$p_2 = p_i TT' e^{-jkl} \left(1 + R^2 e^{-j2kl} + R^4 e^{-j4kl} + \cdots\right) = \frac{TT' e^{-jkl}}{1 - R^2 e^{-j2kl}} p_i \tag{15.11}$$

Inserting the expressions for T, T' and R yields after a little simple algebra:

$$\left|\frac{p_i}{p_2}\right|^2 = 1 + \left[\frac{S_1^2 - S_2^2}{2S_1 S_2} \cdot \sin(kl)\right]^2 \tag{15.12}$$

Figure 15.10 represents the tenfold logarithm of this quantity for some ratios S_2/S_1 of cross-sectional areas as a function of kl. Since the different wave portions interfere with each other the sound insulation of the silencer depends strongly on the frequency. In particular, it vanishes whenever kl is an integral multiple of π, that is, whenever the length l of the chamber is an integral multiple of half the wavelength $\lambda = 2\pi/k$. This is easy to understand since this silencer is a line resonator as described in Section 9.1 with the difference that the line is terminated at both its ends neither rigidly nor with zero impedance

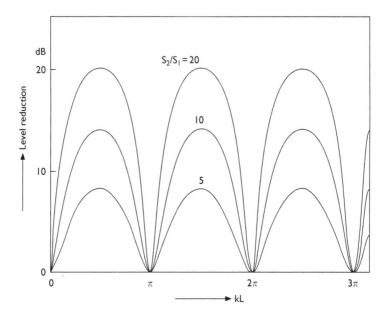

Figure 15.10 Level reduction by a reactive silencer after Figure 15.6b as a function of the frequency parameter kl (l = length of the silencer). Parameter: S_2/S_1.

but with the impedance $Z_0 S_2/S_1$. To moderate the frequency dependence of the attenuation the chamber could be subdivided by a diaphragm inserted asymmetrically in it. Practical mufflers as are used in motorcars or trucks have in general more sophisticated structures and often contain dissipative elements as well.

Sections of pipes with soft or nearly soft walls, that means with walls of vanishing impedance, can also be counted as parts of the family of reactive silencers. Such waveguides or pipes cannot be made for gaseous media. However, for liquids, for example, for water, they can be realised by a hose of compliant material such as rubber. According to Section 8.5 no fundamental wave exists in this case, that is, the lowest cut-off frequency is not zero but has a finite value. For circular cross section with diameter 2a and for water as a wave medium this cut-off frequency is (570/a) Hz. If the angular frequency of the sound wave is below this value the angular wave number k'' after eq. (8.52) is imaginary. Then the intensity-related attenuation constant equals twice the value of k''. However, this formula is only valid if the length of the tube section is considerably larger than its diameter.

15.4.7 Dissipative silencers

A dissipative silencer consists in principle of a section of a duct the walls of which are completely or partially lined with sound-absorbing material

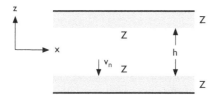

Figure 15.11 Dissipative silencer, schematically. v_n: Normal component of particle velocity, Z: impedance of the wall lining.

(see Fig. 15.11). The reason for the attenuation which a sound wave undergoes is that the normal component of its particle velocity is not zero as it would be in a rigidly walled duct but assumes a finite value. The lining continuously withdraws energy from that component and hence from the sound wave. This consideration tells us, by the way, that the sound wave in the duct cannot be plane but must have a somewhat more complicated structure; otherwise, the only component of the velocity would be the longitudinal one.

To determine the sound field we could solve the wave equation (3.25) taking into account the geometry of the duct and the boundary conditions imposed by the particular wall lining. Doing this one finds infinitely many solutions each of them corresponding to one wave type similar to those described in Section 8.5. However, their angular wave numbers cannot be represented by a closed formula. Of course, they are complex in general which is just what we want since their imaginary parts are proportional to the attenuation per unit length.

Of all these wave types only the fundamental wave is of interest because it can be propagated at all frequencies including the lowest ones. At the same time, this type has the lowest attenuation and hence is well-suited to assess the overall performance of the silencer. Therefore we refrain from this somewhat long-winded approach and restrict ourselves to an elementary, however, less exact derivation of the attenuation constant. For the sake of simplicity it is assumed that the duct has a uniform, locally reacting lining with wall impedance Z.

Let P be the sound power transported by the duct. In the course of propagation it will be diminished according to

$$P = P_0 e^{-mx}$$

Hence $dP/dx = -mP$. On the other hand $-dP/dx$ is the energy absorbed per unit time and length. It is equal to $I_n U$ with U denoting the circumference of the duct, $I_n = \mathrm{Re}\{pv_n^*\}/2$ is the intensity component directed toward the lining. Finally, we express v_n by p_w/Z where p_w is the sound pressure on the surface of the absorbing lining. Thus equating mP to $I_n U$ yields as an intermediate result

$$mP = \frac{U}{2} |p_w|^2 \, \text{Re} \left\{ \frac{1}{Z} \right\}$$

(15.13)

If the impedance is not too small the sound pressure can be expected to be nearly constant over the cross section of the duct. Then we have $p_w \approx p$ and

$$P \approx S \cdot \frac{|p|^2}{2Z_0}$$

(S = cross-sectional area). By combining this expression with eq. (15.13) we obtain:

$$m = \frac{U}{S} \text{Re} \left\{ \frac{Z_0}{Z} \right\}$$

(15.14)

Unfortunately, this derivation does not tell us anything about the range of its validity. However, the more exact derivation mentioned earlier shows that eq. (15.14) is a useful approximation in the frequency range given by

$$\left| \frac{Z_0}{Z} \right| \ll \frac{\omega S}{cU} \ll \left| \frac{Z}{Z_0} \right|$$

(15.15)

This condition is only meaningful if the wall impedance Z is much larger than the characteristic impedance Z_0 of air. Under this preposition the expression $\text{Re}\{Z_0/Z\} = \text{Re}\{1/\zeta\} = \xi/|\zeta|^2$ in eq. (15.14) may be replaced with $\alpha/4$ (see eq. (6.23) with $\vartheta = 0$) where α is the absorption coefficient of the lining. Then we obtain from eq. (15.14) $m = \alpha U/4S$ and for the attenuation per meter $D = 10 \, \text{m} \cdot \log_{10} e = 4.34 \, \text{m}$:

$$D \approx 1.1 \frac{U}{S} \alpha \, \text{dB/m}$$

(15.16)

It is clear that this formula should be used with caution because of the simplifying assumptions which have been made in its derivation. Anyway the preceding discussion holds for infinitely long ducts. Since every real silencer is of finite length, additional losses occur at its entrance and exit.

According to eqs. (15.14) and (15.16), the attenuation is larger the smaller the area and the larger the circumference of its cross section. Therefore the least favourable cross section is the circular one. Particularly good performance show silencers with additional absorbing baffles arranged parallel to the duct walls as depicted in Figure 15.12a. Such baffles can sometimes be used to improve the attenuation of an existing duct.

We conclude this section by glancing at the range of high frequencies. In this case the lining of the wall will only be of slight influence on the lateral

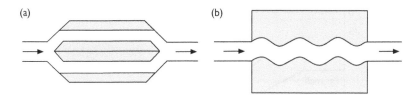

Figure 15.12 Various types of dissipative silencers: (a) with absorbing baffle, (b) with undulated channel.

distribution of the sound pressure, and the wave types will not be very different from those propagating in a rigidly walled duct. The same statement holds for the angular wave number which, for a duct consisting of two parallel rigid plates, is given by eq. (8.46). (In this context m denotes the order of the wave type, that is, an integer.) At high angular frequency and moderate m the root in eq. (8.46) is only slightly smaller than unity; accordingly, the angle φ in Figure 8.12 is very small. On the other hand, at nearly grazing incidence the lining has very low absorption (see Section 6.4), and the same holds for the attenuation in the channel. One can counteract this 'jet formation' by leading the flow through a bent or undulated channel within the absorbing layer. A silencer of this kind is depicted in Figure 15.12b.

15.5   Personal hearing protection

When means of primary or secondary noise control are not feasible or are not sufficient to provide for satisfactory noise protection the last resort is devices to be worn by the endangered persons themselves in order to reduce the harmful or annoying effects of noise. Their purpose is to seal the entrance of the ear channel up to a certain point. The efficiency of such measures is limited by bone conduction which forms a bypass and is not influenced by such devices apart, maybe, from helmets.

Probably the best-known protection is earplugs of formable materials which are brought in a fitting form by the user himself and are inserted into the ear canal. They are usually made of combinations of cotton and wax or vaseline, or of silicone putty. Likewise, earplugs of PVC or polyurethane foam are in use. Prior to application they are rolled and pressed until they match the shape of the canal after insertion. An alternative is custom-moulded earplugs which are manufactured from impressions of the ear canal. Furthermore, premoulded earplugs are also in use; made from soft and flexible materials, they are available in different sizes and fit more or less into the ear canal.

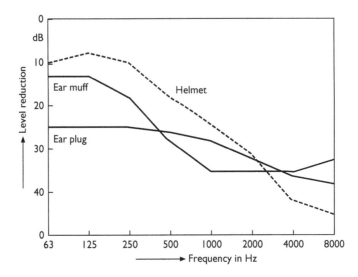

Figure 15.13 Reduction of noise level by various types of hearing protectors.

In very widespread use are earmuffs. These are plastic cups which completely enclose the pinna and which are lined inside with some porous absorption material in order to absorb high-frequency sound and thus to improve their performance. They are kept in place by a headband, much like earphones. It is important that the rim of the cup fits closely to the user's head which is achieved by a cushion filled with plastic foam or some viscous liquid.

Finally, noise protection helmets are employed. They enclose a substantial part of the head and often contain additional cups enclosing the pinnae. Unlike the devices mentioned earlier which only seal the ear channel they impede hearing by bone conduction to some extent.

In Figure 15.13 the level reductions reached with the different hearing protectors are plotted as a function of sound frequency.[2] Since the reproducibility of such measurements is not too high these results should be regarded as averages. This comparison shows that the performance of earplugs is surprisingly good, especially at low frequencies. Noise protection helmets prove to be particularly useful at high frequencies.

2 After H. Berger *et al.* (eds), *The Noise Manual*. Alpha Press, Fairfax VA, 2000.

Chapter 16

Underwater sound and ultrasound

This chapter deals with two special fields of acoustics which are not concerned with audible sound, either because the medium is not air but water, or because the sound frequency is beyond the range accessible to our hearing. In the latter case we speak of ultrasound. Although in our everyday life we are not faced with ultrasound nor water-borne sound both of them are of great practical interest. Thus underwater sound represents the only way to transmit information under water where electromagnetic waves fail because of the electric conductivity of water and the high attenuation caused by it. With the aid of ultrasound the interior constitution of non-transparent bodies and objects can be examined. Furthermore, ultrasound has many applications due to the high acoustical energy densities which can be achieved at elevated frequencies. Between both fields there is some common ground: in underwater sound often waves with ultrasonic frequencies are employed, too, and in ultrasonic engineering the propagation in liquids (and also solids) is more prominent than that in air. Likewise, the methods of sound generation and detection are similar in both fields. This justifies dealing with both of them in one chapter although their applications are different.

16.1 Acoustical detection and localisation of objects (sonar)

The most prominent applications of underwater sound are nowadays summed up under the acronym SONAR which stands for 'Sound Navigation and Ranging'. Sonar engineering is the acoustical counterpart to the better known radar techniques which cannot be employed under water for the reasons mentioned earlier. Both kinds of detection are based on the same idea which is also the basis of the diagnostic ultrasound applications to be described later: a sound transmitter emits an impulsive signal which is partially reflected by some obstacle (often referred to as 'target'). The echo is detected either by the transmitter itself which must be reversible for this purpose, or by a separate receiver arranged close to the transmitter (see Fig. 16.1). From the signal's travel time to the reflecting or scattering object

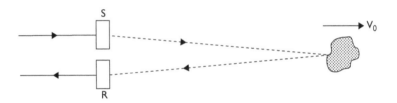

Figure 16.1 Principle of active sonar (S: sound projector, R: receiver, V_0: speed of object with respect to the echo-localising system).

and back to the receiver one can evaluate the distance of the object provided the sound velocity of the medium is known. By employing transmitters and/ or receivers with high directivity and scanning the relevant angular range the location of the object can be determined. However, the resolution and hence the accuracy of target localisation has its natural limits, depending on the sound frequency, on the local and temporal fluctuations of the sound velocity, and on several other factors.

If the target moves with a speed V_0 relative to the observer, the frequency f of each spectral component of the echo signal is altered on account of the Doppler effects (see Section 5.3) by

$$\delta f = \pm 2\frac{V_0}{c}f \qquad (16.1)$$

This means, by measuring the frequency shift the relative speed of an object can be determined too.

Besides the 'active sonar' there is also 'passive sonar', that is, localisation of objects which themselves radiate sound signals as, for instance, ship machinery or ship propellers. In this case only the direction of the target can be detected but not its distance. Both active and passive sonar are widely used, of course, for military purposes, for instance, for the detection of submarines and mines. In addition, there are also many non-military applications. One of the oldest of them is the measurement of sea depth. Here the 'target' is the sea bottom which is often of irregular shape. Part of the incident sound wave may even penetrate the sea bottom and reveal sub-bottom structures by producing echoes from various layers. This is of relevance for sea geology. Furthermore, sonar techniques serve the security of seafaring in that it permits the early recognition of reefs, shallows, icebergs and so on.

Another area of application of sonar is the detecting and tracking of fish and surveying fish populations. It benefits from the large backscattering cross section (see Section 16.3) which many fishes have on account of their gas-filled swimbladder. Therefore virtually all trawlers are equipped with sonar nowadays. However, considerable experience is needed to draw conclusions from received echoes to identify the type and distribution of fish.

16.2   Sound propagation in sea water

In a way, water is better suited as a medium for sound propagation than air since sound attenuation in it is much smaller. In Figure 16.2 the attenuation constant of sea water, expressed in dB/km, is plotted over a logarithmically divided frequency scale. One component is the classical attenuation with its quadratic frequency increase; another, more prominent one is due to the relaxation process attributed to the dissociation of dissolved magnesium sulphate (see Subsection 4.4.2). For comparison it may be noted that the attenuation in air under normal conditions and at 10 kHz is of the order of 100 dB/km (see Fig. 4.12).

The sound velocity in water increases steadily with increasing hydrostatic pressure, that is, with increasing depth. Furthermore, in the interesting range it grows monotonically with the temperature. In shallow water with depth of up to about hundred metres there is enough mixing activity resulting in nearly constant water temperature; therefore the sound speed can be regarded as roughly constant. Matters are quite different in deep water where usually the water temperature shows a pronounced variation with depth which depends on the general climatic conditions, on the season and time of day, but also on the sea state. Mostly, the water is warmest next to the surface. Thus the actual sound speed profile is generally determined by two opposite influences, namely, that of the hydrostatic pressure and of the temperature. This results

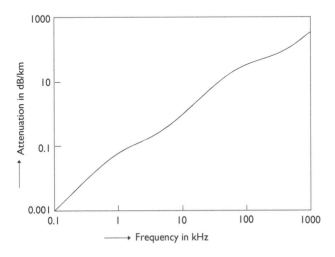

Figure 16.2 Attenuation of sound in sea water.

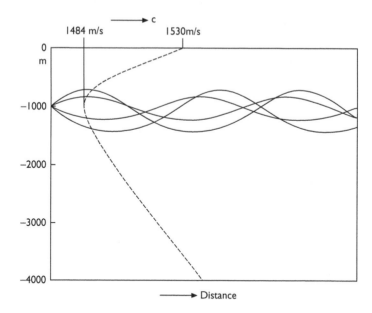

Figure 16.3 Typical sound speed profile in the ocean (broken line) and deep sound channel caused by it.

in a local minimum of sound velocity which is typically about 1000 m below the surface (see Fig. 16.3).

As described in Section 6.2 sound rays in an inhomogeneous medium are curved in general, or more precisely, they are bent towards the side of decreasing sound speed as can be seen from Figures 6.2 or 6.3. Therefore a sound ray emitted horizontally from a transmitter next to the surface will turn downward which may lead to the formation of shadow zones. However, if the sound ray is emitted in a depth near the sound speed minimum into a nearly horizontal direction it oscillates around the horizontal since it is alternately upward and downward curved. Hence the range near the sound speed minimum forms a two-dimensional channel in which the sound waves are confined. Consequently, the sound intensity decreases not proportionally to $1/r^2$ (r = distance) as is typical for spherical wave propagation but only to $1/r$. This corresponds to a level drop of only 3 dB for distance doubling compared to a drop of 6 dB as valid for spherical wave propagation. This particular spreading law, together with the low attenuation of water, explains the enormous distances which can be covered by underwater sound and which may amount at low frequencies to thousands of kilometres. Since the various sound components arriving at a certain point have travelled slightly different distances an impulsive signal will change its shape in the course of its propagation, which means it is subject to dispersion.

In the preceding section, wave propagation in 'deep channels' has been discussed in the ray picture. Alternatively, we could also have adopted a description in terms of wave types or modes as in the discussion of waveguides in Section 8.5 which also show dispersion.

16.3   Strength of echoes

As already mentioned in Section 7.1, the 'acoustic size' of an object is characterised by its scattering cross section the definition of which will be repeated here: let I_0 denote the intensity of a sound wave arriving at an object and P_s the total sound energy it scatters per second, then the scattering cross section of the object is

$$Q_s = \frac{P_s}{I_0} \tag{16.2}$$

For sonar according to Figure 16.1 only that portion of P_s is of interest which is scattered back to the transmitter–receiver. This portion is characterised by the so-called 'backscattering cross section'. It relates the intensity I_{rs} of the scattered wave observed at the location of the transmitter–receiver to the incident intensity I_0:

$$Q_b = 4\pi \, r_s^2 \frac{I_{rs}}{I_0} \tag{16.3}$$

with r_s denoting the distance of the scattering object. This equation means that the ratio of both intensities equals the ratio of the backscattering cross section to the surface area of a sphere with radius r_s. On the other hand, the intensity in a spherically spreading wave is

$$I_0 = \frac{P_0}{4\pi \, r_s^2} \tag{16.4}$$

with P_0 = power output of the sound source. With this expression we obtain

$$I_{rs} = \frac{P_0 Q_b}{\left(4\pi r_s^2\right)^2} \tag{16.5}$$

If the sonar transmitter produces directive sound beams which is usually the case then an additional factor γ_t will turn up in eq. (16.5), the gain of the transmitter (see Section 5.4). Likewise, the directionality of the receiver is accounted for, if necessary, by a further factor γ_r in eq. (16.5). The backscattering cross section is frequency dependent in general; it can be determined for a particular body by calculating the scattered sound field it produces

Table 16.1 Limiting values of the scattering and backscattering cross section of the sphere and circular disk (radius a) normalised by the visual cross section πa^2

Object		Normalised	
		Scattering cross section Q_s	*Backscattering cross section* Q_r
Rigid sphere	$ka \ll 1$	$7/9\,(ka)^4$	$25/9(ka)^4$
	$ka \gg 1$	2	1
Soft sphere	$ka \ll 1$	4	4
	$ka \gg 1$	2	1
Rigid circular disk, normal sound incidence	$ka \ll 1$	$16/27\pi^2(ka)^4$	$16/9\pi^2(ka)^6$
	$ka \gg 1$	2	1

when the body is exposed to a plane wave field. Table 16.1 lists a few limiting cases. Since the backscattering cross section of a more complicated target depends not only on the frequency but also on its orientation with respect to the sonar system, its calculation is quite involved. And still more difficult if not insoluble is the 'inverse' problem, namely, to derive information on the kind, size, orientation, etc. of the target from the small section of the scattered field which is accessible to sonar.

16.4 Ambient noise, reverberation

The detection and localisation of targets in the ocean is afflicted with many more elements of uncertainty. One of them has already been mentioned, namely, the production of erraneous target directions on account of bent sound rays. Another problem is that refraction changes not only the direction from which an echo seems to arrive but also the 'density' of received sound rays and hence the intensity of sound beams, that is, the strength of an echo. Such errors can be corrected if the profile of the sound speed and its temporal variations are known as precisely as possible. Therefore instruments have been developed which permit rapid measurement of this profile. They consist, in principle, of a small sound transmitter and receiver combined with a weight and can be towed underwater by a ship. Both transducers are part of a transmission path the length of which can be increased by a set of mirrors. The local sound velocity is determined from the measured transit time of a signal.

Furthermore, in the sea there is always a certain 'noise level' which adds an unwanted background to the echo signals to be detected and hence interferes with the operation of underwater sound devices. Ambient noise has quite diverse origins. One of them is thermal noise caused by the random motion of molecules. Since this is a fundamental phenomenon it represents the absolute lower limit for the intensity of detectable echo signals. However, in the frequency range below 50 kHz noise due to air bubbles created by surface waves prevails. Its strength depends, of course, on the sea state and the wind force. Furthermore, rain drumming on the surface generates ambient noise within the water. Another source of ambient noise is marine life which, however, plays a role particularly in shallow water. And finally, a significant component of ambient noise may be due to ships. By suitable frequency filtering and more sophisticated methods of signal processing the signal-to-noise ratio may be considerably increased.

But also the operation of a sonar device itself produces unwanted noise since some portions of the projected sound signal are scattered back towards the source by numerous inhomogeneities in the volume and by the rough boundaries. Thus an impulsive sonar signal produces a tail made up of many tiny echoes called reverberation which interferes with the echo from the target. Underwater reverberation is caused by air bubbles, fish and other living organisms, furthermore, by surface waves and irregularities of the sea bottom.

For a quantitative treatment of reverberation we assume that the scattering objects are randomly distributed with an average density of N objects per unit volume, each of them with the backscattering cross section Q_b. Suppose a sound impulse with duration Δt is sent into the medium at the time $t = 0$. After t seconds it will have reached all scattering objects located in a spherical shell with radius $r_s = ct$ and thickness $\Delta r_s = c\Delta t$ (see Fig. 16.4). Their number is $N \cdot 4\pi r_s^2 \cdot \Delta r_s$. The echo signals created by them arrive at the sonar

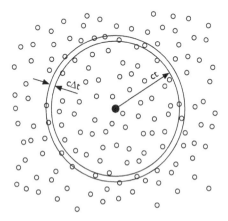

Figure 16.4 Scattering and volume reverberation in the ocean.

system after 2t seconds; because of the random locations of the scatterers they are mutually incoherent and can be added energetically. Hence the average intensity of reverberation is:

$$I_s = 4\pi r_s^2 N \Delta r_s \cdot \frac{Q_b}{4\pi r_s^2} \cdot I_0 = \frac{N Q_b \Delta r_s}{4\pi r_s^2} P_0$$

with the latter expression according to eq. (16.4). In this equation we replace r_s with ct and Δr_s with $c\Delta t$, and also denote the total transit time $2t$ with t_t. Finally, we account for the directivities of the transmitter and the receiver by introducing the gains γ_t and γ_t, respectively, as in the preceding section. Then

$$I_s = \frac{\gamma_t \gamma_r N Q_b \Delta t}{\pi c t^2} P_0 \tag{16.6}$$

Similar reasoning can be applied to 'boundary reverberation' caused by the roughness of the water surface and by irregularities of the bottom. In this case the intensity drops proportionally to $1/t$ (instead of $1/t^2$). In any case the law of energy decay is quite different from what we are used to from room acoustics which reflects the different mechanism of its generation. Furthermore, it should be noted that eq. (16.6) and the corresponding equation for the boundary reverberation yields only a rough picture of the decay; in reality, the decay shows strong and irregular fluctuations.

16.5 Transducer arrays

The most important components of any sonar system are the projectors of underwater signals and the receivers for detecting the echoes induced by these signals. Usually, electroacoustic transducers are used for both these purposes. Nowadays, these are mainly piezoelectric although projectors based on the more traditional magnetostrictive transducer principle are also in use. Also, impulsive underwater signals are occasionally generated by underwater explosions or with hydrodynamic sources. Since most electroacoustic transducers are reversible, that is, since they can convert electric signals into acoustical ones and vice versa, one and the same transducer can be employed for both purposes. In many cases, however, it is more practical to use separate transducers as transmitters and receivers. A more detailed description of the principles underlying electroacoustic transducers is given in the subsequent chapters.

The frequencies of practically applied underwater sound range from a few Hertz to about 1 MHz, depending on the kind of application. Most commonly used sound frequencies are in the range from 5 to 30 kHz where the attenuation is still moderate while the resolution is sufficient for many purposes.

In underwater sound, several transducers of the same kind are usually combined to form a transducer array. Such an array supplies higher power output than a single transducer when operated as a transmitter, or has increased sensitivity when used as a receiver. The main advantage of an array, however, is its directivity. In the simplest case the transducers are arranged equidistantly along a straight line. The properties of such arrays are described in Section 5.6. They concentrate the radiated sound into a plane perpendicular to their extension. A two-dimensional array radiates predominantly in the direction perpendicular to the plane where the transducers are placed. Its directivity function is obtained by an obvious extension of eq. (5.25). (This kind of array represents in a way the transition from the single source to the vibrating piston (see Section 5.8) which is better approximated the smaller the mutual distances of the transducers compared to the wavelength.)

Besides that, circular or cylindrical arrays are also in use. In any case, the total dimension of an array must be several wavelengths if high directivity is to be achieved. Hence it may turn out to be very unpractical or even impossible to sweep it mechanically when a certain angular range is to be scanned. (It should be realised that such arrays are often mounted besides or underneath a ship.) This difficulty can be overcome by electronically changing the directional characteristics, in particular, the direction of maximum radiation (or sensitivity). If, for example, the main lobe of a linear array is to be swept by an angle α_0, the signal feeding the nth array element must be delayed by the time $(n - 1)\tau$ with

$$\tau = \frac{d}{c} \sin \alpha_0 \tag{16.7}$$

as shown in Figure 16.5a. The integers n go from 1 to N, the total number of transducers. The altered directivity function is obtained by replacing $\sin \alpha$ with $\sin \alpha - \sin \alpha_0$ in the derivation of eq. (5.25). Hence we obtain instead of that equation:

$$|R(\alpha)| = \left| \frac{\sin \left[Nkd/2 \, (\sin \alpha - \sin \alpha_0) \right]}{N \sin \left[kd/2 \, (\sin \alpha - \sin \alpha_0) \right]} \right| \tag{16.8}$$

In Figure 16.5b this quantity is represented as a polar diagram for such a 'phased array' consisting of six elements ($N = 6$) and for $kd = 2$; the sweeping angle 30°. It should be noted that the polar diagram is not turned as a whole but is altered in detail. The same method can also be applied to receiving arrays consisting of underwater microphones or, as they are called, hydrophones. Their electrical output signals are added after delaying them properly. Of course, electronic sweeping of the main lobe of the directional characteristics is not restricted to linear arrays but can be applied to any kind of array.

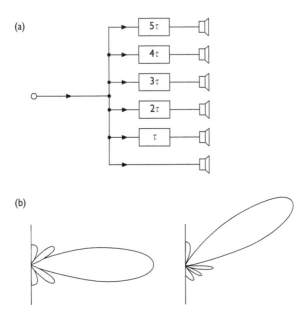

Figure 16.5 Sweeping the main lobe of a linear array with electrical delay units: (a) principle, (b) directional factor (magnitude) of an array consisting of six point sources. Left side: original directional characteristics, right side: main lobe swept by 30°.

The signals used in sonar systems consist in the simplest case of short wave trains of constant frequency. However, the reliability of sonar detection is improved by increasing the frequency bandwidth of the signal since not only the strength but also the shape of a signal is altered by backscattering. Or expressed in a different way, by reflection or scattering of a signal its frequency spectrum is changed in a way which is characteristic for the object to be detected. The bandwidth of the signal can be increased by employing frequency modulation or by transmitting relatively long, more sophisticated signals which after receiving can be swept together into a short impulse by a so-called correlation filter. Generally, in sonar technology signal processing plays a more important role than in any other branch of acoustics. Since this is beyond the scope of this book we shall not go into it in more detail.

16.6   General remarks on ultrasound

Now we turn towards the second subject of this chapter, namely, ultrasound. As already mentioned in the introduction the term describes all sounds with frequencies above the upper limit of the human hearing. Although this limit differs from person to person and also varies in the course of life, a frequency of roughly 20 kHz is a reasonable value. Accordingly, the ultrasonic range lies above 20 kHz.

The propagation of ultrasound follows basically the same laws as that of sound of any frequency. However, the weighting of the various phenomena influencing the propagation is somewhat shifted and this more so the higher the frequency. Thus, because of the smaller wavelengths, diffraction by obstacles does not play the prominent role it has in the range of audible sound. Therefore, it is often said that the propagation of ultrasound, particularly of somewhat elevated frequencies, is 'quasi-optical' which means that the ray concept familiar in optics is applied with more justification in ultrasonics than in the audio range. Thus we can speak of 'illuminating' an object or a region by an ultrasonic beam.

On the other hand, sound attenuation which is often neglected in the audible range becomes more prominent at ultrasonic frequencies. Concerning the physical mechanisms of sound attenuation we refer to Section 4.4. As a general rule, attenuation is more prominent in gases than in liquids, and it is higher in liquids than in solids. For this reason, the use of ultrasound in air is rather limited; in most ultrasonic applications the wave medium is liquid or solid.

In these applications one distinguishes 'diagnostic' methods which do not require very intense sound waves from those which are based on the high sound intensities as are generated relatively easily in the ultrasonic range. In the former applications ultrasound serves as a carrier of information, mainly to learn about the interior state or structure of non-transparent bodies. Here, non-destructive flaw detection in metals and other materials as well as medical diagnostics is in the foreground of interest. On the other hand, high intensity ultrasound is employed to achieve certain changes in materials and objects. In the first line of such applications is ultrasonic cleaning as well as joining with ultrasound.

16.7   Generation and detection of ultrasound

Today, technical ultrasound is generated almost exclusively by electrical means. In the foreground of interest there is the piezoelectric sound generator, which is easy to operate and can be adapted to quite different requirements. A more thorough description of it is found in Chapters 17 and 19. At present we just mention that its essential component is mostly a disk or layer of piezoelectric material arranged between two metal electrodes (see Fig. 17.2). When an alternating electrical voltage is applied to these electrodes the disk reacts to the variations of the electrical field strength by varying its thickness which leads to sound emission into the environment.

Since the piezoelectric effect is reversible it can also be used for detecting ultrasound signals: a sound wave impinging on a piezoelectric disk gives rise to variations of its thickness which are associated with alternating electrical charges on the electrodes, due to the piezoelectric effect. One and the same piezoelectric element can be used for both generating and receiving ultrasonic signals, a fact which is often exploited in technical applications. In addition, various sorts of ultrasound microphones have been developed. Since they are mostly used in liquid media they are usually referred to as hydrophones. More will be said on hydrophones in Chapter 18.

If the acoustical wavelength in the piezoelectric layer is smaller than its thickness d, the layer must be regarded as a waveguide. Then the strains created by the applied voltage travel within the piezoelectric in the form of elastic waves which are repeatedly reflected from its end faces. The super-position of all these waves leads to a standing wave similar to that in an air-filled tube which is closed at both its ends. This standing wave is particularly pronounced if the thickness of the piezoelectric layer equals an integral number of half-wavelengths. In this case we have excited a normal mode of the layer. The corresponding frequencies – the eigenfrequencies or resonance frequencies – are given by eq. (9.1). However, with an arrangement shown in Figure 17.2 only normal modes of odd order can be excited if both faces of the disk have the same mechanical load. Hence, the resonance frequencies of a piezoelectric thickness transducer are

$$f_n = (2n + 1) \cdot \frac{c_L}{2d} \quad (n = 0, 1, 2, \ldots) \tag{16.9}$$

(c_L = speed of longitudinal waves). For high intensity applications these transducer resonances are generally desired because of the high power yield; usually the fundamental resonance ($n = 0$) is used. However, for generating broadband sound signals, for instance, short impulses, these resonances are suppressed to some degree by attaching a block of damping material to one side of the disk. Typical transducers such as are used for non-destructive testing of materials are shown in Figure 16.8.

Piezoelectric transduction is not the only way to detect ultrasound; mechanical, thermal or optical effects are also employed for this purpose although to a minor extent. Thus a limited sound beam traversing an otherwise undisturbed liquid exerts a constant pressure on an obstacle called the radiation pressure as already described in Section 4.5. If the obstacle or target is a sound-absorbing plate perpendicular to the axis of the sound beam the radiation pressure is numerically equal to the energy density in the liquid; if the plate reflects the incident sound, the radiation pressure is twice as high. Therefore the energy density and hence the intensity of the sound beam can be determined by measuring the radiation force acting on the target. Instruments constructed for this purpose are called radiation balances. Figure 16.6 shows one particular example of such a balance.

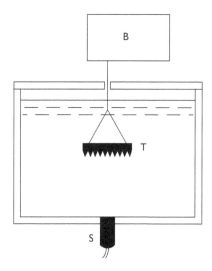

Figure 16.6 Ultrasonic radiation balance (S: ultrasound source, T: target, B: balance).

Thermal sensors of ultrasound determine the sound intensity from the rise of temperature of a sound-absorbing body which is placed in the sound field. Optical methods are based on the fact that a harmonic (progressive or standing) ultrasound wave acts as an optical diffraction grating due to the regular density changes caused by it. Comparing the light intensities in the various diffraction orders permits an absolute determination of the sound intensity. Furthermore, the sound-induced variations of density can be used to visualise extended sound fields by schlieren optical methods.

16.8   Diagnostic applications of ultrasound

The most important applications of low-intensity ultrasound are non-destructive testing of materials, tools, machine components, etc., and medical sonography. In both cases the goal is the examination of a medium and the inhomogeneities hidden in it.

Apart from special applications the method almost exclusively employed is the impulse echo method. The principle underlying this procedure was already shown in Figure 16.1. Since the wavelengths are relatively short in the ultrasonic range, the formation of highly directive sound beams does not present any difficulties. Very often the reflectivity of a boundary in the test object is so weak that a substantial part of the incident sound energy can enter the region behind and detect further boundaries or obstacles. Then the received signal will exhibit several or even many echoes. If the test object is a plate, a tube, a container and so forth the strong echo produced by the rear

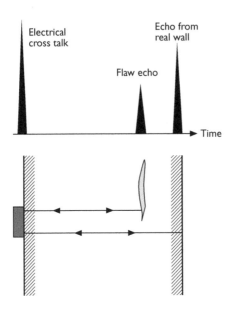

Figure 16.7 Non-destructive testing of materials with the impulse echo method.

wall can be employed to determine the thickness of the object provided the sound speed of the material is known. Inversely, a rear wall echo facilitates the localisation of a flaw in a body even without knowing the sound velocity of it (see Fig. 16.7).

As before, the strength of an echo is determined by the backscattering cross section of the object which creates it (see Section 16.3). Since this quantity increases strongly with frequency, the centre frequency of the signal determines the size of the smallest detectable object. On the other hand, increasing the frequency means also increasing the attenuation of the ultrasound signal. Therefore in practical applications the choice of test frequency is a compromise between competing requirements. In technical material testing the applied frequencies are mostly between 1 and 10 MHz. About the same holds for medical sonography (see Subsection 16.8.2) although sometimes, namely, for the examination of small or thin organs (eye, skin), ultrasound of much higher frequencies are employed.

16.8.1   Non-destructive testing of materials

A typical transducer for testing materials is presented in Figure 16.8a. It consists essentially of a disk of piezoelectric material the front side of which is usually covered with a thin protective layer, and its rear side is in contact with damping material. The latter should combine high interior losses with

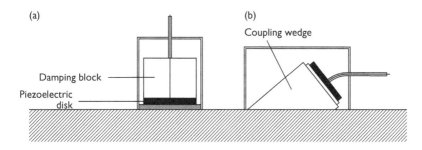

Figure 16.8 Piezoelectric probes for non-destructive testing: (a) probe for perpendicular sound incidence, (b) angle probe for oblique sound incidence.

a characteristic impedance close to that of the piezoelectric material. For non-perpendicular irradiation of test signals into a workpiece as is employed, for instance, for the inspection of welds, angle probes are in use which are fitted out with a wedge (see Fig. 16.8b). Of course, the user has to make allowance for the refraction of the waves entering the test material and also for wave type conversion as described in Section 10.2. At a sufficiently oblique angle a purely transverse wave will penetrate into the test specimen which is sometimes advantageous.

To inspect some material or workpiece the probe is lightly pressed onto the surface of the specimen after applying a liquid layer (water or oil) to it to ensure good acoustical contact. Alternatively, the test specimen can be immersed in a water tank where the transducer is arranged. Sometimes, the ultrasonic wave is coupled to the test specimen by a water jet. In any case the surface or part of it must be systematically scanned, either manually or automatically.

In principle, non-destructive testing with ultrasound can be applied to virtually all materials, with, however, varying success, depending on the interior structure of the material. Generally, the sound waves in metals are attenuated by scattering from inhomogeneities (see Subsection 4.4.3), for instance, of graphite inclusions in cast iron. At the same time, the scattered sound portions form some background noise or 'reverberation' similar to that described in Section 16.4. Fortunately, most types of steel are well-suited for ultrasonic testing, more or less the same holds for light metals like aluminium and magnesium and their alloys. Much more problematic is the inspection of copper alloys such as brass or bronze and particularly of cast iron. Flaw detection in concrete or artificial stones is possible at very low frequencies only because of the coarse structure of this materials.

Ultrasound is used for the inspection of raw materials and semi-finished products just as for flaw detection in finished workpieces, in the latter case prior to the first use of machine components or assemblies as well as in the course of maintenance periods. In the foreground of interest are particularly important or heavily stressed components. Just as examples we mention sheet metal, rods, axles and tubes, containers of all kind, weld seams, railway wheels and tracks and so forth, a list which could be continued ad infinitum.

16.8.2   Ultrasonic imaging in medicine (sonography)

In material testing the echo signals are usually displayed in the form of an oscillogram (see Fig. 16.7). In medical sonography this kind of presentation called 'A-scan' is rarely applied nowadays because the examining doctor prefers a true, pictural survey over a somewhat more extended region or organ. Therefore the scanning of a region with sound rays is carried out electronically ('B-scan'). For this purpose either linear transducer arrays are used consisting of 60–240 piezoelectric elements arranged side by side, each of them a few wavelengths wide (see Fig. 16.9a). The active part of this array consists of a group of elements which are connected in parallel and which have a directivity as described in Section 5.6. After each completed probing cycle one element at each side of the active group is switched on or off in such a way that the whole active region of the transducer is laterally shifted by one unit. At the same time, the trace on the oscilloscope undergoes a small lateral offset. Another commonly used device is the sector scanner which contains a rotating head carrying several transducer elements. It scans a sector of the region under test or can even produce a panorama display.

In any case the echo signals control the brightness of the luminous spot of a monitor. By suitable synchronisation a close correspondence of the examined body region with the ultrasonic display is achieved. The sequence of probing cycles is high enough to ensure a real-time representation. Hence, also time-variable processes such as the motion of the heart valves can be displayed. As an alternative the frequency shift caused by the Doppler effect after eq. (16.1) can be exploited. This latter method is also well-suited in determining the flow velocity in blood vessels since each blood corpuscle generates a tiny echo.

Sonography is particularly useful in the examination of soft tissues which do not show big differences in their characteristic impedances. Therefore, the ultrasound beam can easily traverse the boundary between different organs and can reach a high penetration depth. The attenuation is of the order of 1 dB/cm at 1 MHz in such tissue; it can be compensated at least partially by electronic time gain control. In general, biological tissue presents itself by irregular patterns of speckles. These patterns are by no means displays of the tissue structure itself but are caused by interferences of numerous weak echo components produced in the tissue. Nevertheless, they contain information on the tissue; furthermore, they mark the boundaries between different tissues or organs and facilitates the assessment of their position

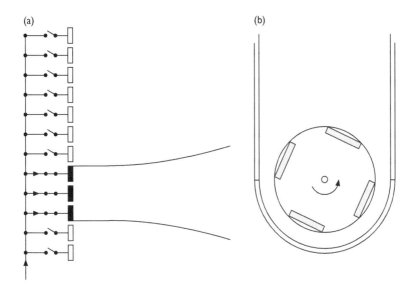

Figure 16.9 Transducers for B-scan sonography: (a) linear array, (b) sector scanner.

and size. Air-filled organs such as the lungs and regions behind them cannot be examined with ultrasound since their surface reflects the incident sound completely. A similar statement holds for bones. It is of particular advantage in sonography that it does not employ ionising radiation and thus avoids the health risks associated with X-rays; the applied sound intensity energy can be kept so low as to exclude tissue damages due to excessive mechanical strain or intolerable heat production.

On account of its efficiency, its flexibility and its simple operation sonography is applied in nearly all branches of medicine, for instance, in internal medicine, in gynaecology and obstetrics, in cardiology, ophthalmology, urology and many more fields.

16.9   Applications of high intensity ultrasound

16.9.1   Cavitation

One of the most remarkable effects of strong ultrasound waves in liquids is cavitation and the phenomena associated with it. Cavitation means the formation of voids or cavities in regions where negative pressures occur. We encountered it already in Subsection 15.2.2, however, in a different context. In ultrasonic sound fields it is the negative phase of the sound pressure which may give rise to cavitation.

Strictly speaking, the tensile strength of physically pure liquids is too high as to be overcome by the negative pressures encountered in common ultrasonic fields. In real liquids, however, numerous microscopically small solid particles are suspended which may stabilise small amounts of gas and thus act as nuclei for the onset of cavitation. As a consequence, the cavitation threshold, that is, the minimum sound pressure amplitude needed to produce cavitation, is dramatically reduced. At frequencies below 30 kHz its order of magnitude is 1 bar (corresponding to an intensity of about 0.3 W/cm^2 in water), and at higher frequencies it grows monotonically with frequency.

In contrast to the stable gas bubbles which we are familiar with these cavitation voids contain only small amounts of gas. Under the influence of the sound field they perform either strongly non-linear pulsations, or they implode as soon as the negative pressure vanishes which created them. Such a collapse starts very slowly at first, then the inward motion of the bubble wall becomes faster and reaches in the final state of the implosion extremely high velocities. During this process the remaining gas is being highly compressed, and pressure peaks of 10 000 bar may be generated. Since this compression is nearly adiabatic the temperature of the gas may become high enough to emit a short light impulse. This is probably the reason for the faint light emission originating from strong ultrasound fields in liquids which is known as sonoluminescence. Furthermore, certain chemical reactions can be initiated or accelerated in cavitation fields. This is the basis of sonochemistry. In any case, cavitation effects a strong temporal and spatial concentration of energy which is exploited in several applications.

16.9.2 Ultrasonic cleaning

To clean an object with ultrasound, it is immersed in a vessel or tank filled with a cleaning liquid that is exposed to an intense ultrasound field. The cleaning process is brought about by cavitation produced on the contaminated surface which provides the nuclei needed for cavitation inception. On the one hand, strong and short pressure impulses emerging from imploding cavities act on the surface and loosen insoluble dirt particles. On the other hand, strong local flows occur in the direct vicinity of the cavitation bubbles because they do not move synchronously. These currents remove the dirt particles from the surface and provide for a quick exchange of cleaning liquid.

Ultrasonic cleaning is carried out in cleaning tanks of quite different sizes which are made of stainless steel or of plastics (see Fig. 16.10). The sound field is mostly generated by piezoelectric compound transducers as described in Section 19.7 which insonify the liquid from the bottom or from one wall. The sound frequency is usually between 20 and 50 kHz. The cleaning liquid is either aqueous (alkaline or acid) or organic, depending on the kind of contamination.

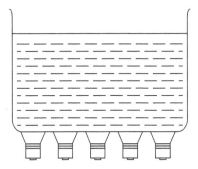

Figure 16.10 Ultrasonic cleaning tank.

Ultrasonic cleaning proves particularly useful whenever the highest degrees of cleanness is required, or when the objects to be cleaned are mechanically delicate or very small, or if they have irregularly shaped surfaces inaccessible to brushes, etc. Examples are products of fine mechanics and precision engineering, medical instruments, optical lenses, jewellery, television screens, electronic circuits, radioactively contaminated objects and many other items.

16.9.3   Ultrasonic joining

Another application of high intensity ultrasound firmly established in industrial production is joining, mainly welding and bonding of plastic materials and parts. The welding process is achieved by thermal plastification or liquification due to the sound energy dissipated in the work material. Therefore this method is well-suited for the treatment of thermoplasts such as polystyrene and its co-polymers, for polycarbonate and many more materials, but not duroplasts.

For producing a weld, the two parts are pressed together between the 'anvil' and the actual welding tool, the 'sonotrode' (see Fig. 16.11a). The latter serves at the same time for introducing the vibrational energy which is generated by an efficient ultrasonic vibrator, typically, a compound transducer. It is fed to the sonotrode via a velocity transformer (see Section 19.7) in such a way that the sonotrode vibrates perpendicularly to the surface of the bond. The frequency is 20–30 kHz. The welding process is initiated by local plastification of the material at some isolated contact points where the highest energy concentration occurs. Since sound absorption in plastics increases with rising temperature in general, those parts of the material already plastified will be heated faster and faster until extended areas will become liquid and finally both components are joined to each other. The entire welding

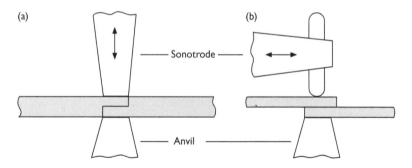

Figure 16.11 Ultrasound welding: (a) of plastics, (b) of metals.

process takes only a fraction of a second. In any case, the heat is produced exactly where it is needed which is a particular advantage of ultrasonic welding. Therefore ultrasonic bonding of plastics is applied to the production of a countless number of products in nearly all branches of the plastics-processing industry.

Ultrasonics can also be used as well for joining metals to each other or with non-metals. In this application, however, the sonotrode oscillates not perpendicular to but parallel with respect to the joint. In this way, a tangential motion of both mating surfaces relative to each other is effected (see Fig. 16.11b). By this action, the yield strength of the material is exceeded at isolated contact spots and the surfaces are levelled down by plastic deformation, and the final joint is achieved by molecular attraction forces. Thus, this kind of welding is not or not predominantly a thermal process as in welding of plastics. Best suited for ultrasonic welding is copper as well as aluminium and its alloys, either with themselves or with other metals. Joints of metals with semiconductor materials, with glass or ceramic materials are also possible.

16.9.4 Drilling and cutting

For drilling holes into a workpiece, a tool shaped according to the desired hole – similarly as with ultrasonic welding of plastics – is set into vigorous vibrations perpendicular to the surface of the workpiece. As in bonding, these vibrations are generated by a power transducer in combination with a velocity transformer. Between the drilling tool and the workpiece an aqueous suspension of an abrasive (silicon carbide, boron carbide, diamond powder) is applied as shown in Figure 16.12. When the vibrating tool approaches the surface, an alternating, transverse flow in the abrasive slurry underneath the tool is created. Additionally, strong cavitation is produced in the liquid. Both effects set the grains of the abrasive in fast motion by which the material

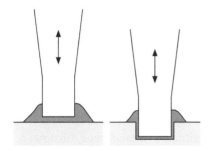

Figure 16.12 Ultrasonic drilling.

under the tool is eroded. Thus, the process in reality is some kind of grinding. By continuously lowering the tool a dip in the surface is gradually produced. This motion must, however, be sufficiently slow to avoid direct contact between tool and workpiece. With hollow tools small disks can be cut out of plates. Thin slices are cut from bar of stock by using a thin steel lamella as a tool.

Usually, the tool is soldered to the tip of the velocity transformer; it need not to be made of a particularly hard material. To drill large holes hollow tools are advantageous as they must erode less material. They have the additional advantage that fresh abrasive can be supplied continuously through the tool to the cutting region. As in ultrasonic welding it is important that the transforming piece including the drilling or cutting tool is exactly tuned to the frequency at which the transducer operates.

A particular advantage of ultrasonic drilling is that its application is not restricted to producing circular holes and that it is particularly well-suited for machining hard or brittle materials such as glass, ceramics, hard metal or gems.

16.10 Generation of high and highest ultrasound frequencies

Before dealing with the methods of generating ultrasound of very high frequencies the question of an absolute upper limit of all acoustical phenomena mentioned shall be discussed as referred to already in the Introduction.

First of all, it should be realised that sound waves of extremely high frequencies, if they exist at all, can be observed only in solids because in liquids and gases the attenuation would be much too high. Even under this condition we restrict the discussion to sound in perfect crystals which show still relatively moderate attenuation because their elementary constituents (atoms, molecules, ions) are arranged in a regular lattice.

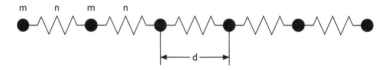

Figure 16.13 Model of a one-dimensional crystal.

Figure 16.13 shows a one-dimensional crystal lattice, imagined as a straight chain of equidistant point masses, m, with mutual distances d. The cohesive forces between the masses are idealised as springs with compliance n. If a longitudinal wave travels through this 'crystal' all masses vibrate with equal amplitudes but with mutual phase differences

$$\Delta\varphi = k_L d = 2\pi \frac{fd}{c_L}$$

This phase difference grows monotonically with increasing frequency. Sound propagation stops when adjacent masses vibrate exactly in opposite phases, then the progressive wave has become a standing wave. This is the case at the frequency

$$f_{max} = \frac{c_L}{2d} \tag{16.10}$$

(A somewhat less crude estimate would yield a factor π instead of 2 in the denominator which, however, is not relevant in this context). With $c_L \approx 5000\,\text{m/s}$ and a typical 'atomic distance' of $d \approx 2.5 \cdot 10^{-10}\,\text{m}$ one can estimate the order of the upper frequency limit to be

$$f_{max} \approx 10^{13}\,\text{Hz} = 10\,\text{THz}$$

and it is an interesting question as to how close this limit is approached by modern experimental methods.

With piezoelectric thickness or shear transducers as described in Section 16.7 ultrasound can be generated and detected at quite high frequencies. Thus, very thin quartz disks can be excited in their fundamental thickness mode (n = 0 in eq. (16.9)) at frequencies up to about 100 MHz. Still higher frequencies are reached with thin foils of polyvinylidenfluoride (PVDF) which is a piezoelectric high polymer. Thus the fundamental mode of a customary PVDF foil with 10 μm in thickness is about 200 MHz. Furthermore, very thin layers of certain piezoelectric materials can be fabricated by vapour deposition or sputtering. In this case, certain monocrystalline such as sapphire must be used as substrates which enforce a certain orientation within the layer. Best suited for this process are piezoelectric materials like cadmium sulphide, zinc sulphide or zinc oxide; they permit the fabrication of thickness transducers with fundamental resonances as high as about 3 GHz.

To arrive at significantly higher frequencies one has to abandon the concept of a thickness resonator and to make a virtue of necessity, so-to-speak: instead of producing sound with a piezoelectric element in the form of a thin disk, a little piezoelectric rod, for instance, of quartz is employed as a sound generator, as an acoustic transmission line and as a sound detector; its length is in the order of 1 cm. One of its ends is placed in that part of an electrical coaxial resonator where the electrical field is strongest and the field lines dive nearly at right angle into the rod (see Fig. 16.14). Provided this face is sufficiently plane and perpendicular to the so-called crystallographic X-axis of the quartz it will emit a plane longitudinal wave which travels along the rod and is detected at the other end of it with a similar resonator. To keep the attenuation as low as possible the whole set-up must be operated at liquid helium temperature (4.2° K). With this method frequencies of up to about 100 GHz are reached. Its limit is set by the difficulty of fulfilling the high demands in the preparation of the rod's end faces. Therefore quite different ways must be gone if sound with still higher frequencies is to be generated.

To explain them we must return to the crystal lattice mentioned at the beginning of this section. This lattice is never completely at rest. Instead, its constituents, that is, atoms, ions, etc., perform erratic oscillations around their rest positions the magnitudes of which increase with rising temperature. The energy of all these oscillations is identical to the heat content of the body.

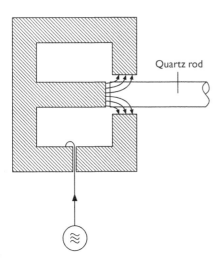

Figure 16.14 Generation of monofrequent ultrasound up to about 100 GHz after Bömmel and Dransfeld.

According to the concepts of quantum theory the vibrational energy stored in a solid body cannot be continuously increased or reduced but only stepwise by integral multiples of a finite amount which is called a vibrational quantum, a sound quantum or a phonon. Its energy is related to the frequency of vibration by

$$E = h \cdot f \tag{16.11}$$

where h is Planck's constant ($h = 0.6624 \cdot 10^{-33}$ Ws^2). This situation corresponds to that of a cavity containing electromagnetic radiation the energy of which can only be altered by supplying or removing energy quanta of finite size, the so-called light quanta or photons. On the other hand, vibrational energy can be supplied to a solid body by heating it. If we heat, for instance, a thin metal film deposited on the solid by a vigorous, very short electrical current impulse it radiates a burst of longitudinal and transverse phonons into the substrate which propagate with sound velocity through the body. This fact can be experimentally verified provided the solid is a crystal which is virtually free of imperfections and which is cooled to liquid helium temperatures. However, the sound energy is distributed over a wide frequency range reaching the upper limit given by eq. (16.10).

Even more interesting is a method invented by Eisenmenger and Dayem[1] to generate and to detect monofrequent sound quanta (although not sinusoidal sound waves). It is based on the idea that in a superconducting metal the conduction electrons are not single particles but exist in the form of so-called Cooper pairs. These electron pairs can travel nearly unimpeded through the metal lattice; this fact is the reason for the high electrical conductivity of superconductors. To break up such a Cooper pair into two single electrons a small but finite energy 2Δ is needed. Conversely, if two electrons 'recombine' to form a Cooper pair this energy will be released, predominantly in the form of sound quanta with frequency

$$f = \frac{2\Delta}{h} \tag{16.12}$$

according to eq. (16.11). These facts are illustrated by the energy band scheme shown in Figure 16.15 which is similar to that of a semiconductor. In the upper band there are – if at all – only unpaired electrons, whilst the lower one contains only Cooper pairs. For generating sound one 'just' has to find a way of injecting unpaired electrons from outside into the superconductor.

This can be achieved by combining two superconductors of the same kind which are separated by a thin insulating layer. By applying an electrical voltage $U > 2\Delta/e$ (e = elementary charge) their band schemes can be shifted

1 W. Eisenmenger and A. H. Dayem, Quantum generation and detection of incoherent phonons in superconductors. *Physics Review Letters* **18**(4) (1967) 125.

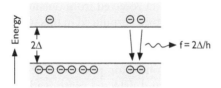

Figure 16.15 Energy band scheme of a superconductor.

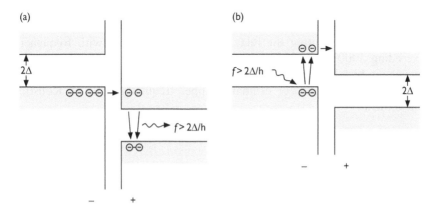

Figure 16.16 Generation and detection of monochromatic phonons in the high gigahertz range: (a) generation, (b) detection.

Source: W. Eisenmenger and A. H. Dayem, Quantum generation and detection of incoherent phonons in superconductors. *Physics Review Letters* **18(4)** (1967).

against each other as shown in Figure 16.16a. Then the electrical field draws the Cooper pairs to the right side; they can pass the insulating barrier on account of the tunnel effect known from quantum theory. At the right side they find themselves as singles in the upper band of the right superconductor since the energy supplied by the bias voltage is sufficient to break up the electron binding. To recombine, that is, for reaching the lower band, they must lose their excess energy which occurs in two steps: at first they 'fall' either directly or via intermediate steps to the lower edge of the upper band ('relaxation radiation'). The sound quanta produced thereby have frequencies between 0 and $(eU - 2\Delta)/h$. Afterwards the electrons recombine to Cooper pairs. In this latter process phonons with the frequency $2\Delta/h$ are generated.

To detect sound quanta a similar arrangement is used, however, with the difference that now the bias is smaller than $2\Delta/e$ (see Fig. 16.14b). Now the cooper pairs cannot tunnel through the insulating layer. However, they can be broken up by sound quanta received from outside provided their energy is at least 2Δ. The single electrons produced in this way are drawn across the insulating layer and give rise to an electrical current which is proportional to the rate of their production and hence to the intensity of the received sound.

By using tin as a superconductor phonons with frequencies of up to 280 GHz can be generated and detected, with lead the corresponding frequency is 650 GHz. The superconducting metal is deposited in the form of two crosswise strips deposited onto the end faces of a small sapphire rod serving as a waveguide with low attenuation. The insulating layer is only 10–50 Å thick ($1 \text{ Å} = 10^{-10}$ m) and consists of the oxide of the metal used. It is self-evident that the whole arrangement must be cooled down to the temperature of liquid helium.

Today, by exploiting the relaxation radiation sound with frequencies exceeding 3000 GHz can be generated. It is very remarkable that the frequencies which have been made accessible by these experimental techniques are not too far from our estimated upper frequency limit of acoustical phenomena.

Chapter 17

Electroacoustic transducers

Today, acoustical signals can be analysed and processed in nearly every desired way; furthermore, we can transmit them over long distances or store them in several ways and retrieve them at any time. However, in order to do this in an efficient way the signals must first be converted into electrical signals; then, after processing, transmitting or storing them they must be reconverted into sound. For both these steps electromechanical transducers are needed, or electroacoustic transducers as they are mostly called in this particular context. Likewise, the sound waves used in underwater sound or ultrasound are almost exclusively generated with electroacoustic transducers. The same holds for the detection or reception of such waves.

This chapter deals with some basic properties of electroacoustic transducers. Furthermore, the various transducer principles will be described here. In a way the chapter may be regarded as a preparation for the subsequent chapters which are devoted mainly to transducers for the audio range, that is, to microphones and loudspeakers.

An electroacoustic transducer is a system in the sense of Section 2.10. This means it has two ports, one of them being the input terminal to which the signal to be converted is applied, while the second one is the output terminal which yields the result of the conversion. First of all it is assumed that this system operates linearly, that is, that the principle of linear superposition is valid. Of course, there are no real transducers which completely meet this requirement. Moreover, we suppose that the transducers we consider are reversible, which means that they can be used for the conversion of mechanical vibrations or signals into electrical ones as well as for the reverse operation. In any case, the mechanical port of the transducer consists of some plate or pin, etc. to which the mechanical vibrations are applied or where such vibrations appear as the output signal. However, in the sense of the force–voltage analogy as explained in Section 2.7 we can equally well represent the mechanical port by two 'poles' to which a 'voltage' F is applied and into which a 'current' v flows (see Fig. 17.1). In what follows we count any

Figure 17.1 Electromechanical transducer, schematically (F: force, v: velocity, U: voltage, I: electrical current).

mechanical or electrical power flows as positive if they are directed towards the transducer port.

As a basic requirement we should demand that the conversion process does not alter the signal to any noticeable extent. In the ideal case for any input signal $s_1(t)$ the corresponding output signal $s_2(t)$ should be given by the simple relationship

$$s_2(t) = K \cdot s_1(t - \Delta t) \qquad (17.1)$$

with constant K. The time interval Δt in the argument of s_1 is to indicate that we allow for some delay of the output signal provided it is sufficiently small.

Equation (17.1) is a very restrictive requirement. Even if a transducer does not contain any non-linear component in the sense of Section 2.11 it will change the shape of a signal. To illustrate this we consider the Fourier transform of eq. (17.1):

$$S_2(\omega) = K \cdot S_1(\omega)e^{j\omega \Delta t} \qquad (17.2)$$

with $S_1(\omega)$ and $S_2(\omega)$ denoting the complex spectra of the time signals $s_1(t)$ and $s_2(t)$. Hence, a transducer satisfying eq. (17.1) conserves the frequency spectrum of a signal apart from the simple phase factor indicating the time delay. Any deviation from this ideal behaviour causes what is called a 'linear distortion'. If we take into account that our hearing is not very sensitive to changes of the phase spectrum we arrive at the less stringent requirement:

$$|S_2(\omega)| = K \cdot |S_1(\omega)| \qquad (17.3)$$

which means that the transfer function $G(\omega)$ (see eq. (2.65)) of the transducer must have constant magnitude. Even in this relaxed form it cannot be exactly fulfilled. At best we can expect that a real transducer meets condition (17.2) or (17.3) within a limited frequency range.

The majority of electroacoustic transducers employ electrical or magnetic fields to link mechanical and electrical quantities. In fact, it is an everyday experience that such fields are associated with forces. So every child knows the attraction which a magnet exerts on small iron particles; likewise, it is a common experience that a hair comb after use is able to attract little scraps of paper. Depending on the kind of field we distinguish transducers with electrical fields, which we shall call 'E-transducers' hereafter, and such with magnetic fields or 'M-transducers'. However, the basic relationship between mechanical and electrical quantities is not always linear. In these cases particular means to linearise the transducers are needed which will be described in due course.

17.1 Piezoelectric transducer

Many solid materials are electrically polarised when they are deformed which manifests itself as an electrical charge on its surface. This property is called piezoelectricity. The piezoelectric effect is reversible: when a body of such a material is exposed to an electrical field its dimensions will undergo certain changes.

As a simple but important example of a piezoelectric transducer we consider in Figure 17.2 a thin disk of thickness d which has been cut into a suitable orientation, say, of a quartz crystal. We suppose that both its faces are covered with thin metal layers serving as electrodes. If an electrical voltage U is applied to them a virtually homogeneous electrical field will be established in the disk. This gives rise to a small change of thickness which is proportional to the field strength U/d and which can be imagined as a consequence of an elastic tensile stress σ created by the electric field:

$$\sigma = e \cdot \frac{U}{d} \tag{17.4}$$

Herein e is a material constant, namely, the piezoelectric constant. For quartz its value is $e = 0.159 \ N/Vm$. Multiplied by the area S of the disk

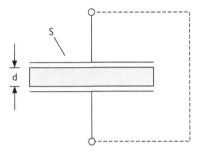

Figure 17.2 Piezoelectric disk with electrodes.

face this equation reads

$$F_{v=0} = \frac{eS}{d} \cdot U = NU \tag{17.5}$$

The subscript $v = 0$ of the force symbol F is to indicate that this full force appears only when the disk is 'clamped', that is, when it is hindered from performing any motion. If, on the contrary, the motion of the disk is not impeded from outside then the piezoelectrically generated force is completely compensated by elastic forces of the transducer material, hence the disk appears force-free. The factor N relating the applied voltage to the force $F_{v=0}$ is the so-called transducer constant:

$$N = \frac{eS}{d} \tag{17.6}$$

Its unit is N/V or As/m (A = Ampere) which is the same since

$$1\,VAs = 1\,Nm \tag{17.7}$$

Reversely, if a relative change of thickness ξ/d is imposed by some exterior forces the disk material will become electrically polarised. At short-circuited electrodes (dashed lines in Fig. 17.2) the polarisation influences a charge Q on the electrodes with density, that is, the dielectric displacement

$$D = \frac{Q}{S} = -e \cdot \frac{\xi}{d} \tag{17.8}$$

By differentiating this equation with respect to time we obtain a relationship linking the electrical current $I = dQ/dt$ to the velocity $v = d\xi/dt$ by which the thickness of the disk is changed:

$$I_{U=0} = -\frac{eS}{d} \cdot v = -N \cdot v \tag{17.9}$$

Again, the current I has a subscript which now indicates that the electrical port of the transducer is short-circuited. That the same constant N appears in both eqs. (17.5) and (17.9) is due to the reversibility of the transducer.

According to these equations the piezoelectric transducer – and this holds for every E-transducer – can be modelled as an electrical transformer with the transformation ratio $1:N$, provided we adopt the force–voltage analogy, also known as the impedance analogy.

The change of thickness is just one of several different deformations which a piezoelectric disk can experience in an electric field. Another one is a change of lateral dimensions which is usually referred to as the 'transverse piezo-effect'. Furthermore, under suitable circumstances the electrical field can give rise to a shear deformation. Which of these changes will occur depends on the kind of material and on the orientation of the electric field with respect to certain characteristic directions within the material. In piezoelectric crystals such as quartz or Seignette salt (potassium sodium tartrate) these directions are the crystallographic axes. For each of the mentioned effects the piezoelectric constant is different.

That which has been described so far is not the only way to relate electrical with mechanical quantities. Another one which is also very common is by using the 'piezoelectric modulus', which connects the strains, for instance, the relative thickness change of a disk, with the electrical field strength and the dielectric displacement with the mechanical stress.

The piezoelectric transducer can be represented by an equivalent electrical circuit, which can be thought of as the entrails, so-to-speak, of the box shown in Figure 17.1. As mentioned before, its core is a transformer; however, a few other circuit elements must be added. At first, a disk as sketched in Figure 17.2 with the electrodes on both its faces is an electric capacitor with capacitance C_0. In the circuit it is connected in parallel with the poles of the electrical port. Second, the disk reacts, independently on its electrical properties, to any thickness changes with an elastic force as mentioned earlier. Hence it acts also as a spring (however, a very stiff one). Its compliance, ξ/F, is, according to eq. (10.6):

$$n_0 = \frac{d}{SY} \qquad\qquad (17.10)$$

($Y =$ Young's modulus). In the equivalent circuit of Figure 17.3a this spring is represented by a capacitor which is subject to the full velocity v appearing at the output. In this circuit the transformer can even be left out if we note that an ideal electrical transformer with the voltage-transformation ratio $1 : N$ transforms an impedance Z into Z/N^2. Hence, its effect is accounted for by replacing n_0 with n_0N^2, v with Nv and F with F/N. The equivalent circuit modified in this way is shown in Figure 17.3b.

It is evident from these equivalent circuits that a disk has a larger electrical capacitance when its surfaces are force-free ($F = 0$) than that for clamped surfaces. Conversely, a piezoelectric disk is less stiff when its electrodes are short-circuited ($U = 0$) than with its electrical port open ($I = 0$).

The electrical equivalent circuit describes the transducer's function correctly as long as all quantities vary at a sufficiently slow rate, that is, at low frequencies. Then the deformation of the transducer material and also its dielectric polarisation can be regarded as constant throughout its thickness. Only under this condition the piezoelectric disk can be represented with lumped elements. At higher frequencies any elastic disturbance travels across the disk in the form of an elastic wave, leading to standing waves and resonances as described in Section 16.7. In Section 19.7 another equivalent circuit will be presented which accounts for these effects.

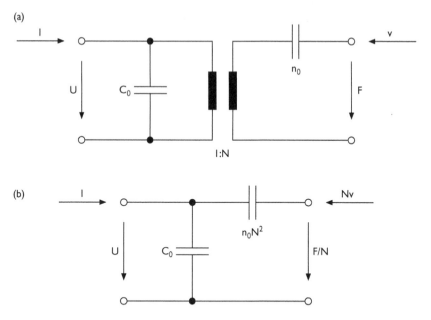

Figure 17.3 Electrical equivalent circuit of the piezoelectric transducer: (a) with transformer, (b) simplified version.

In the audio range the piezoelectric transducer has to compete with several other kinds of transducers. However, in the ultrasonic range it is the most commonly used transducer, both for generation and detection of ultrasound.

Piezoelectricity is a relatively common property of materials, although the strength of the effect varies very much. It was first discovered in quartz. Nowadays, this material has lost its importance for electroacoustic applications. For other uses, however, quartz is still of considerable interest: on account of its small elastic losses mechanical resonators consisting of quartz elements have a high Q-factor. Therefore quartz crystals are used in quartz filters, or to stabilise the frequency of electrical oscillators. Another common application of vibrating quartz elements are quartz watches. Examples of other piezoelectric crystals are tourmaline, Seignette salt or lithium sulphate. Most piezoelectric transducers for practical applications, however, consist of ceramics made of certain ferroelectric materials such as barium titanate, lead zirconate titanate (PZT) or lead metaniobate. Transducers of these materials can be fabricated in nearly any shape, that is, not only as disks or rods, but also as spherical or cylindrical shells, for instance. Since these materials are made up of irregularly oriented crystallites, they must be electrically polarised prior to their use. The same holds for foils of piezoelectric

Table 17.1 Properties of piezoelectric materials

Material	Sound velocity (longitudinal) c_L m/s	Density ρ_0 g/cm³	Relative permittivity ε	Piezoelectric constant e As/m²
Quartz (x-cut)	5700	2.65	4.6	0.17
Lithium sulphate	5470	2.06	9	0.66
Lead zirconate titanate (PZT-5A)	4350	7.75	1700	15.8
Lead metaniobate	3300	6.0	225	3.2
Polyvinylidene fluoride (PVDF)	2200	1.78	10	0.14

high polymers such as polyvinylidene fluoride (PVDF). Table (17.1) lists the properties of some piezoelectric materials.

17.2   Electrostatic transducer

If we leave out the piezoelectric material from the parallel-plate capacitor shown in Figure 17.2 we arrive at the electrostatic transducer, also known as dielectric or capacitive transducer. Its function as a sound generator is based upon the attractive forces between electrical charges of opposite sign. The reverse effect exists too: varying the distance of the electrodes of a charged capacitor changes its capacity and hence its voltage, or its charge, or both of them, depending on the electrical load of the electrodes. Any change of electrical charge is linked to a charging or discharging current.

Unfortunately, the relation between the electrical field strength and the forces connected to it is non-linear: the electrodes of a parallel-plate capacitor charged to a voltage U attract each other with the force:

$$F_{tot} = \frac{C_0}{2d} \cdot U^2 \qquad (17.11)$$

where C_0 is its capacitance and d the distance of the electrodes. Therefore the transducer must be linearised. For this purpose a bias voltage U_0 is superimposed on the signal voltage $U_\approx$. Supposing that the latter is sufficiently small compared to the bias we can write

$$U^2 \approx U_0^2 + 2U_0U_\approx$$

Inserting this into eq. (17.11) yields:

$$F_{tot} = \frac{C_0}{2d} \cdot U_0^2 + \frac{C_0U_0}{d} \cdot U_\approx \qquad (17.12)$$

Hence, the total force F_{tot} is composed of a constant part which must be balanced by a suitable suspension of the electrodes, and by the alternating force represented by the second term:

$$F_{v=0} = \frac{C_0 U_0}{d} \cdot U_\approx = N U_\approx \qquad (17.13)$$

As in eq. (17.5) the subscript $v = 0$ indicates that this force refers to the clamped condition.

Reversely, if we change the electrode distance by a small amount ξ the capacitance will also undergo a slight change δC. Since the capacitance is inversely proportional to the distance we have

$$\frac{C_0 + \delta C}{C_0} = \frac{d}{d + \xi} \approx 1 - \frac{\xi}{d}$$

or

$$\frac{\delta C}{C_0} \approx -\frac{\xi}{d} \quad \text{for } \xi \ll d \qquad (17.14)$$

Now we imagine that the electrodes are short-circuited with respect to alternating currents. This must be achieved with a large external capacitor, otherwise a constant bias voltage across the electrodes could not be maintained. Then the variation δC in eq. (17.14) is associated with a variation of the electrical charge

$$\delta Q = U_0 \delta C \approx -\frac{C_0 U_0}{d} \cdot \xi \qquad (17.15)$$

By differentiating this relation with respect to time we obtain the short-circuit current; on the right ξ must be replaced with the velocity v:

$$I_{U=0} = -\frac{C_0 U_0}{d} \cdot v = -N \cdot v \qquad (17.16)$$

It may be illustrative to compare the transducer constant N of an electrostatic transducer with that of a piezoelectric transducer of equal size. From eqs. (17.6) and (17.13) (second part) it follows with $C_0 = \varepsilon_0 S/d$ that

$$\frac{N_{piezo}}{N_{cap}} = \frac{e \cdot d}{\varepsilon_0 U_0}$$

where $\varepsilon_0 = 8.854 \cdot 10^{-12}$ As/Vm is the permittivity of vacuum. Suppose the piezoelectric constant is $e = 15.8$As/m^2 (lead zirconate titanate), and the field strength due to the bias is $U_0/d = 5000$V/cm $= 0.5 \cdot 10^6$ V/m. Then

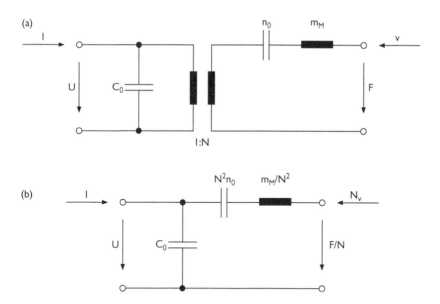

Figure 17.4 Electrical equivalent circuit of the electrostatic transducer: (a) with transformer, (b) simplified.

the ratio of transducer constants is as large as $3.57 \cdot 10^6$! We shall return to this point in Section 17.6.

The electrical equivalent circuit of the electrostatic transducer is sketched in Figure 17.4. (For the sake of clarity the supply of bias voltage U_0 is omitted.) It is similar to that of the piezoelectric transducer, apart from an inductance m_M which represents the mass of the movable components. In real transducers one electrode is a thin, electrically conducting membrane. Generally, the compliance n_0 consists of two components. One of them, n_M, is due to the stretched membrane, the other one, n_A, is the compliance of the air enclosed between both electrodes which cannot escape when the distance is rapidly reduced and hence will be compressed and decompressed. Since both 'springs' are actuated by the same elongation ξ but with different forces F_M und F_A with $F_M + F_A = F_{v=0}$ we have

$$\frac{1}{n_0} = \frac{F_M + F_L}{\xi} = \frac{1}{n_M} + \frac{1}{n_A} \qquad (17.17)$$

Thus, the resulting compliance is calculated in the same way as the capacitance of two capacitors connected in series. The compliance of a closed air cushion is obtained from eq. (6.28a) by dividing that expression by the area S of the plates since we are presently not referring to pressures but to forces. Therefore

$$n_A = \frac{d}{\rho_0 c^2 S} \tag{17.18}$$

Earlier, the transducer equations (17.13) and (17.16) have been derived under the premise that the alternating voltage is small compared to the constant bias U_0 and that the elongations of the membrane are small to the distance d of both electrodes. These conditions are no severe restriction when the transducer is used as a sound receiver. They limit, however, its application as a sound generator, where often large amplitudes are required, especially at low sound frequencies. For this reason the electrostatic sound generator has gained some importance only where small amplitudes are sufficient, namely, in earphones.

17.3   Dynamic transducer

In the transducers to be described in the following three sections mechanical and electrical quantities are linked by magnetic fields, hence we call them M-transducers. The basis of the dynamic transducer is the electromotive force, also known as Lorentz force, which an electrical conductor in a magnetic field experiences when an electrical current is passed through it. The reverse effect, namely, the conversion of a mechanical variation into a corresponding electrical one, is due to induction, that is, to the generation of an electrical voltage in a conductor which is moved across a magnetic field. Both effects are basically linear provided the magnetic field is homogeneous.

Figure 17.5a shows a straight conductor of the length l, that is, of a piece of wire or a rod which is placed in a homogeneous magnetic field with the flux density B and is arranged perpendicular to the flux lines. (At present we ignore any connections to the conductor.) Let I denote an electrical current passed through the conductor, then a force will act on it which is perpendicular to both the current and the field lines. If the conductor is clamped, this force is

$$F_{v=0} = Bl \cdot I \tag{17.19}$$

The factor connecting the current I and the force F is the transducer constant now denoted with M:

$$M = Bl \tag{17.20}$$

It has the dimension N/A or Vs/m.

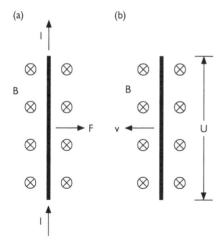

Figure 17.5 Principle of the dynamic transducer: (a) Lorentz force acting on a moving conductor, (b) voltage induced in a moving conductor.

If the current is switched off and instead the conductor is moved perpendicularly to its length and to the magnetic field with velocity v (see Fig. 17.5b), an open-circuit voltage

$$U_{I=0} = -Bl \cdot v \tag{17.21}$$

will be induced in it.

The contents of eqs. (17.18) and (17.20) cannot be represented by an electrical transformer unless we change to the force–current analogy. To keep the force–voltage analogy which we preferred so far, we have to 'invent' a new element, the so-called gyrator which has the peculiarity of relating an input current I with a force $M \cdot I$ at the mechanical port and, reversely, a velocity appearing at the mechanical side to the electrical voltage $M \cdot v$. This gyrator is the nucleus of the equivalent circuit of the dynamical transducer shown in Figure 17.6a. On its electrical side we have to add two elements accounting for the resistance R and the inductance L_0 of the conductor. Similarly, two more elements on the mechanical side represent the mass m_M of the moving parts and the compliance n_M of their suspension.

Similarly as in Figure 17.3b we can do without the gyrator by including its transforming properties into the mechanical circuit elements. This is achieved by eqs. (17.19) and (17.21): a gyrator terminated with the mechanical impedance $F/v = Z_m$ is equivalent to a circuit element with the electrical impedance M^2/Z_m (see Fig. 17.7), that is, the gyrator 'transforms' an impedance into its reciprocal. It may be helpful to verify this statement by

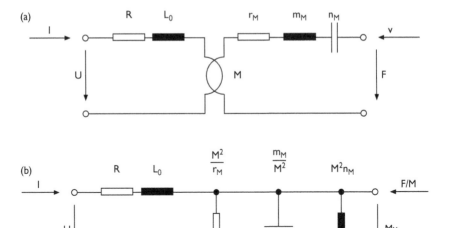

Figure 17.6 Electrical equivalent circuit of the dynamic transducer: (a) with gyrator, (b) simplified version.

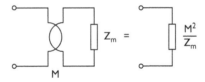

Figure 17.7 Impedance transformation by a gyrator.

considering the dimensions: Z_m has the dimension Ns/m, and the dimension of $M = Bl$ is Vs/m. From this we obtain the dimension of M^2/Z_m:

$$\frac{V^2 s^2}{m^2} \cdot \frac{m}{Ns} = \frac{V^2 s}{Nm} = \frac{V}{A}$$

the latter with eq. (17.7). Generally, the gyrator converts a circuit into the dual circuit: currents and voltages will be interchanged, elements connected in series will become connected in parallel, a capacity is changed into an inductivity etc. With these explanations, it should not be too difficult to understand the equivalent circuit shown in Figure 17.6b. Here, the mass of the conductor is represented by a capacitor with capacitance m_M/M^2 while the compliance of the spring which keeps the movable part in position appears as an inductance $n_M M^2$. Again, it is recommended to verify these assertions by checking the dimensions. Incidentally, the equivalent circuits shown in Figure 17.6 have a strange consequence: loading a dynamic loud-speaker system with a mass m results in a capacitance m/M^2 appearing at the electrical contacts of the system. Suppose, for example, the loud-speaker system has a transducer constant of 1 N/A, then a weight of 100 g fixed to the moving coil of the loudspeaker produces a capacitance of 0.1 Farad at its electrical input!

In most dynamic transducers the conductor consists of a wire wound in the form of a cylindrical coil. It is arranged in the cylindrical gap of a magnetic circuit which produces a radial flux perpendicular to the wire (see Figs. 18.9 and 19.1)

17.4 Magnetic transducer

The magnetic transducer (in the narrow sense) consists basically, as depicted in Figure 17.8, of a magnet and a movable part – the armature – made of magnetically soft iron. This part and the magnetic pole next to it are separated by a narrow air gap with width d. Furthermore, either the magnet or the armature carries a coil to which an electrical current can be applied, or which may yield an electrical voltage. Provided the air gap is sufficiently narrow the magnetic field in it can be assumed as nearly homogeneous.

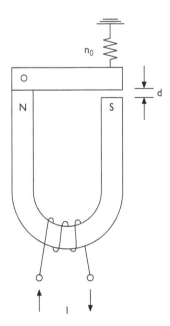

Figure 17.8 Magnetic transducer, schematic.

Under these premises the armature is attracted by the force

$$F_{tot} = \frac{\mu_0}{2} SH^2 \tag{17.22}$$

Herein S is the area of the air gap and H is the magnetic field strength in it; $\mu_0 = 1,2566 \cdot 10^{-6} Vs/Am$ is the permeability of vacuum. As with the electrostatic transducer the force is proportional to the square of the electrical quantity – here of the magnetic field strength. To linearise the transducer function the total magnetic field is made up of a constant component H_0 due to the permanent magnet, and a variable component $H_\approx \ll H_0$ produced by an electrical current I passed through the coil. Under the premise that the magnetic circuit has constant cross section S and that

$$d \gg \frac{L_1}{\mu_1} + \frac{L_2}{\mu_2} \tag{17.23}$$

the latter component is

$$H_\approx = \frac{wI}{d} \tag{17.24}$$

In these expressions L_1 and L_2 are the lengths of the field lines within the magnet and the armature; μ_1 and μ_2 are the relative permeabilities of those elements, w is the number of turns of the coil and I is the electrical current. Inserting $H = H_0 + H_\approx = H_0 + wI/d$ into eq. (17.22) yields

$$F_{tot} = \frac{\mu_0}{2} SH_0^2 + \mu_0 S \frac{wH_0}{d} \cdot I$$

This equation is similar to eq. (17.12); again, the attractive force is composed of a constant part and the alternating component

$$F_{v=0} = \frac{w\Phi_0}{d} \cdot I \tag{17.25}$$

where we introduced the magnetic flux

$$\Phi_0 = \mu_0 SH_0 \tag{17.26}$$

produced by the magnet. The constant force is balanced by the spring shown in Figure 17.8. According to eq. (17.25) the transducer constant of the magnetic transducer is

$$M = \frac{w\Phi_0}{d} \tag{17.27}$$

For using the transducer in the reverse direction we note that – under the condition (17.23) – the magnetic flux is inversely proportional to the width of the air gap. Suppose the width is changed from d to $d + \xi$ with $\xi \ll d$, then

$$\frac{\Phi}{\Phi_0} = \frac{d}{d+\xi} \approx 1 - \frac{\xi}{d}$$

or, after differentiation with respect to time:

$$\frac{d\Phi}{dt} = -\frac{\Phi_0}{d} \cdot v$$

The change of magnetic flux induces an electrical open-circuit voltage in the coil:

$$U_{I=0} = w\frac{d\Phi}{dt} = -\frac{w\Phi_0}{d} \cdot v = -Mv \qquad (17.28)$$

The electrical equivalent circuit of the magnetic transducer is the same as that of the dynamic transducer (see Fig. 17.6).

As with the electrostatic transducer, the condition $\xi \ll d$ limits the range of linear operation. The fact that the transducer constant depends itself on the actual width of air gap, strictly speaking, gives rise to non-linear distortions unless the displacement ξ is very small. This condition is particularly restrictive if the transducer is to be used as a sound source. Nevertheless, in the early days of broadcasting the magnetic loudspeaker was widely used, mainly due to its high efficiency.

17.5 Magnetostrictive transducer

At first glance, the magnetostrictive effect which forms the basis of the transducer to be described in this section has some formal similarity with the piezoelectric effect: if a ferromagnetic body is magnetised a change of its dimensions is observed. Thus, a rod of iron, cobalt or nickel (see Fig. 17.9) alters its length when an electric current is flowing through a coil surrounding the rod which magnetises its material. As with the piezoelectric effect this change can be conceived as being caused by an interior tensile stress

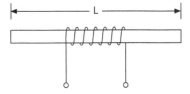

Figure 17.9 Magnetostrictive transducer, schematic.

which is the primary consequence of the magnetic polarisation. There is, however, an important difference between both effects: while the piezoelectric effect is basically linear up to very strong electrical fields, the change in length induced by magnetostriction is generally not proportional to the field strength. Hence linear operation of this transducer requires again, as with the magnetic transducer, a constant bias field. One way to achieve this is to add a strong constant current to the signal current. It is only under this condition that there is a relation analogous to eq. (17.4)

$$\sigma = e_m \cdot H \tag{17.29}$$

It should be noted that the 'piezomagnetic constant' e_m depends not only on the kind of material and its history but also on the magnetic bias. Multiplying this equation with the cross-sectional area S of the rod and inserting the relation $H = wI/L$ valid for slender rods (L = length of the rod, w = number of turns) leads to:

$$F_{v=0} = \frac{w e_m S}{L} \cdot I \tag{17.30}$$

Accordingly, the transducer constant is

$$M = \frac{w e_m S}{L} \tag{17.31}$$

It is only in very long and thin rods or in closed rings that the demagnetisation caused by the free magnetic poles can be neglected. Otherwise, it must be accounted for by a proper factor in eq. (17.31).

The magnetostrictive effect is reversible: changing the length of a magnetised rod is accompanied by a change of the magnetic flux which induces a corresponding voltage in the coil. With the precautions mentioned earlier the following relation holds:

$$U_{I=0} = -\frac{w e_m S}{L} \cdot v \tag{17.32}$$

Again, the equivalent circuit of the magnetostrictive transducer agrees with that of both transducers described before (see Fig. 17.6).

What has been said in Sections 16.7 and 17.1 holds also for the magnetostriction transducer: if the dimensions of the ferromagnetic body, here the length of the rod shown in Figure 17.9, are not small compared to the acoustical wavelength of the material any deformation will travel along the rod as a longitudinal wave which is repeatedly reflected from its ends, thus forming standing waves, that is, one-dimensional normal modes.

The magnetostriction effect is strongest in nickel which was for a long time the most important material for magnetostrictive transducers. Besides, there are ferrites which are also well-suited for this purpose. To avoid demagnetisation the ferromagnetic material is shaped as a closed core. If nickel is employed the core must be built up of laminations isolated from each other – as with an electrical transformer – in order to keep losses by eddy currents as small as possible.

The magnetostrictive transducer had its particular merits as a robust sound projector in underwater sound. Likewise, it played an important role in the generation of intense ultrasound with frequencies up to about 50 kHz. In these applications the resonances due to the finite dimensions of the transducer core are considered as beneficial since one is not so much interested in a wide frequency bandwidth but in high acoustical power output. Nowadays, magnetostrictive transducers have been widely superseded by piezoelectric ones.

17.6 The coupling factor

A quantity well-suited for comparing the different transducer principles is the coupling factor which can be regarded as a measure of the efficiency of electromechanical transduction. It depends not only on the transducer constant N or M but accounts as well for the inevitable storage of energy in some constructive elements which are not involved in the transduction process.

Suppose we apply a direct voltage to an E-transducer. In any case the capacitor C_0 in the equivalent circuit of Figure 17.3b will be charged. If the mechanical port is force-free ($F = 0$), the capacitor $N^2 n_0$ will be charged up to the same voltage. Physically, this corresponds to storing the potential energy W_{mech} in the elastic element of the transducer (piezoelectric crystal, stretched membrane, air cushion). Now the coupling factor compares the energy W_{mech} to the total energy supplied by the electrical source:

$$k^2 = \frac{W_{mech}}{W_{el} + W_{mech}} \tag{17.33}$$

The electrical energy stored in a capacitor with capacitance C is $W = CU^2/2$ where U is the voltage across its terminals. Applying this formula to both capacitors in Figure 17.3b yields

$$k^2 = \frac{n_0 N^2}{C_0 + n_0 N^2} \tag{17.34}$$

The denominator of this expression is the capacitance $C_{F=0}$ measured at the electrical port of the transducer when the mechanical port is short-circuited, that is, when it is force-free:

$$C_{F=0} = \varepsilon_0 \varepsilon_{F=0} \frac{S}{d} \qquad (17.35)$$

Now we apply these relations to the piezoelectric and the electrostatic transducer. The compliance of the piezoelectric disk can be inserted from eq. (17.10). After introducing the transducer constant $N = eS/d$ into the nominator of eq. (17.34) we obtain:

$$k_{piezo}^2 = \frac{e^2}{\varepsilon_0 \varepsilon_{F=0} Y} \qquad (17.36)$$

Hence the coupling factor of this transducer is a material constant.

The transducer constant of the electrostatic transducer is $N = C_0 U_0/d$ (see eq. (17.13)). As shown by the example presented in Section 17.2 it is usually much smaller than that of an equally sized piezoelectric transducer. Since this difference cannot be compensated for by a larger compliance n_0 the second term in the denominator of eq. (17.34) can be neglected against the first one. As the relevant spring element we consider the air layer between both electrodes, and its compliance is given by eq. (17.18). Herewith and with $C_0 = \varepsilon_0 S/d$ the square of the coupling factor of a capacitive transducer becomes:

$$k_{cap}^2 = \frac{\varepsilon_0 U_0^2}{\rho_0 c^2 d^2} \qquad (17.37)$$

To compare both coupling factors we return to the example presented in Section 17.2. Setting $Y = 1.3 \cdot 10^{11}$ N/m^2 for the Young's modulus of the piezoelectric material and $\varepsilon_{F=0} = 1700$ for its relative dielectric constant yields

$$k_{piezo} = 0.36$$

For the capacitive transducer, on the other hand, we obtain with $c = 340$ m/s and $U_0/d = 0.5 \cdot 10^6$ V/m:

$$k_{cap} = 0.0038$$

We must not infer, however, from this large difference that the electrostatic transducer is in every respect inferior to the piezoelectric one. In fact, the efficiency of transduction is not the only aspect in selecting a transducer for a particular task.

To calculate the coupling factor of M-transducers we start from Figure 17.6b neglecting all loss elements. Now we regard the constant driving current I as the input variable. The energy supplied by it is stored in the inductance L_0 of the conductor and – at low frequencies – in the inductance $M^2 n_M$ representing the compliance of its suspension. In analogy to eq. (17.34) the coupling factor is found to be

$$k^2 = \frac{n_M M^2}{L_0 + n_M M^2} \tag{17.39}$$

As an example we consider a dynamic loudspeaker with a transducer constant M = 1 N/A. We assume that the elastic suspension of the diaphragm and the moving coil has a compliance of $5 \cdot 10^{-4}$ m/N; the inductance L of the moving coil be 0.5 mH. With these data eq. (17.39) yields:

$$k_{dyn} = 0.707$$

Thus, the dynamic transducer effects a very strong coupling between electrical and mechanical quantities.

17.7 Two-port equations and reciprocity relations

Figure 17.10a and b show once more the equivalent circuits of the E-transducer and the M-transducer in a somewhat modified form. In both circuits the impedances appearing on the mechanical side are combined in a mechanical impedance

$$Z_m = j\omega m_0 + \frac{1}{j\omega n_0}$$

Similarly, Figure 17.10b introduces an electrical impedance

$$Z_e = R_0 + j\omega L_0$$

From Figure 17.10a the four-pole equations of the E-transducer are easily derived. We choose that version which relates the voltage U and the force F

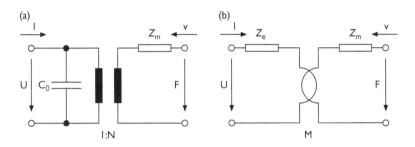

Figure 17.10 Electrical equivalent circuit: (a) of the E-transducer, (b) of the M-transducer. Z_e: electrical impedance, Z_m: mechanical impedance.

on the left side with the current I and the velocity v on the right one:

$$U = \frac{1}{j\omega C_0} I - \frac{N}{j\omega C_0} v \tag{17.40a}$$

$$F = -\frac{N}{j\omega C_0} I + \left(Z_m + \frac{N^2}{j\omega C_0} \right) v \tag{17.40b}$$

The corresponding equations of the M-transducer as obtained from Figure 17.10b read

$$U = Z_e I - Mv \tag{17.41a}$$

$$F = MI + Z_m v \tag{17.41b}$$

In both pairs of equations (17.40) and (17.41) the second coefficient of the first equation is equal to the first coefficient of the second equation, except possibly for the sign. That means

$$\frac{F_{v=0}}{I} = \pm \frac{U_{I=0}}{v} \tag{17.42}$$

whereby the upper sign holds for E-transducers, the lower one for M-transducers. The meaning of this relation which represents one particular form of the so-called reciprocity relations is the following: if an electrical current is sent into the electrical side of the transducer, a certain force $F_{v=0}$ is observed at its mechanical port when the transducer is clamped. Now the transducer is operated in the reverse direction; the mechanical port is now its input side and is set in motion with a velocity v giving rise to an open-circuit voltage $U_{I=0}$ at the left port. In both cases the ratio of output quantity to the input quantity is the same, apart, possibly, from the sign.

   Further versions of the reciprocity relations may also be derived from the four-pole equations (17.40) and (17.41), for example

$$\frac{v_{F=0}}{I} = \mp \frac{U_{I=0}}{F} \tag{17.43}$$

Again, the denominator of these fractions is regarded as the input quantity while the numerator is the quantity appearing at the output, and as before the upper sign refers to the E-transducer, the lower one to the M-transducer.

   The reciprocity relations are the basis of a very flexible and precise method for the calibration of microphones and other transducers which will be described in Section 18.9.

Chapter 18

Microphones

Microphones in the widest sense are electroacoustical sound receivers, that is, devices which convert acoustical or mechanical vibrations into electrical signals. In most cases they are electroacoustical transducers in the sense of Chapter 17 with a mechanical port to which an oscillation or an alternating force is applied, and an electrical port which yields an electrical voltage or current which resembles as much as possible the input signal.

The main subject of this chapter is microphones for airborne sound and for the range of audible sounds. They are used to record sound signals such as speech or music. Furthermore, they serve to measure sound field quantities, in particular, the sound pressure. In addition there are also microphones for water-borne sound – so-called 'hydrophones' which are employed in underwater sound and in ultrasonics. They are mostly designed for different frequency ranges than that of audio sound and will be treated in Section 18.8. Furthermore, this chapter contains a section devoted to sensors of structure-borne sound. Their typical field of use is in building acoustics and noise control.

For the performance of a microphone several characteristic features are important. One of them is sensitivity. For pressure microphones this is defined as the ratio of the open-circuit voltage to the applied sound pressure; its dimension is V/Pa or mV/Pa. Another one is the frequency dependence of the sensitivity which may be associated to 'linear distortions'. At best the sensitivity can be expected to be constant within a certain limited frequency range. And finally, the directivity of a microphone is a third feature which is of significance in practical applications. On the other hand, non-linear distortions are of lesser significance since the moving parts of microphones usually vibrate only at very small amplitudes.

18.1 Principles of microphones for airborne sound

Regarded from the mechanical point of view most microphones for airborne sound consist of a thin diaphragm which forms one side (or a part of it) of a small box or capsule as sketched in Figure 18.1a. We shall discuss first the

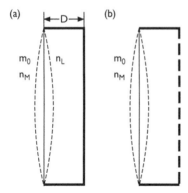

Figure 18.1 Capsule of a microphone, schematic: (a) closed, (b) open.

behaviour of such an arrangement in the sound field under the assumption that in the frequency range of interest its dimensions are small compared with all acoustical wavelengths.

At first we suppose that the box is tightly closed, and that the enclosed air is at atmospheric pressure. Let S denote the area of the diaphragm and p a deviation from the static pressure p_0, for instance, that occurring in a sound wave. Then the force acting on the diaphragm is $F = Sp$. Under its influence the diaphragm will be deformed inward or outward depending on the sign of the pressure p. So, the air in the capsule will be compressed or expanded, and it responds like a spring with some compliance n_A. An additional restoring force is due to the tension or to the bending stiffness (if there is any) of the diaphragm. Combining the corresponding compliance n_M with n_A according to eq. (17.17) results in the total compliance n_0. A second important element is the mass m_0 of the diaphragm. (The variation of the diaphragm deflection over the area S can be accounted for by considering a suitable average.) Together with n_0 it forms a resonance system of the kind described in Section 2.5 with the resonance frequency

$$\omega_0 = \frac{1}{\sqrt{m_0 n_0}}$$
(18.1)

Furthermore, we represent all the mechanical losses caused by radiation etc. by a 'frictional resistance' r. Then the mechanical input impedance of the capsule is

$$Z_m = \frac{F}{j\omega\xi} = r + j\left(\omega m_0 - \frac{1}{\omega n_0}\right)$$
(18.2)

Here ξ is the (average) elongation of the diaphragm. If the sound frequency is significantly lower than the resonance frequency, that is, if $\omega \ll \omega_0$, the loss term r and the mass term $j\omega m_0$ in eq. (18.2) can be neglected. Then the motion of the diaphragm is then 'stiffness controlled', and its elongation is simply

$$\xi = n_0 S \cdot p \qquad (18.3)$$

In this frequency range the diaphragm and the capsule function as a 'pressure receiver'.

The situation is different if the rear side of the box is perforated as shown in Figure 18.1b. Then the inner side of the diaphragm is exposed to the same pressure as the rear side of the capsule; the total force acting on the diaphragm is proportional to the pressure difference Δp between both sides of the box:

$$F = S\Delta p \approx S\frac{\partial p}{\partial x} \cdot D \qquad (18.4)$$

Here D is the thickness of the capsule and $\partial p / \partial x$ is the x-component of the pressure gradient in the sound field. (The second version of eq. (18.4) is valid since we assumed the depth of the capsule as small compared with the wavelength.) Another consequence of the perforation is that the air layer loses its restoring force since it is no longer enclosed in the capsule. So, we have $n_A = \infty$ and hence $n_0 = n_M$ after eq. (17.17). This means that under otherwise equal conditions, the diaphragm has become less stiff. Now its displacement is given by

$$\xi = n_M F = n_M SD \cdot \frac{\partial p}{\partial x} \qquad (18.5)$$

again under the assumption $\omega \ll \omega_0$. Hence, the system diaphragm plus capsule is now sensitive to the gradient of the sound pressure. Suppose the sound field consists of a single plane wave arriving at the diaphragm under an angle θ, that is, according to eq. (6.5)

$$p(x, y) = \hat{p}e^{-jk(x\cos\theta + y\sin\theta)} \qquad (18.6)$$

Then the gradient of the sound pressure $\partial p / \partial x$ at the location of the capsule $(x = y = 0)$ is

$$\left(\frac{\partial p}{\partial x}\right)_0 = -jkp \cdot \cos\theta \qquad (18.7)$$

By inserting this into eq. (18.5) we obtain

$$\xi = -jkn_M SD \cdot p \cdot \cos\theta \qquad (18.8)$$

This relation shows the essential features of a gradient receiver: the increase of the membrane deflection with the angular frequency $\omega = ck$, and its characteristic dependence on the angle under which the sound wave arrives. The latter corresponds to that of a dipole source (see Section 5.5). Hence we can conceive a gradient capsule as well as a dipole receiver. The frequency dependence of the pressure sensitivity vanishes, by the way, if the openings in the rear wall are very long and narrow. Then the motion of the diaphragm is controlled by the flow resistance r which the air experiences when it is forced through the openings. Hence, the first term in eq. (18.2) is the dominant one and $\xi = F/j\omega r$. Combined with eqs. (18.4) and (18.7) this leads to:

$$\xi = -\frac{SD}{cr} \cdot p \cdot \cos\theta \qquad (18.9)$$

It should be noted that the mechanical input impedance of some microphone capsules is considerably more complicated than that in eq. (18.2). An example will be discussed in Section 18.4. In general, it can be stated that the linear distortions of a microphone – including phase distortions – will be smaller the less complicated its mechanical construction and the simpler, therefore, the mathematical expression representing its mechanical input impedance.

The preceding discussion is to give some insights into the mechanical behaviour of electroacoustic sound receivers. To become a microphone such a device must be combined with an electroacoustic transducer which senses the membrane motion. Then it will act as a pressure receiver, a pressure gradient receiver or a receiver detecting the particle velocity in a sound wave, depending on whether the transducer is sensitive to the elongation ξ or to the velocity $j\omega\xi$ of the membrane.

18.2 Condensor microphone

Probably the simplest way to convert acoustic signals into electrical ones is by arranging a metal plate inside the capsule, parallel to the diaphragm (see Fig. 18.4). Together with the membrane it forms a parallel-plate capacitor, and the whole device acts as an electrostatic transducer as described in Section 17.2. However, in practical use the condensor microphone as it is called is not operated with the electrical terminals short-circuited as was assumed with eq. (17.15) but under open-circuit conditions. Its complete equivalent circuit, valid for low frequencies ($\omega \ll \omega_0$), is presented in Figure 18.2. The input velocity v is transformed into an electrical current Nv giving rise to the voltage across the capacitance C_0

$$U_{I=0} = \frac{Nv}{j\omega C_0} = U_0 \frac{\xi}{d} \qquad (18.10)$$

$CN = C_0 U_0/d$ which appears at the electrical terminal; d is the distance of the back electrode from the diaphragm. The bias voltage is supplied via a high resistance R, and the capacitor C_s is to keep direct current away from the electrical terminal.

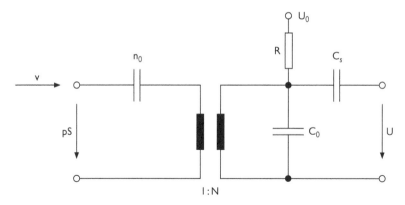

Figure 18.2 Electrical equivalent circuit of the condensor microphone (C_s: separating capacitor, U_0: bias voltage).

Basically, the condensor microphone detects the elongation ξ of the diaphragm. In closed box conditions (see Fig.18.1a) this is proportional to the pressure; inserting ξ from eq. (18.3) leads to

$$U_{I=0} = \frac{n_0 U_0 S}{d} \cdot p \qquad (18.11)$$

Hence, below its resonance frequency the condensor microphone is a pressure receiver with frequency-independent sensitivity.

At very low frequencies the open-circuit conditions can be impaired by the resistor R through which a constant polarising voltage U_0 is supplied. The condition $R \gg 1/\omega C_0$ defines the lower frequency limit of the microphone:

$$\omega_1 = \frac{1}{RC_0} \qquad (18.12)$$

The total frequency response of a condensor microphone with closed capsule is represented in Figure 18.3. Below the limiting frequency ω_1 defined in eq. (18.12) the sensitivity grows proportionally with the frequency, corresponding to a rise of 6 dB per octave. It then remains constant until the resonance frequency after eq. (18.1); for still higher frequencies μ it falls in proportion to $1/\omega^2$, or, expressed logarithmically, at 12 dB per octave. On this general characteristic, the effect of incomplete diffraction may be superimposed: a microphone which is very small compared with the acoustical wavelength is, so-to-speak, not present for the sound waves because the latter is completely diffracted around the capsule. However, with increasing frequency the microphone becomes acoustically, that is, in relation to the wavelength, larger and larger and generates an increasingly stronger

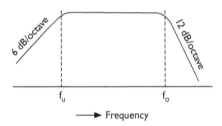

Figure 18.3 Frequency response of a condensor microphone.

secondary sound field. At relatively low frequencies, backscattering in the direction of the incident wave dominates as is shown in Figure 7.4 (left diagram) for a spherical obstacle. The sound pressure of this secondary field adds to that of the incident wave. As a consequence, the resulting pressure level in front of the (nearly) rigid, circular diaphragm of a condensor microphone may exceed that of the primary wave at frontal incidence by more than 10 dB. (Another consequence of the diffraction is some directivity of the microphone, see Section 18.6.)

If the rear wall of the capsule is perforated we have to combine eq. (18.10) with eq. (18.5):

$$U_{I=0} = \frac{n_M U_0 SD}{d} \cdot \frac{\partial p}{\partial x} \tag{18.13}$$

In this case the condensor microphone is a pressure gradient sensor.

If the motion of the diaphragm is not controlled by its stiffness but instead by some frictional resistance r we have to insert eq. (18.9) into eq. (18.10) and obtain:

$$U_{I=0} = -\frac{U_0 SD}{crd} \cdot p \cdot \cos \vartheta = -\frac{U_0 \rho_0 SD}{rd} \cdot v_x \tag{18.14}$$

The latter version follows since $p \cdot \cos \theta / \rho_0 c$ is the component v_x of the particle velocity perpendicular to the diaphragm. Accordingly, the microphone responds now to the particle velocity in an incident sound wave.

Condensor microphones are built in many different forms. Figure 18.4 shows a typical example. The diaphragm is made of steel, nickel or aluminium, and its thickness is about 10–20 μm. Hence, its bending stiffness is negligibly small, and the elastic restoring force stems solely from its mechanical tension which, on the other hand, is so high that the contribution of the air cushion between both electrodes to the total stiffness $1/n_0$ (see eq. (17.16)) is relatively small. If needed the stiffness of the air layer can be further reduced by grooves, or by blind or even clear holes in the fixed back electrode without increasing the distance d of both electrodes. The latter is usually of the same

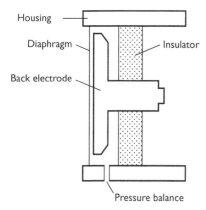

Figure 18.4 Condensor microphone, schematic.

order as the thickness of the diaphragm and determines the sensitivity of the microphone. The bias voltage U_0 is mostly in the range 50 to 100 V. Since the capacitance of the microphone is of the order of 100 pF the resistor R must amount to several hundred megohms to meet the open-circuit condition in eq. (18.12) at frequencies as low as, say, 20 Hz. This makes the microphone susceptible to external stray fields; therefore careful electrical screening is imperative. To avoid any reduction of the sensitivity by the additional capacity of connecting leads a preamplifier with high input impedance is usually mounted in the housing of the microphone which matches it to a longer cable. Furthermore, the capsule is provided with a fine bore by which the internal pressure is matched to the varying atmospheric pressure. Because of its high flow resistance it cannot impair the performance of the microphone.

The fabrication of such a microphone makes great demands on mechanical precision. On the other hand the condensor microphone maintains a leading position in acoustical measuring techniques and for high-quality sound recordings, due to its stability and its simple structure. Its sensitivity is of the order of 10 mV/Pa.

Another version which operates without an externally applied bias voltage is the electret microphone invented by *G. Sessler*. Here the polarising field U_0/d in the earlier formulae is created by an 'electret', that is, by a dielectric material carrying a permanent electrical polarisation very similar to the magnetic polarisation of a magnet. If the space between two electrodes of a parallel-plate capacitor is partially filled with the electret material, a constant electric field will be produced in the empty space as shown in Figure 18.5.

Usually, electret materials are suitable polymers which must be polarised before being used. This is achieved by exposing the material at elevated temperature to a strong electrical field which is maintained during the

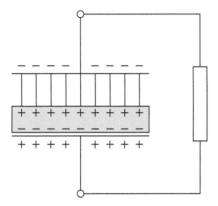

Figure 18.5 Electret in a parallel-plate capacitor.

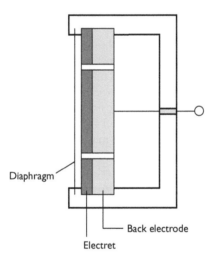

Figure 18.6 Electret microphone (backplate electret).

subsequent cooling process. Another way is to shoot charged particles into the polymer which will be trapped inside. The inhomogeneous distribution of charges too corresponds to a dielectric polarisation.

Two main types of electret microphones are in use: in one of them a thin metallised electret foil forms the diaphragm of the microphone. Alternatively, the electret material can be placed on the backplate of the condensor microphone as depicted in Figure 18.6. Electret microphones are available in a great variety of sizes and forms. Their main advantages are their relatively low price and their small size. For these reasons they are widely used wherever sounds are to be recorded.

It should be mentioned that a condensor microphone can also be operated in quite a different way, namely, as a capacitor with pressure-dependent capacitance which controls the frequency of an electrical high-frequency oscillator. The result is a frequency-modulated electrical signal from which the original audio signal is obtained by demodulation. This kind of operation has the advantage that no bias voltage is needed and that, as a consequence, the frequency range of the microphone has no lower limit. It is self-evident that in this case the microphone is not a reversible transducer but just a pressure-sensitive circuit element.

18.3   Piezoelectric microphones

In a piezoelectric microphone the diaphragm in Figure 18.1 is in mechanical contact with a body of piezoelectric material which performs the transduction of its deflection into an electrical signal.

Like the electrostatic transducer, the piezoelectric transducer as described in Section 17.1 is an elongation sensor as long as its active element is small compared with the acoustical wavelength. Reading the equivalent circuit depicted in Figure 17.3a in 'reverse' direction, that is, from right to the left, we see that a velocity v of the moving part results in an electrical current Nv which produces the open-circuit voltage

$$U_{I=0} = \frac{N}{j\omega C_0} \cdot v = \frac{N}{C_0} \cdot \xi \tag{18.15}$$

across the capacitor C_0. Here ξ means – as in Section 18.1 – the elongation of the diaphragm in Figure 18.1a. Combined with eq. (18.3) this leads to:

$$U_{I=0} = \frac{n_0 SN}{C_0} \cdot p \tag{18.16}$$

Because of the higher dielectric constant of the piezoelectric material the open-circuit condition is not as restrictive as it is for a condensor microphone. This holds in particular if the piezoelement consists of a ceramic material. Nevertheless, the cable connecting the microphone to an amplifier must not be too long, otherwise C_0 would be increased by the cable capacitance resulting in a lower sensitivity.

In spite of the similarities between the piezoelectric and the electrostatic transducer principle there is also an important difference between both types of microphones: due to the very principle of the piezoelectric transducer a

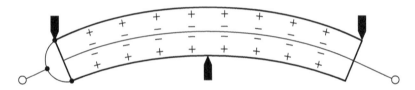

Figure 18.7 Bimorph piezoelectric element.

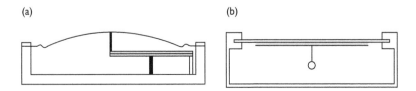

Figure 18.8 Piezoelectric microphones.

solid body, namely, a piezoelectric element, has to be deformed by the sound pressure. Hence, n_0 is not the compliance of a stretched thin diaphragm, eventually reduced by an air cushion behind, but is determined by the elastic properties of the piezoelectric element and the way it is stressed by the sound wave. To keep n_0 sufficiently low so-called bimorph elements are used as sensors which respond to bending forces. A bimorph element consists of two equally oriented thin piezoelectric plates of equal orientation or polarisation which are metallised on both sides and which are glued together (see Fig. 18.7). If the element is bent one of the components will be stretched while the other one is compressed. Because of the transverse piezo effect electrical charges of opposite sign are produced on the surfaces; these charges are added by connecting the plates in electrical opposition. In one type of piezoelectric microphone a little pin conveys the vibrations of the diaphragm to such a bimorph element as depicted in Figure 18.8a. As an alternative, the diaphragm itself can be realised by a bimorph element. And finally, a thin disk of piezoelectric material can be rigidly connected to a metal membrane as shown in Figure 18.8b; again, it is the transverse piezoelectric effect which produces electrical surface charges when the membrane is bent.

It is evident that in all these cases the diaphragm cannot be too thin and hence has a larger mass than the membrane of a condensor microphone. Even more important is its bending stiffness which, according to eq. (10.10), increases with the third power of its thickness. Hence the diaphragm of a piezoelectric microphone is necessarily less mobile. Furthermore, like any waveguide of finite extension, it has normal modes and resonances.

Principally, this holds also for the membrane of a condensor microphone. However, here the resonances are so strongly damped that they do not manifest themselves. In contrast, the resonances of a piezoelectric microphone are principally more pronounced because – according to eq. (2.23) – the higher mass is associated with a higher Q-factor. The sensitivity of a piezoelectric microphone is comparable to that of a condensor microphone.

18.4 Dynamic microphones

The dynamic transducer principle is another useful basis of microphone construction. One widely used version of a dynamic microphone is the moving-coil microphone. Figure 18.9 presents a cross-sectional view of it. Essentially, it consists of a small coil within the cylindrical air gap of a permanent magnet which produces a radial magnetic field in it. The coil is attached to the diaphragm which is itself relatively stiff and obtains its mobility from its corrugated support.

According to eq. (17.21) the output voltage of this microphone is, as with all dynamic transducers, proportional to the velocity of the moving coil. Therefore, if the pressure sensitivity is to be frequency independent, the motion of the diaphragm should be resistance-controlled, that is, with a mechanical input impedance just consisting of a resistance r (see eq. (18.2)). In this case we have

$$U_{I=0} = Mv = MrS \cdot p \tag{18.17}$$

In principle this could be achieved by designing the system as a heavily damped resonator with such high losses that the resistance term r is the

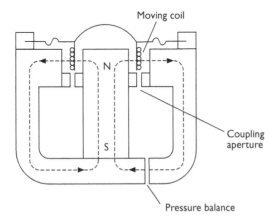

Figure 18.9 Moving-coil microphone.

prevailing one in eq. (18.2). However, the sensitivity of such a microphone would be very low. A reasonably flat frequency response without sacrificing too much sensitivity is achieved by coupling the air volume immediately behind the diaphragm by a narrow aperture to another chamber which, together with the coupling aperture, forms a resonant cavity (see Subsection 8.3.3), damped by the losses occurring in the aperture. Sometimes a moving-coil microphone contains still further resonant cavities. It is the task of the designer to choose all these components in such a way that the mechanical impedance Z_m of the system does not vary significantly within a given frequency range. In this way excellent microphones for speech and music recordings are fabricated. In general, the inductance of the moving coil can be neglected against its electrical resistance which is typically 200 Ohm. Therefore the microphone can be connected to a cable without dramatic loss of sensitivity.

Another version of the dynamic microphone which was widely used in the past is the ribbon microphone. It comes rather close to the moving conductor which was shown in Figure 17.5b to explain the dynamic transducer principle. In this microphone the electrical conductor is a light corrugated metal ribbon about 3 cm long which is loosely suspended between the poles of a permanent magnet. The corrugation stiffens the

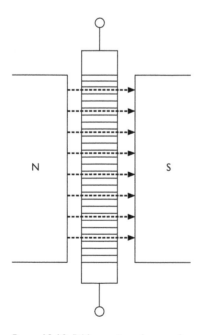

Figure 18.10 Ribbon microphone, schematic.

ribbon in a lateral direction in Figure 18.10 but not in the vertical direction. Hence the ribbon can easily respond to external forces. The field lines run, as indicated in the figure, in a horizontal direction. The ribbon is exposed on both its front and back side to the sound field. So, its response is proportional to the pressure gradient. Because of its low tension it resonates at a very low frequency. At higher frequencies, it is mass-controlled, that is, its mechanical impedance after eq. (18.2) is $Z_m \approx j\omega m_0$ (m_0 = mass of the ribbon) and the open-circuit voltage appearing at the electrical terminal is

$$U_{I=0} = M \cdot \frac{F}{j\omega m_0} = M \cdot \frac{K}{j\omega m_0} \cdot \frac{\partial p}{\partial x} \tag{18.18}$$

(K is some constant factor.) On the other hand, the pressure gradient is related, after eq. (3.15), to the corresponding component of the particle velocity of the sound wave:

$$\frac{\partial p}{\partial x} = -j\omega \rho_0 v_x$$

so that finally

$$U_{I=0} = -\frac{M}{j\omega m_0} K \cdot j\omega \rho_0 v_x = -\frac{MK\rho_0}{m_0} \cdot v_x \tag{18.19}$$

Hence, the ribbon microphone in the described form is a velocity microphone with the dipole characteristics typical of a gradient receiver. Since the electrical resistance of the ribbon and also the induced voltage are very small usually an electrical transformer is connected to ends of the ribbon to match the microphone to the impedance of a cable which is typically 200 Ohm. With this, a sensitivity of about 1 mV/Pa is achieved.

18.5 Carbon microphone

For a long time, the carbon microphone was the standard microphone in telephone communication techniques. This is because of its high sensitivity and its mechanical robustness, properties which were deemed more important in its practical operation than high fidelity. Nowadays, however, it has been replaced with other microphone types; only in old telephone sets may it be found.

The carbon microphone consists basically of a chamber or capsule which is partly filled with loose carbon granules arranged between two electrodes. On one side of the chamber it is closed with a metal membrane. An external pressure on the membrane is transferred to the interior of the chamber, thus improving the electrical contact between the granules. Hence an impinging sound wave modulates the resistance of the cell. By driving a constant electrical current through the microphone a voltage which is proportional to the resistance of the microphone will appear at its terminals. Obviously, the carbon microphone is not a reversible transducer but just a pressure-sensitive electrical resistor.

Besides the linear distortions which are caused by pronounced resonances of the membrane the carbon microphone is afflicted with non-linear distortions which are ascribed to the very principle of its operation. Therefore all attempts to create high-quality sensors on this basis have ultimately failed.

18.6   Microphone directivity

If a microphone is to respond to the sound pressure then it must have angle-independent sensitivity since the sound pressure itself does not contain any information on the directional structure of a sound field. This requires that the microphone is very small compared to the wavelength, a condition which is easily fulfilled at very low frequencies but not at higher ones. We can safely assume that most nominal pressure microphones have a noticeable directivity at elevated frequencies merely on the grounds of their size, that is, that their sensitivity depends more or less on the direction of sound incidence.

In many practical situations the directivity of a microphone is quite desirable. In sound recording it enables us to favour one particular sound source and to suppress sounds from others. This holds in particular if speech or other sounds are to be picked up in noisy environments. It holds as well for sound recording in very reverberant rooms: as is well-known (see Section 13.6) too much reverberation may impair the intelligibility of speech or the transparency of music. By using directional microphones the reverberation can be suppressed up to a certain degree, since many components of the reverberant sound field arrive from directions for which the microphone is not very sensitive.

As with sound sources, the directivity of microphones is described by a directivity factor R (θ, ϕ) (see Section 5.4), which is proportional to the sensitivity as a function of some suitably defined angles θ and ϕ. It is usually normalised by setting R $= 1$ for the direction of maximum sensitivity. Likewise, the characteristic figures derived from this function such as the half-width of the main lobe, or the gain after eq. (5.16a), can be applied as well to microphones.

As discussed in Chapter 5 every sound source can be thought of as being composed of several or many point sources, or even of a continuous distribution of infinitesimal point sources. For each of these elementary sources the reciprocity principle mentioned in Section 5.2 is valid which states that the ratio of the sound pressure received in some observation point and the volume velocity of a point source remains unaltered if the position of the source and receiving point are interchanged. From this principle it can be inferred that a sound receiver consisting of several or an infinity of elementary 'point receivers' has the same directivity factor as an equally shaped and structured sound source. This holds, for instance, for microphone arrays or microphones with an extended diaphragm.

One type of directional microphones was already mentioned in this chapter, namely, those which respond to the pressure gradient in a sound wave. Because of the shape of their directional diagram represented in polar coordinates (see Fig. 5.6b) they are often referred to as 'figure-of-eight' microphones. By combining such a microphone with an omnidirectional one and adding both outputs a directional microphones with the general directional factor

$$R(\theta) = \frac{A + B \cos \theta}{A + B} \qquad (18.20)$$

can be realised with A and B denoting frequency-independent constants. Thus, B = 0 corresponds to an omnidirectional microphone and A = 0 to a gradient microphone. The characteristics determined by A = B is represented by a cardioid in polar coordinates as shown in Figure 18.11a. Further, well-known combinations are those corresponding to the 'supercardioid' with A/B = 0.5736 and the 'hypercardioid' with A/B = 0.342 (see Fig. 18.11b and c).

Of course, the combination of two separate high-quality microphones in the described way is a very costly way of achieving directivity. Another, less expensive way is to incorporate an acoustical delay in a microphone. Figure 18.12 shows a microphone with nearly cardioid characteristics which is brought about by lateral apertures in its housing. Sound arriving frontally reaches the diaphragm the direct way and via a detour which is twice the distance of the apertures from the diaphragm. However, if the sound arrives from the rear side, that is, from the right in the figure, both sound portions reach the diaphragm nearly simultaneously and cancel each other. The

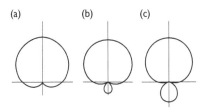

Figure 18.11 Directional microphone with (a) cardioid characteristics, (b) 'super-cardioid' and (c) 'hypercardioid'.

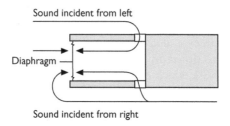

Figure 18.12 Directional microphone with one capsule.

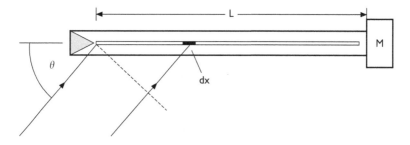

Figure 18.13 Line microphone, schematically (M microphone capsule).

difference between such a 'delay type' microphone and the gradient microphone shown in Figure 18.1b is that the distance D in the latter should be very small compared to the wavelength whereas in the microphone of Figure 18.12 the lateral openings must not be too close to the diaphragm in order to produce noticeable delay effects.

Still more prominent is the role an acoustical delay plays in the line microphone shown in Figure 18.13. Its essential part is a metal tube with a slit along its whole length L; the actual pressure sensor is a condensor or dynamic microphone attached to one of its ends. Suppose a sound wave arrives at the microphone under an angle θ with respect to the tube axis $(y = 0)$. The sound pressure in this wave is given by eq. (18.6), then the sound pressure along the slit is $p(x) = \hat{p}\exp(-jkx\cos\theta)$. It will set the air in the slit into oscillations which are perpendicular to the tube wall. Hence every length element dx of the slit becomes the origin of two waves travelling in the tube in opposite directions. One of them will be absorbed by the porous wedge at the left end of the tube $(x = 0)$ while the other one travels towards the microphone M. Since the distance of the slit element dx from the microphone is $L - x$, this portion is associated with another phase factor $\exp[-jk(L - x)]$.

Hence its contribution to the sound pressure at the microphone membrane is

$$dp_m = Ce^{jkx(1-\cos\theta)}dx$$

with C denoting a constant. The directivity factor $R(\theta)$ is obtained by integrating this expression over the tube length L; the result, normalised in the described way, reads:

$$R(\theta) = \frac{\sin\left[(kL/2)(1-\cos\theta)\right]}{(kL/2)(1-\cos\theta)} \tag{18.21}$$

It is only for frontal sound incidence, that is, for $\theta = 0$, that all sound portions arrive with equal phases at M, hence this is the direction of maximum sensitivity.

Figure 18.14 represents two polar plots of the directional characteristics $|R(\theta)|$ calculated for two different ratios L/λ. They should be compared with the directional diagrams of an oscillating piston, shown in Figure 5.15. Again, these diagrams contain several satellite maxima the number of which increase with increasing ratio L/λ, separated by zeros.

After what has been said earlier, the principle of the linear array (see Section 5.6) can also be applied to the construction of directional microphones. Since the handling of such an arrangement is rather awkward it is only rarely used in airborne sound. However, in underwater sound it finds widespread application. This holds as well for two-dimensional microphone (or rather hydrophone) arrays.

The counterpart of the oscillating piston as described in Section 5.8 is best realised by a concave mirror of metal or other reflecting material with a microphone located in its focus. If the mirror is large compared to the wavelength its directivity is about the same as that of an equally sized piston.

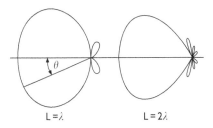

$L=\lambda$ $L=2\lambda$

Figure 18.14 Directional characteristics of a line microphone, for two different length/wavelength ratios.

18.7 Hydrophones

Hydrophones are microphones for detecting sound in liquid media. They find wide application especially in underwater sound techniques and are important components of most sonar systems (see Section 16.5). In ultra-sonics they are mainly used for measuring purposes as, for instance, for the examination and the surveillance of the sound field produced by ultrasonic sound projectors.

Most hydrophones are based on the piezoelectric transducer principle. In contrast to microphones for airborne noise they do not need a diaphragm since the characteristic impedance of piezoelectric materials is much closer to that of liquids although still higher than it. The piezoelectric high polymer polyvinylidene fluoride is even almost perfectly matched to water. Therefore hydrophones are constructed in such a way that the sound waves interact directly with the transducer material.

As pointed out at the beginning of Section 18.6 a hydrophone must be small compared to all wavelengths in the frequency of interest range if it is to be used as a pressure receiver. This condition ensures that the hydrophone has no directivity, that its active part is free of resonances and that it does not disturb the sound field. If this condition is fulfilled its function can be described by the low-frequency equivalent circuit of Figure 17.3 from which eq. (18.16) follows. However, this relation is strictly valid only if the pressure is applied to the end faces of a piezoelectric disk or stack which can be ensured by suitable design. If, on the contrary, the piezoelectric element is exposed on all sides to the sound field an effect is only obtained if the piezoelectric material is sensitive to omnidirectional pressure and hence to volume variations which is the case, for instance, for lithium sulphate.

Figure 18.15 shows two constructions of piezoelectric hydrophones. In Figure 18.15a the active element is a small cylinder of piezoelectric material which is inside in contact with some absorptive material. Constructions with one or several piezoelectric disks are also in use (see Fig. 18.15b). Particularly, high-frequency bandwidth is achieved with the needle hydrophone; it consists of a metal needle the tip of which is covered by a thin film of polyvinylidene fluoride (PVDF), which of course must be polarised before use.

18.8 Vibration pickups

Vibration pickups are microphones for detecting structure-borne sound. Of course, such sensors can only be applied to the surface of a solid body. They play an important role in noise control of machines but are also used for the close examination of sound transmission through partitions and floors.

In principle, all components of the vibrations of a solid body are accessible to measurement. However, since only the component perpendicular to the

(a)

(b)

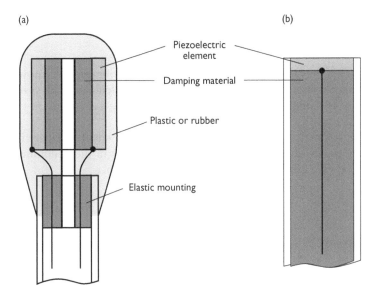

Piezoelectric
element

Damping material

Plastic or rubber

Elastic mounting

Figure 18.15 Hydrophones.

surface causes sound radiation into the environment it is this component which is of most interest.

The basic problem in vibration measurement is the lack of a reference point which is known to be at rest. If such a point were available the task of a vibration sensor would be just to detect the elongation of some object point relative to that reference point. Fortunately, every mass is a reference not for the elongation but for the acceleration and hence for the inertial force it experiences. In this respect the mass is different from all other mechanical 'circuit elements': the force acting on a spring depends on the difference of elongations between both its ends, and the same holds for a resistance element representing mechanical losses. In contrast, the inertial force developed by a mass depends merely on its acceleration relative to any unaccelerated system (inertial system in the language of relativity), for instance, relative to some point on the earth's surface. (The rotation of the earth can be left out of consideration in this context.) Hence, the measurement of vibrations can be carried out with any damped mass-spring system which is attached to the object to be examined and the response of which can be calculated.

The behaviour of such a system which is depicted in Figure 2.9 has already been discussed at the end of Section 2.7. We denote with $\eta = v/j\omega$ the displacement of the solid surface to be examined and with $\eta_m = v_m/j\omega$ that of the mass m. Then we obtain from eq. (2.32) (first version) the only observable quantity, namely, the difference $\eta - \eta_m$ of both displacements:

$$\eta - \eta_m = -\frac{(\omega/\omega_0)^2}{1 + j\omega r n - (\omega/\omega_0)^2} \cdot \eta \qquad (18.22)$$

If the data of the resonator and the frequency of the vibration are known then the displacement η of the object under test can be calculated. This procedure is particularly simple if one of the following limiting cases is given:

1 The resonance frequency of the resonator is small compared to all frequencies of interest, $\omega_0 \ll \omega$. Then all terms in the denominator of eq. (18.22) except $(\omega/\omega_0)^2$ can be neglected, hence

$$\eta - \eta_m \approx \eta \qquad (18.22a)$$

This means that the mass remains virtually at rest and the measurable difference $\eta - \eta_m$ equals the elongation of the object. This is the principle of a seismograph.

2 If, on the contrary, the resonance frequency of the measuring system is much higher than all frequencies to be considered ($\omega_0 \gg \omega$) then all frequency-dependent terms in the denominator of eq. (18.22) can be neglected and we obtain:

$$\eta - \eta_m \approx \frac{1}{\omega_0^2} \cdot (-\omega^2 \eta) \qquad (18.22b)$$

Since $-\omega^2 \eta$ is the acceleration of the object to which the resonator is attached such a vibration pickup is called an accelerometer.

The vibration pickups mostly used in acoustical measuring techniques are of the latter type. The difference elongation $\eta - \eta_m$ is detected either by a piezoelectric bimorph element as described in Section 18.3, or a piezoelectric disk is used which, at the same time, acts as the spring of the resonator. Figure 18.16 shows schematically the practical construction of a piezoelectric accelerometer.

Another detector of structure-borne sound is the strain gauge. It is not sensitive to the displacement or acceleration of an object but to its strain caused by elastic forces. Basically, it consists of a wire cemented to the surface under test. If the latter is strained, the same holds for the wire, thus its length will be extended, and lateral contraction reduces its thickness (see Subsection 10.3.1). Both changes cause an increase of its electrical resistance which is detected in an electrical bridge circuit. In a strain gauge the wire is folded into a long zig zag shape and embedded in a plastic strip which is rigidly glued to the surface under test by means of a special adhesive. As an alternative, particular semiconducting materials have been developed the

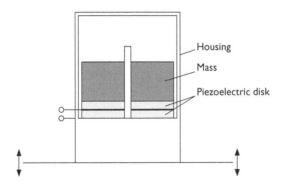

Figure 18.16 Piezoelectric accelerometer.

electrical conductivity of which depends on the mechanical stress they are subjected to. With such materials a great variety of strain gauges can be fabricated.

Finally, remote measurements of vibrations can be performed with optical methods, for instance, with a so-called laser vibrometer.

18.9 Microphone calibration

For sound recordings it is sufficient to have some knowledge of the sensitivity of the microphone which can be obtained by comparison with a microphone of known sensitivity. However, if the microphone is to be used for absolute measurements or as a standard for comparisons, exact knowledge of its sensitivity is indispensable. This means it must be calibrated.

One way of calibrating a microphone is by exposing it to a sound field the properties of which are known on account of the nature of its generation. A very common way of doing this is by using a small, thick-walled chamber which can be fitted to the microphone under test and in which a well-defined pressure variation is created by means of a small reciprocating piston (see Fig. 18.17). We denote with ξ its displacement from its resting position and with S its area, then the change of the chamber volume is ξS. Under the condition of very small chamber dimensions (as usual in comparison with the wavelength) a volume reduction $-\xi S$ leads to a relative density variation $\tilde{\rho}/\rho_0 = \xi S/V_0$ which in turn is associated with a pressure variation after eq. (3.13):

$$p = \frac{\rho_0 c^2 S}{V_0} \cdot \xi \tag{18.23}$$

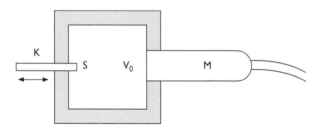

Figure 18.17 Microphone calibration with a compression chamber (M: microphone, P: oscillating piston).

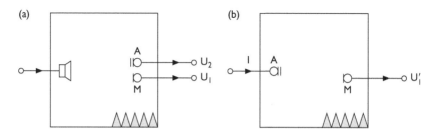

Figure 18.18 Microphone calibration using the reciprocity method (M: microphone to be calibrated, A: auxiliary transducer): (a) comparing both microphones, (b) auxiliary transducer as sound source.

The displacement of the piston can be measured, for instance, with a microscope, or known from the very way it has been produced. Incidentally, the earlier equation yields immediately the compliance n_A of the chamber volume, that is, the ratio of the elongation ξ and the force pS exerted by the piston:

$$n_A = \frac{V_0}{\rho_0 c^2 S^2} \tag{18.24}$$

A very exact and versatile method of calibration is the reciprocity method. What is needed for this procedure is some sound source, typically a loudspeaker, and an auxiliary transducer which is known to be reversible; this means it can be used both as a microphone and as a loudspeaker. The calibration is carried out in a defined environment, for instance, in an anechoic chamber or a suitably constructed pressure chamber (see earlier). The procedure comprises two steps: at first (see Fig. 18.18a), both the microphone M to be calibrated and the auxiliary transducer A are exposed to the sound field produced by the loudspeaker. From the open-circuit output voltages U_M and U_A of both transducers the ratio of their sensitivities σ and σ' is obtained:

$$\frac{\sigma}{\sigma'} = \frac{U_M}{U_A} \tag{18.25}$$

In the second step (see Fig. 18.18b) the original sound source is replaced with the transducer A which is now operated as a sound source and is fed with a current I. The sound pressure it produces at the location of the microphone M is proportional to the volume velocity $Q_{p=0}$ of the sound source, the factor of proportionality be denoted with R_e. Then the output voltage U'_M of the microphone in this part of the procedure is

$$U'_M = \sigma \cdot p = \sigma \cdot R_e Q_{p=0} \tag{18.26}$$

Now we apply the reciprocity relation of reversible transducers to the auxiliary transducer A using the version of eq. (17.43) which we multiply with the active area of the transducer A. (In general, this will be the area S_m of its membrane.) This means the velocity $v_{F=0}$ turns into the volume velocity $Q_{p=0}$ and the force F into the sound pressure p'. Finally, we replace $U_{I=0}$ in eq. (17.43) with U'. Then

$$\frac{Q_{p=0}}{I} = \mp \frac{U'}{p'} = \mp \sigma' \tag{18.27}$$

Inserting $Q_{p=0}$ from eq. (18.27) into eq. (18.26) yields

$$U'_M = \sigma \sigma' R_e I = \sigma^2 \cdot \frac{U_A}{U_M} R_e I \tag{18.28}$$

where the signs have been omitted. Therefore the sensitivity of the microphone to be calibrated can be calculated from measured quantities:

$$\sigma = \sqrt{\frac{U_M U'_M}{U_A R_e I}} \tag{18.29}$$

The constant R_e is sometimes called the 'reciprocity parameter'. For the field of a spherical wave, it follows from eq. (5.6):

$$R_e = \frac{p(r, t)}{Q} = \frac{j\omega\rho_0}{4\pi r} e^{-jkr} \tag{18.30}$$

In practice only its absolute value $\omega\rho_0/4\pi r$ is used. For the compression chamber we obtain from eq. (18.23)

$$R_e = \frac{\rho_0 c^2}{j\omega V_0} \qquad\qquad (18.31)$$

The index $p = 0$ in eq. (18.27) indicates that the auxiliary transducer A should be free of any mechanical load, strictly speaking. However, the mechanical impedance of a real transducer is mostly so high that its reaction to the medium can be neglected.

Chapter 19

Loudspeakers and other electroacoustic sound sources

The subject of this chapter is sound generation by means of electroacoustic transducers and, in particular, the generation of airborne sound in the audio range. The sound sources employed for this purpose are called loudspeakers if the sound is projected into free air. In contrast, we speak of earphones or headphones if the sound source is arranged next to the listener's ear. In the concluding section we shall deal with electroacoustic sound sources used in underwater sound and in ultrasonics.

Loudspeakers are the most widely used electroacoustic devices. We find them in every radio or television set, in every stereo system, and even our motor car is equipped with several loudspeakers. Likewise, earphones are also in common use taking into account the various types of telephones. In any case a smooth frequency response over a wide frequency range is an important quality criterion. Or to put it somewhat more precisely, it is usually required that the sound pressure amplitude in the far field of a loudspeaker (or at the ear channel), relative to the voltage or current of the electrical input signal, does not show noticeable frequency dependence within a sufficiently broad frequency band. To meet this requirement is principally more difficult in loudspeakers than earphones which explains the enormous differences in loudspeaker quality. Another important aspect is the output power attainable with a loudspeaker, of course, under the condition that the non-linear distortions are within tolerable limits.

In any case we have to distinguish between two basic components of an electroacoustic source: the size and shape of the sound emitting surface, called the diaphragm or 'membrane', and the electroacoustic driver system which sets this surface into vibration. The former is decisive for the structure of the generated sound field, for instance, for the directional distribution of the sound energy. In contrast, the construction of the transducer determines the electrical properties of the sound transmitter but is also responsible for the occurrence of linear and non-linear distortions.

If one wants to project sound from a limited source, say from a loudspeaker diaphragm into an open space, one is faced with a fundamental problem. Its essential point is contained already in eq. (5.6) according to which the sound pressure in the field of a point source with constant volume velocity grows proportionally to the frequency, hence the radiated power grows with the square of the frequency. Since we can conceive a more complicated sound source as a combination of several or many point sources this frequency dependence is found in all expressions for the radiated power or the radiation resistance. In particular, after eq. (5.45) the sound power radiated from a circular piston in an infinite baffle is in the low-frequency limit (ka ≪ 1 with a = piston radius):

$$P_r \approx \frac{\rho_0 S^2 \hat{v}_0^2}{4\pi c} \cdot \omega^2 \qquad (19.1)$$

In this expression $\hat{v}_0$ denotes the velocity amplitude of the piston while S is its area. Although most loudspeaker diaphragms are neither plane nor rigid, especially when it comes to somewhat higher frequencies, we can adopt the laws of the radiating piston for common loudspeakers without too severe errors.

To maintain a frequency-independent sound pressure at some point of the far field the acceleration $j\omega v_0$ of the membrane must be kept constant throughout the whole frequency range. This is tantamount to saying that its velocity v_0 is inversely proportional to the frequency, or that the membrane displacement is inversely proportional to the square of the frequency. Obviously, this would lead to intolerable vibration amplitudes when the frequency tends towards zero. The situation is still worse for a sound source with dipole characteristics since the sound pressure in a point of its field grows – again with the dipole strength $\hat{Q}d$ kept constant – with the square and its total power output with the fourth power of the frequency (see eq. (5.23)).

The relationship between membrane motion and sound pressure which is so unfavourable for loudspeaker construction is a characteristic feature of sound radiation into three-dimensional space. If our goal were to generate a frequency-independent sound pressure within a pressure chamber, that is, in a zero-dimensional space, so-to-speak, we would have to maintain just constant displacement amplitude of the oscillating piston, according to eq. (18.23). If the space is one-dimensional, that is, confined in a rigid-walled tube with the sound field generated by a piston (see Fig. 4.3), the latter must vibrate with constant velocity if the sound pressure is to be frequency independent. The same holds for the generation of a plane wave by means of a plane surface of infinite extension. As soon as the sound radiating surface is finite, however, the boundary will emit a diffraction wave as described in Subsection 5.8.1 which alters the sound field all the more with a smaller sound source.

19.1   Dynamic loudspeaker

The loudspeaker which is almost exclusively used today is based on the dynamic transducer principle as described in Section 17.3. Its basic structure is shown in Figure 19.1. As in the moving-coil microphone (see Section 18.4) the conductor interacting with the magnetic field is a cylindrical coil of wire arranged concentrically in the annular gap of a strong permanent magnet. Its poles are shaped in such a way that a radial magnetic field is produced in this gap. The coil, often called 'voice coil', is either self-supporting or wound on a light cylinder. When an electrical current flows through the coil an axially directed electromotive force is exerted on it and conveyed to the diaphragm or membrane which is rigidly attached to the coil. To increase its stiffness the diaphragm is generally shaped as a cone (see Fig. 19.2a). Therefore the loudspeaker shown in Figure 19.1 is also known as a 'cone loudspeaker'. Some loudspeakers, however, have a diaphragm with double curvature as shown in Figure 19.2b.

At its outer perimeter the diaphragm is attached to the frame of the loud-speaker by a flexible surround to its rim; the coil is centred by another flexible element, the so-called spider. Both elements have a high compliance for deflections in an axial direction but impede motions perpendicular to the axis. We denote with n_M the compliance of the suspension of the diaphragm

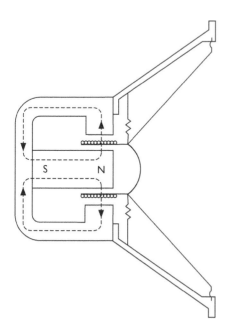

Figure 19.1 Dynamic loudspeaker.

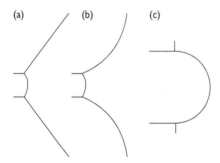

Figure 19.2 Various forms of diaphragms: (a) cone, (b) double-curvature diaphragm and (c) dome-shaped diaphragm.

including the coil, and with m_M its mass. Then the resonance frequency of the mechanical system is:

$$\omega_0 = \frac{1}{\sqrt{m_M n_M}} \tag{19.2}$$

For frequencies above ω_0 the system is mass-controlled, that is, its mechanical behaviour is determined by the mass m_M, to which, strictly speaking, the 'radiation mass' after eq. (5.46) should be added. Hence, if an electrical current I with the frequency $\omega \gg \omega_0$ is forced through the coil the diaphragm vibrates with the velocity

$$v_0 = \frac{Bl}{j\omega m_M} \cdot I \tag{19.3}$$

As earlier, B denotes the magnetic flux density in the gap of the magnet and l is the total length of the wire wound on the coil. This is exactly the frequency dependence which compensates that of P_r in eq. (19.1), hence the sound pressure in the generated sound field is proportional to the driving current. The resonance frequency after eq. (19.2) marks the lower limit of the useful frequency range. On the other hand, the range in which eq. (19.1) is valid reaches up to about $\omega = 2a/c$ corresponding to ka = 2, according to Figure 5.16. At higher frequencies the radiation resistance R_r and hence the acoustic output power P_r at constant velocity v_0 does no longer increase with the square of the frequency but approaches a constant value. Therefore the total acoustic power falls off with $1/\omega^2$ if the current is kept constant. This tendency is partially compensated by the increasing directivity of the loudspeaker setting in at ka < 2 as may be seen from the polar plots in Figure 5.15. For receiving points on the middle axis of the loudspeaker this compensation is perfect. In fact, eq. (5.40) along with eq. (19.3) leads to a frequency-independent sound pressure for $\theta = 0$ since

$$\lim_{x \to 0} \frac{2J_1(x)}{x} = 1$$

In points off the axis, however, the increasing directivity causes a loss of high-frequency components.

The condition that the loudspeaker is driven with frequencies-constant electrical current is not very stringent because for frequencies below about 1000 Hz the inductance of the voice coil can be neglected in comparison to its resistance which is typically 4 or 8 ohms. Therefore the loudspeaker can be fed without problem from a constant voltage amplifier. This has an additional advantage: the mechanical losses of a loudspeaker as caused by the viscosity of the air in the gap, by elastic losses in the suspension system and by radiation are only moderate, thus the Q-factor of the loudspeaker resonance is high if it is operated with constant current. If, on the contrary, the loudspeaker is driven from a low-impedance source maintaining constant voltage, the electrical port of the loudspeaker is virtually short-circuited. Hence the system experiences additional damping by the current induced in the voice coil when it moves in the magnetic field.

The properties of a rigid piston as a sound projector apply only to a common loudspeaker if the latter is set – as assumed for the piston – into an infinite rigid plane. Without that the loudspeaker would be much less effective: since both sides of the diaphragm emit sound in opposite phases the loudspeaker would act at low frequencies as a dipole source as mentioned in the introduction to this chapter, or put in a different way, the pressure differences between the front and the rear side of the diaphragm would partially cancel each other. We encountered this phenomenon, named 'acoustical short-circuit', already in Sections 5.5 and 10.3.4. In Section 19.4 we shall deal with several ways of avoiding this undesired effect.

Generally, the diaphragm of loudspeakers is composed of paper, sometimes of plastics or aluminium. At medium and high frequencies it vibrates no longer uniformly since flexural resonances may be excited on it which impair the frequency response of the loudspeaker. They can be prevented to some degree by increasing the stiffness of the membrane material in its central part. This has the additional advantage that at elevated frequencies only the central part of the diaphragm is active which reduces the directivity of the loudspeaker. Nevertheless, it is not easy to achieve a smooth loudspeaker response over a wide frequency range. One way of improving this is by applying motional feedback. In this method a control signal is derived from a separate coil closely attached to the voice coil. Hence this signal is proportional to the velocity of the voice coil and is used to control the amplifier in a negative feedback loop. Even more efficient is to pass the feeding signal through an electrical filter which has a transfer function inverse to that of the loudspeaker and hence eliminates all linear distortions. It has been shown that phase distortions may be disregarded in this process since our hearing is not very sensitive to them.

Of course, it is much easier to optimise a loudspeaker for a restricted frequency range. Therefore the whole audio range is often subdivided into several frequency bands, for instance, in a low-frequency, a mid-frequency and a high-frequency band, which are served by separate loudspeakers fed from suitable cross-over networks. Loudspeakers for the high-frequency range, so-called tweeters, have often a stiff dome-shaped diaphragm of plastic with a diameter of a few centimetres which is driven at its periphery (see Fig. 19.2c). Another advantage of separating low from high frequencies is prevention of so-called Doppler distortions: a diaphragm vibrating simultaneously at low and high frequency can be conceived as a moving sound source with respect to the high-frequency components; hence the high-frequency sound will become frequency modulated on account of the Doppler effect (see Section 5.3). Further non-linear distortions are due to the properties of the suspension system which becomes less compliant at high elongations. Moreover, the moving coil may reach into a region of reduced flux density B, so the transducer constant itself depends on the elongation. By suitable design of the air gap and the voice coil distortions of this kind can be diminished but not altogether eliminated.

The transient behaviour of the dynamic loudspeaker is mainly determined by its mechanical resonance. If a slightly damped loudspeaker is excited with a very short impulsive force or voltage its membrane reacts with a deflection similar to that of Figure 2.1b. This long ringing must be shortened by additional damping, that is, by reducing the Q-factor of the system. A particularly favourable condition is met if the system is critically damped corresponding to a Q-factor of 0.5 (see Section 2.6).

19.2 Electrostatic or condensor loudspeaker

The diaphragm of a condensor loudspeaker consists of a thin and light foil of metal or a metallised plastic. To keep the resonance frequency of the system sufficiently low the stiffness of the diaphragm must be extremely small. This is achieved on the one hand by stretching the membrane as loosely as possible, on the other by perforating its back electrode and hence avoiding any additional stiffness caused by an air layer between both electrodes. The mechanical impedance of the membrane is so low that the mechanical properties of the electrostatic loudspeaker are mainly determined by its radiation. Moreover, the system has a relatively low Q-factor which is favourable with respect to its transient response.

At low frequencies the radiation impedance of the loudspeaker consists mainly of its reactive component, $j\omega m_r$ with m_r denoting the 'radiation mass' (see Subsection 5.8.3). Hence the velocity of the membrane is

$$v_0 = \frac{N}{j\omega m_r} \cdot U \qquad (19.4)$$

(N = transducer constant.) According to eq. (19.1) this results in a frequency-independent power output. At high frequencies, however, the reactive component of the radiation impedance Z_r can be neglected, hence Z_r is real and

$$Z_r \approx R_r \approx SZ_0 \qquad (19.5)$$

Therefore

$$v_0 = \frac{N}{R_r} \cdot U \qquad (19.6)$$

Again the total power radiated $P_r = |v_0|^2 R_r / 2$ proves to be constant.

The electrostatic loudspeaker has the disadvantage that only small membrane displacements can be generated without producing unacceptably strong distortions. Thus, to achieve sufficiently high sound pressures particularly at low frequencies such a loudspeaker must have a large area. Figure 19.3 shows a symmetrical electrostatic loudspeaker. The membrane is located in the middle between two perforated electrodes which are fed by voltages of opposite sign. With this design certain non-linear distortions can be suppressed.

To show this we go back to eq. (17.12), replacing d in this formula and also in the capacitance $C_0 = \varepsilon_0 S / d$ with $d \pm \xi$ where ξ is the displacement of the membrane and the different signs refer to both sides of the loudspeaker. Then the force on one side of the system is:

$$F_\approx = \frac{\varepsilon_0 S U_0}{(d \pm \xi)^2} \cdot U_\approx \qquad (19.7)$$

It acts on the membrane with the impedance Z'_m including the radiation impedance with which it is loaded. Therefore the left side of this equation

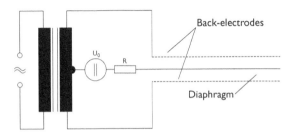

Figure 19.3 Electrostatic loudspeaker.

can be expressed by $j\omega\xi Z'_m$. Then we obtain from the earlier formula:

$$d^2\xi \pm 2d\xi^2 + \xi^3 = \frac{\varepsilon_0 S U_0}{j\omega Z'_m} \cdot U_\approx \tag{19.8}$$

If these expressions for both parts of the loudspeaker are added the terms with ξ^2 cancel and so do the distortions caused by it.

Because of its favourable transient behaviour, the electrostatic loudspeaker is highly appreciated by some connoisseurs. Nevertheless, it represents no real alternative to the much more robust and less costly dynamic loudspeaker.

19.3   Magnetic loudspeaker

Today, the magnetic loudspeaker is of historical interest only, because it is inferior to the dynamic loudspeaker when it comes to distortions. Nevertheless, in the beginning of the broadcasting era and even in the thirties it was in widespread use because of its simple construction and its high efficiency. Particularly, the latter property was a great advantage regarding the then moderate power output of electrical amplifiers.

Figure 19.4 shows schematically the design of a magnetic loudspeaker as was contained in many cheap radios at that time. Its essential components are a permanent magnet and a small bar of soft iron, the armature, which is kept in front of the magnetic poles by a flexible spring. The driving current is fed to a small coil which is wound around the armature and polarises it

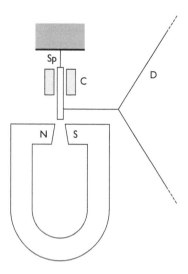

Figure 19.4 Magnetic loudspeaker (Sp: spring, C: coil, D: diaphragm).

magnetically. Depending on the polarity of this current the armature is drawn either towards the left or the right side. This motion is conveyed to a conical paper membrane by a little pin. As in the electrostatic loudspeaker shown in Figure 19.3 the symmetrical design of the loudspeaker enables certain distortions to be suppressed and so larger amplitudes to be generated.

19.4   Improvement of loudspeaker efficiency

The contents of this section apply to dynamic loudspeakers which need special measures to prevent the sound radiated from the rear of the diaphragm from interfering with that from the front and thus avoiding what has been called 'acoustical short-circuit' (see Section 19.1).

For the theoretical model of a rigid piston as treated in Section 5.8 this is achieved by assuming the piston to be surrounded by a rigid plane of infinite extension. We could come close to this condition by inserting the actual loudspeaker system flush into the wall of a room which can be imagined to be repeatedly mirrored by the adjacent side walls and hence extended to infinity. We arrive at a more practical solution by contenting ourselves with a baffle, that is, a panel with finite dimensions (see Fig. 19.5). However, this measure will not completely eliminate the sound from the rear since a part of it is diffracted around the edge of the baffle and will interfere with the sound originating from the front. This leads to fluctuations of the frequency response which may be reduced by placing the loudspeaker asymmetrically into the baffle panel. If the (mean) distance of the loudspeaker system from the edge is less than about one-quarter of the wavelength the acoustical short-circuit will again be significant. Thus the main effect of a baffle is just to shift the cancellation of sound components towards lower frequencies.

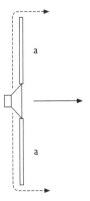

Figure 19.5 Loudspeaker mounted in a baffle.

By suitably folding a rectangular baffle we arrive at an open cabinet which is handier than a large baffle but has the same drawbacks. Furthermore, standing waves will be set up in it which impair the quality of sound reproduction.

19.4.1   The closed loudspeaker cabinet

The most common way of avoiding an acoustical short-circuit is by mounting the loudspeaker system into one wall of an otherwise closed enclosure (see Fig. 19.6). However, it should be noted that the loudspeaker interacts with the air enclosed in the box. Thus the air is alternately compressed and rarefied by the motion of the diaphragm, hence at low frequencies it reacts like a spring increasing the stiffness of the system. Accordingly, the compliance n_M in eq. (19.2) has to be replaced with n_0 which is given by eq. (17.17):

$$\frac{1}{n_0} = \frac{1}{n_M} + \frac{1}{n_A}$$

with

$$n_A = \frac{V_0}{\rho_0 c^2 S^2} = \frac{V_0}{\kappa p_0 S^2}$$

(see eq. (18.24)). In this expression V_0 is the volume of the enclosure and S the area of the membrane, p_0 is the atmospheric pressure and $\kappa = 1.4$ the adiabatic exponent of air. With a small enclosure the shift of the resonance frequency due to the enclosed air may be quite considerable. It can be counteracted by increasing the compliance of the membrane suspension, then the total compliance n_0 is mainly determined by the second term in the earlier equation. This has the additional advantage that aging effects are less noticeable.

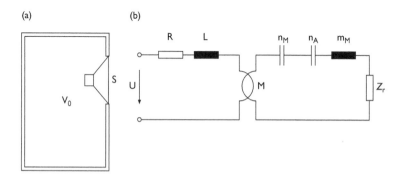

Figure 19.6 Closed box: (a) section, (b) electrical equivalent circuit (Z_r radiation impedance).

At higher frequencies cavity resonances of the box will be excited (see Chapter 9), leading to fluctuations of the frequency response. The simplest way to suppress them is by filling the whole enclosure with damping material such as glass wool or cotton wool. As a desirable side effect, the compressions of the enclosed air will be no longer adiabatic but isothermal; as a consequence the constant κ in the earlier equation (second version) can be omitted which increases the compliance of the cabinet by the factor 1.4.

Although a closed cabinet prevents the acoustical short-circuit it influences the radiative properties of a loudspeaker in quite a different way than a baffle panel. In any case, the sound emitted by the diaphragm will be diffracted around the cabinet thus reducing the power output compared to that in eq. (19.1). Furthermore, the radiated power depends also on the position of the loudspeaker in a room. If it is placed in front of a room wall which reflects the sound it is mirrored by the wall; consequently, in a way there are two sound sources (see also Section 13.1). At very low frequencies their contributions arrive at some point with equal phase, hence the sound pressure is twice that in the free field corresponding to a level rise by 6 dB. If the loudspeaker is placed in a corner the sound pressure is raised by a factor of four corresponding to a level increase by 12 dB. At higher frequencies the changes caused by the room are much more involved as detailed in Section 9.5.

19.4.2 The bass-reflex cabinet

In a way it is unsatisfactory to waste the sound energy radiated from the rear side of a loudspeaker diaphragm. The bass-reflex cabinet is a method which uses that sound portion to extend the range of efficient sound radiation towards lower frequencies.

Figure 19.7a represents a section of a bass-reflex cabinet. It differs from the cabinet shown in Figure 19.6 in having a vent in the front wall of the cabinet including a short tube inside. This port emits sound when the air confined in it is set into oscillations. From outside it is loaded with the radiation impedance Z_r'. Furthermore, the air confined within the tube – including the end corrections – represents a mass m' which is moved by the pressure variations inside the enclosure.

The equivalent circuit of the bass-reflex cabinet is shown in Figure 19.7b. The mass m_M of the diaphragm and the compliance n_M of its suspension are connected in series with each other and with its radiation impedance Z_r since these elements are subject to the same velocity. This, however, does not hold for compliance n_A of the air volume V_0 since it carries only the difference 'current' $v_M - v'$ (v_M and v' are the velocities of the diaphragm and the air in the vent, respectively). The branch connected to the right side of the driving gyrator M has three resonances, namely, two series resonances

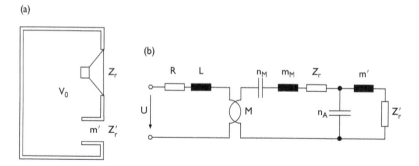

Figure 19.7 Bass-reflex box: (a) section, (b) electrical equivalent circuit (Z_r, Z_r' radiation impedances of the diaphragm and the vent).

and one parallel resonance occurring at the angular frequency determined by m′ and n_A:

$$\omega_0' = \frac{1}{\sqrt{m'n_A}} \tag{19.9}$$

lying between both other resonance frequencies. By properly designing the vent the resonance frequency ω_0' should be made to coincide with the resonance ω_0 of the loudspeaker system itself given by eq. (19.2). Then the lower of both series resonances will improve the efficiency of radiation and thus extend the useful range towards low frequencies. In reality, matters are a little more involved than shown in the equivalent circuit since both masses m_M and m′ are coupled to each other not only by the air within the cabinet but also by the sound fields produced by them.

At very low frequencies the air confined in the vent moves in the opposite direction to the motion of the diaphragm: if the membrane is slowly pressed inside the displaced air simply escapes through the vent. However, with increasing frequency the resonance system consisting of the elements n_A and m′ causes a phase shift which lessens the counteraction of the diaphragm and the vent. At frequencies well above the resonance frequency ω_0' both elements radiate with equal phase.

One disadvantage of the bass-reflex box is the more pronounced transients caused by increasing the number of energy stores. They may be perceived particularly at low frequencies.

19.4.3 Horn loudspeakers

As already noted in the introduction to this chapter a piston oscillating in a rigid-walled tube (see Fig. 4.3) with frequency-constant velocity generates a sound wave the pressure and hence the power of which are frequency independent. Therefore a suggestion may be to exploit this fact for sound radiation into the space by using a tube with gradually growing diameter, that is, by a horn. In fact, if the piston is combined with an exponential horn (see Subsection 8.4.2) the radiation resistance and hence the generated power remains constant as long as the increase of the lateral dimensions per wavelength is small, that is, the frequency is sufficiently high. With falling frequency, the radiated power is diminished, very gradually at first, but faster and faster when approaching the cut-off frequency $\omega_c = c\varepsilon$ where ε denotes the flare constant as defined by eq. (8.34). The mathematical expression of this behaviour is eq. (8.43) which shows the frequency dependence of the radiation resistance and the content of which is represented in Figure 8.11. Below the cut-off frequency ω_c the radiation impedance becomes imaginary which indicates that no sound is emitted any more into the horn.

In real horn loudspeakers the diaphragm is driven by specially designed dynamical systems. The radiation resistance which loads the diaphragm of the driving system can be further increased by arranging for a 'compression chamber' between the diaphragm and the throat of the horn. This is more correctly described as a constriction of cross section (see Fig. 19.8a). Let S_M denote the area of the diaphragm and S_0 the cross-sectional area at the throat while v_M and v_0 are the corresponding velocities. Then the conservation of mass requires

$$\frac{v_0}{v_M} = \frac{S_M}{S_0} \tag{19.10}$$

hence this compression chamber can be conceived as an impedance-matching device (see also Section 8.3). In most horn loudspeakers it is incorporated in the driving unit. In this way, loudspeakers of quite high efficiency are constructed. However, it may happen that the velocities occurring in the

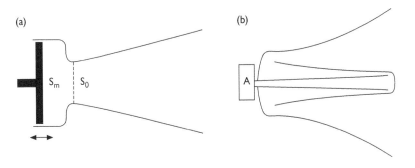

Figure 19.8 Horn loudspeaker: (a) with 'compression chamber', (b) folded horn.

throat of such a horn are no longer very small compared to the sound speed and hence may give rise to non-linear distortions.

The earlier statements on the function of horn loudspeakers apply to infinitely long horns. For horns of finite length the onset of radiation at the cut-off frequency is not as marked as in Figure 8.11; a small amount of sound energy is still emitted into the horn in the range $\omega < \omega_c$. Furthermore, standing waves associated with resonances will be set up in the horn since its wide end – its 'mouth' – generally reflects some of the arriving sound. Fortunately, these reflections can be minimised by choosing a proper length. To estimate this length L, consider a horn with circular cross section and with a given flare constant ε, its diameter at its mouth to be $2R_m$. We assume that the latter is loaded with the radiation resistance of an equally sized piston. According to Figure 5.16 this radiation resistance approaches $\pi R_m^2 Z_0$ for $kR_m \gg 1$. In this case, the horn is nearly matched to the free space and reflections from its mouth will be negligible. In order to meet this condition in the whole range above the limiting frequency $\omega_c = c\varepsilon$ we set:

$$\frac{\omega_c}{c} R_m = 1 \quad \text{or} \quad \varepsilon R_m = 1$$

On the other hand the shape of the horn is given by $R(x) = R_0 \cdot \exp(\varepsilon x)$ with R_0 denoting the radius at the throat. Accordingly, $dR/dx = \varepsilon R(x)$ and in particular

$$\left(\frac{dR}{dx}\right)_{x=L} = \varepsilon R_m \tag{19.11}$$

Since this product is 1, eq. (19.11) simply requires that the angle between the wall of the horn at $x = L$ and its axis should be about 45°. Then the optimum length L is obtained from $R_m = R_0 \cdot \exp(\varepsilon L)$:

$$L = \frac{1}{\varepsilon} \log_e \left(\frac{R_m}{R_0}\right) = \frac{1}{\varepsilon} \log_e \left(\frac{1}{\varepsilon R_0}\right) \tag{19.12}$$

the latter expression results from $\varepsilon R_m = 1$. Suppose the cut-off frequency $f_c = \omega_c / 2\pi$ of a horn is 100 Hz corresponding to a flare constant $\varepsilon = 1.85\,\text{m}^{-1}$. Then the diameter of the mouth after eq. (19.11) should be $2R_m = 1.08$ m. If we assume a throat radius of 1 cm eq. (19.12) tells us that the length of the horn must be at least 2.16 m! Of course, a horn can be folded in many ways to get a more handy loudspeaker. Figure 19.8b shows an example.

Of course, the earlier considerations are only qualitative since the wavefronts in widely opened horns are not plane but are nearly spherical. For exact calculations of horn shapes the curvature of wavefronts must be taken into account.

Horn loudspeakers are very commonly applied in sound reinforcement systems. The horn shapes used often differ from exponential in order to adapt their directivity to the purpose for which they are used (see next section). If just speech is to be reinforced as, for instance, for announcements in railway stations or airports it is sufficient to choose the limiting frequency as high as 200–300 Hz. The high efficiency of horn loudspeakers is of particular advantage for portable megaphones. They consist of a folded horn loud-speaker which is fed by a battery-operated amplifier and is combined with a microphone.

19.5 Loudspeaker directivity

Most loudspeakers project the sound they produce preferably in a certain direction except at very low frequencies. This directivity is often desirable especially in sound reinforcement applications when large audiences are to be supplied with sound as, for instance, in sports arenas or in large halls. One benefit of loudspeaker directivity is that the power output of amplifiers and loudspeakers can be kept smaller when the sound energy is directed towards the place where it is needed, namely, the audience area. Another advantage is the reduced excitation of reverberation which may impair speech intelligibility. And finally, well-designed directional loudspeakers may help to repress acoustic feedback as is caused when portions of the loudspeaker signal are projected towards the microphone which is originally intended to pick up the sound to be reinforced.

For piston-like loudspeakers eq. (5.40) or (5.41) illustrated by Figure 5.15 gives at least some guidance for the directional characteristics which can be expected. With horn loudspeakers matters are more difficult since no closed formula are available for calculating or estimating the directivity, which depends on the shape and the length of the horn as well as the size and the shape of its mouth. Thus it has to be determined experimentally. Many horns have rectangular cross sections with different expansions in both directions, for instance, exponential in vertical direction but with flat side walls. By properly combining curvatures a nearly frequency-independent directivity can be achieved (constant directivity horns).

Another common design is the multicellular horn consisting of several individual horns created by subdividing the cross section with partitions. Furthermore, linear loudspeaker arrays, sometimes also called line or column systems are widely used. These are realisations of the linear array described in Section 5.6 and consist of several equal loudspeaker systems arranged at equal distances along a line (see Fig. 19.9) and which are fed with the same electrical signal. Like single loudspeakers, the loudspeakers of a line system are usually mounted in an enclosure to avoid acoustical short-circuit. Usually, each loudspeaker of a line system has some directivity of its own unless the frequency is so low that it can be regarded as point source. Thus, to obtain the overall directivity of a line system the array function R after eq. (5.25) must be multiplied with R_0, the directional factor of the single

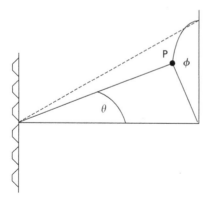

Figure 19.9 Loudspeaker array.

loudspeaker. At this point it must be noted that both directional factors, R and R_0, have rotational symmetry, but with respect to different axes: R is symmetrical with respect to the line axis while R_0 has the loudspeaker axis as axis of symmetry. Therefore it is useful to characterise a particular direction not by the elevation angle α as in Figure 5.6 but by its complement $\alpha' = 90° - \alpha$, that is, by the angle between that direction and the line axis. Then:

$$\sin \alpha = \cos \alpha' = \sin \theta \cos \phi$$

ϕ is the azimuth angle of the considered direction, measured from the vertical line in Figure 19.9, and θ denotes the polar angle with respect to the horizontal axis. Hence the overall directional factor is:

$$R(\theta, \phi) = R_0(\theta) \cdot \frac{\sin\left[(Nkd/2)\cos\phi\sin\theta\right]}{N\sin\left[(kd/2)\cos\phi\sin\theta\right]} \tag{19.13}$$

This formula, however, overlooks that the actual directivity of a line system is also influenced by the loudspeaker enclosure and, perhaps, by the interaction between the single loudspeakers.

In Section 16.5 it has been shown how the directivity of arrays can be altered by electronic 'beam steering' – a method which is commonly applied in underwater sound. To apply it to loudspeaker arrays the input signals must be suitably delayed against each other. Since electrical delay units are relatively cheap today this techniques can be used as well in sound reinforcement. Thus, for instance, it is no longer necessary to mount loudspeaker arrays vertically; instead, they can be placed underneath the ceiling, giving them the desired directivity by proper delays. In this way they are less visible and the stage area looks more pleasant.

19.6 Earphones

Since earphones are placed immediately at the ear the problems concerning sound radiation into the open space are not present. Earphones can be operated with low electrical energy which is an important advantage when it comes, for instance, to portable telephones. Another advantage is that the reproduction is not influenced by the environment, neither by reverberation of a room nor by noise produced in it. Conversely, persons using earphones do not disturb their environment.

The essential component of any earphone is a diaphragm with a diameter of a few centimetres which is set in motion by an electroacoustic transducer. Today, the magnetic earphone which was very common in the past is only used as in-the-ear receivers in hearing aids and other battery-operated devices where the high efficiency of the magnetic transducer is an important benefit. The majority of high-quality earphones are dynamic; only a small fraction of them are electrostatic earphones. The latter are highly appreciated by critical listeners because of their high fidelity qualities. However, they require higher technical expenditure since they need a relatively high signal voltage and a polarising voltage. In telephones piezoelectric earphones are widely used nowadays which are similar in their construction to piezoelectric microphones (see Section 18.3).

Regarding their general design one has to distinguish open and closed earphones. Amongst the latter are those which are inserted into the entrance of the ear canal and seal it at the same time. Other earphones enclose the listener's pinna (circum-aural earphones), the contact with the head is achieved by a soft but tight rim. A smaller modification of this is the supra-aural earphone which is placed onto the pinna either directly or with an intermediate sealing cushion. In open earphones a definite distance between the ear and the system is maintained by a sound transparent cushion. The telephone does without such a cushion. Figure 19.10 offers an overview over the various types of earphones.

The acoustical situation is best defined for the closed earphone. Here the region between the diaphragm and the ear acts at least at low frequencies as a compression chamber: the diaphragm produces small volume changes and hence pressure variations in the chamber. These are, according to eq. (18.23), proportional to the displacement ξ of the diaphragm. Suppose an earphone has a chamber volume $V_0 = 8\,\mathrm{cm}^3$ and a diaphragm with the area $S = 4\,\mathrm{cm}^2$, then after eq. (18.23) a displacement amplitude of less than $0.004\,\mu\mathrm{m}$ is sufficient to generate an effective sound pressure of $0.02\,\mathrm{Pa}$ corresponding to a sound pressure level of $60\,\mathrm{dB}$. It is evident that at such small elongations one need not bother with non-linear distortions.

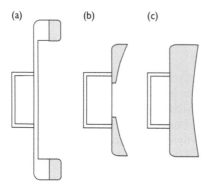

Figure 19.10 Various forms of earphones: (a) circum-aural, (b) supra-aural and (c) open earphone.

Theoretically, the mechanical resonance of the transducer system should be at the upper end of the interesting frequency range. Then we can expect that the displacement of the diaphragm and hence – with the mentioned restrictions – the sound pressure at the ear is proportional to the supplied current, if the transducer is dynamic, or to the voltage if it is electrostatic. However, at low frequencies the diaphragm must vibrate with somewhat larger amplitudes in order to make up for the inevitable leaks of the chamber.

However, an earphone with frequency-independent response provides by no means the same acoustical impression which a listener would have in the free sound field in which the arriving sound waves are diffracted by the head and the pinnae. This effect causes characteristic peaks and valleys in the frequency response. We are used to these distortions from our infancy and perceive them as natural. Moreover, they enable us to localise sound sources because they are different at both ears if the sound arrives from a lateral direction. Hence, the frequency response should agree more or less with the head-related transfer functions (HRTF) mentioned in Section 12.7, strictly speaking. In principle this can be achieved by a filter or equaliser, which, for instance, models the HRTF for frontal sound incidence (free field equalising). It is effected by suitably designed mechanical elements in the earphone, similar to those in dynamic microphones.

With open earphones the aperture of the ear canal is in the near field of the diaphragm, therefore without equalising the sound pressure shows a pronounced frequency dependence. The frequency response of a telephone receiver can only partially be smoothed by equalisation since the receiver is held by hand and hence is not in a defined position relative to the ear.

19.7   Sound transmitters for water-borne sound and for ultrasound

In most transmitters of underwater sound the piezoelectric transducer principle is employed, although magnetostrictive projectors are also in use today. Both of them are well-suited for sound radiation into water: the sound is produced in a solid material which is rather well matched to water because of its high characteristic impedance, therefore diaphragms as used in loudspeakers are not needed. The same holds for the generation of ultrasound since in most applications the wave medium into which the sound is introduced is liquid or solid.

In order to discuss the piezoelectric transducer we go back to Figure 17.2. Application of an alternating electric voltage gives rise to thickness changes of the piezoelectric disk, corresponding to the variations of the voltage. However, at somewhat elevated frequencies we cannot expect the elastic stresses and strains to be constant in space; then standing waves associated with pronounced resonances will be set up within the disk, as described in Section 16.7. Therefore the equivalent circuit sketched in Figure 17.3 is no longer an adequate representation of the transducer. A circuit which accounts for wave propagation in the piezoelectric material is presented in Figure 19.11. Each of the sources which are thought of being located at the surfaces of the disk creates a force $\sigma = eU/d$ per unit area after eq. (17.4), with e denoting the piezoelectric constant while d is the thickness of the disk and U denotes the electrical voltage. On its end faces the disk is loaded with impedances Z_1 and Z_2 and with the input impedance of a transmission line. Z_1 and Z_2 are the characteristic impedances of the media which are in

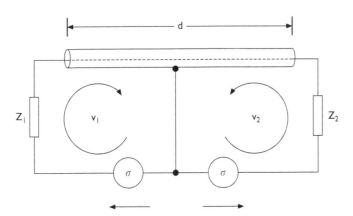

Figure 19.11 Equivalent line-circuit of a piezoelectric thickness transducer. σ: elastic stress; Z_1, Z_2-mechanical load impedances on the faces of the piezoelectric body.

contact with the end faces provided their lateral dimensions are large compared with the wavelength. When d is small compared to the wavelength the transmission line degenerates into a capacitor with the 'capacitance' n_0 which models the elastic compliance of the piezoelectric disk in Figure 17.3. The electrical capacitance C_0 of the parallel-plate capacitor which in reality is connected in parallel to both voltage sources has been omitted. In addition to the one shown in the figure, several different circuits have been developed in the course of time which are equivalent to that in Figure 19.11.

The simplest case is that of symmetric loads, in which $Z_1 = Z_2$, accordingly $v_1 = v_2$. In the middle of the transmission line the particle velocity vanishes because of the symmetry; in other words, at this point there is a velocity node of the standing wave. Thus, each half of the line is terminated with the impedance ∞. After eq. (8.10) with $l = d/2$ and $Z(0) \to \infty$ its input impedance is $-jZ_0 \cdot \cot(kd/2)$ with Z_0 denoting the characteristic impedance of the transducer material. Hence the velocity of an end face is:

$$v = v_1 = v_2 = \frac{\sigma}{Z_1 - jZ_0 \cot(kd/2)} \tag{19.14}$$

and the acoustical power radiated towards one side of the disk is

$$P = \frac{1}{2}|v|^2 \cdot SZ_1 = \frac{1}{2} \frac{SZ_1 |\sigma|^2}{Z_1^2 + Z_0^2 \cot^2(kd/2)} \tag{19.15}$$

The contents of this formula are represented in the diagram Figure 19.12 as a solid line, and the abscissa is the frequency parameter $kd = \omega d/c_L$ where c_L is the speed of longitudinal waves in the transducer material. What leaps to the eye are the regularly distributed maxima which are the more pronounced the more the characteristic impedance of the transducer material differs from that of the surrounding wave medium. They occur whenever kd is an odd integer of π, hence the thickness d of the disk is an odd integer of half the longitudinal wavelength λ_L. Consequently, the maxima indicate thickness resonances of the disk. The corresponding resonance frequencies are

$$f_n = (2n + 1) \cdot \frac{c_L}{2d} \qquad (n = 0, 1, 2, \ldots) \tag{19.16}$$

which agrees with eq. (16.9).

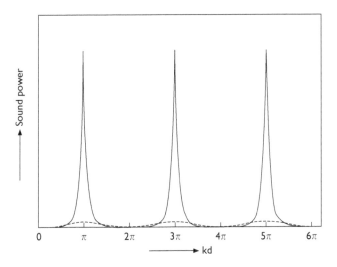

Figure 19.12 Sound power emitted from one face of a piezoelectric disk as a function of the frequency parameter kd for $Z_1/Z_0 = 0.1$. Solid line: symmetrically loaded transducer, broken line: one face loaded with the characteristic impedance Z_0.

From the equivalent circuit of Figure 9.11 the following facts can be inferred which are useful in several applications:

1 If one of both end faces of a piezoelectric disk remains unloaded, that is, if $Z_2 = 0$, for example, then the power output of the other face is the fourfold of the power after eq. (19.15).
2 If the piezoelectric disk is loaded on one of its end faces with its own characteristic impedance ($Z_2 = Z_0$), then the power radiated from the other face is

$$P = \frac{2SZ_1 |\sigma|^2}{(Z_0 + Z_1)^2} \sin^2\left(\frac{kd}{2}\right) \qquad (19.17)$$

It is shown in Figure 19.12 as a dashed line. The resonances have not completely disappeared, since in the equivalent circuit there are still two sources interfering with each other. But they are much less pronounced now, and the bandwidth of the sound source has been enhanced. This is exactly the purpose of the damping block mentioned in Section 16.7. However, the increase of frequency bandwidth must be paid for by a significantly reduced power output.

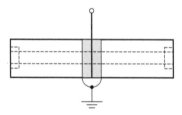

Figure 19.13 Piezoelectric compound transducer.

To tune an ultrasound transmitter to a relatively low resonance frequency (20–50 kHz) very thick layers (or long rods) of piezoelectric material are needed if we follow eq. (19.16). In the compound transducer represented in Figure 19.13 this waste of costly transducer material is avoided by attaching end pieces of metal to the active material. If desired these pieces can be shaped in a suitable way, for instance, to improve sound radiation. For electrical reasons it is practical to employ two piezoelectric disks of opposite polarisation; the components of the transducer are pulled together with a substantial screw. The resonance frequency $\omega_0 = 2\pi f_0$ of the compound transducer is found by solving the equation

$$\tan\left(\frac{\omega_0 d}{2c_L}\right) \cdot \tan\left(\frac{\omega_0 l}{c_L'}\right) = \frac{Z_0}{Z_0'} \tag{19.18}$$

Here d is, as before, the total thickness of the piezoelectric layer and l the length of one end piece; the dashed symbols refer to the material of the latter.

With such compound transducers considerable acoustical powers can be achieved. They are employed in underwater sound as transducers in transmitting arrays, and also in high intensity ultrasound they find widespread application. (See, for instance, Figure 16.10.) If ultrasonic vibrations with high amplitudes are needed as, for instance, in ultrasonic welding or drilling a compound transducer is often combined with a 'velocity transformer'. This is an inverse horn, in effect, that is, a tapered rod including the case of an abruptly changing cross section, on which a standing extensional wave is excited by the transducer attached to the thick side of the horn. Figure 19.14 shows a few examples of transforming horns. In the 'stepped transformer' sketched on the left the amplitudes at both end faces are inversely proportional to their areas; other horn shapes produce different transformation ratios.

The principle of the magnetostrictive sound generator has already been explained in Section 17.5. Evidently, the equivalent circuit of Figure 19.11 applies to this transducer too. However, it would not be very practical to construct a transmitter exactly according to Figure 17.9. To avoid demagnetisation closed cores are used as in electrical transformers in which the

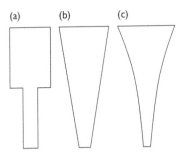

Figure 19.14 Various velocity transformers: (a) stepped transformer, (b) conical transformer and (c) exponential horn.

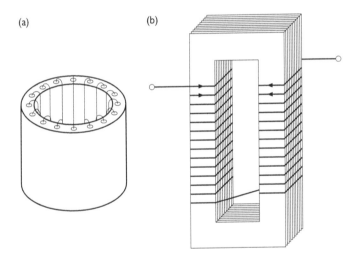

Figure 19.15 Magnetostrictive transducers.

magnetic flux is completely confined inside the ferromagnetic material. Furthermore, the core is built up of nickel laminations which are isolated from each other to keep losses by eddy currents as low as possible. The frequency range attainable with such transducers reaches up to about 50 kHz. Figure 19.15 shows two designs of magnetostrictive sound projectors.

Chapter 20

Electroacoustic systems

Electroacoustic systems for the transmission and reproduction of sound are employed in a great variety of sizes and designs today. They are intended to serve quite different purposes. In the first place we may think of installations to supply sound of sufficient loudness to large audiences. But we should also keep in mind that every car radio or home stereo set is part of such an electroacoustic transmission chain. The input terminal of such a system is usually a microphone M (see Fig. 20.1) which converts acoustical vibrations into a corresponding electrical signal. The end of the chain consists of a transducer which reconverts the electrical signals into sound. This is sometimes an earphone, however, mostly a loudspeaker L is used for reproducing the original signal. If the loudspeaker operates in a room then the latter is also a part of the sound system, strictly speaking. Between the microphone and the loudspeaker there may be a variety of electrical components for processing the signal. In any case there are amplifiers A, and quite often the spectrum of the signal is altered by a filter F. Furthermore, the signal can be stored in various ways (St) and is reproduced at some later time and somewhere else. Or the signal is modulated on a high-frequency carrier and transferred by means of sending and receiving antenna over large distances. Today, the transmission chain often includes an artificial satellite.

Likewise, the requirements which must be met by a sound system differ widely. In the simplest case the signal to be transmitted is just speech for which the quality of transmission need not meet the highest standards. This holds for the telephone or for public address systems in railway stations, airports and so forth. For other devices as, for example, for hearing aids the demand of extreme miniaturisation is to the fore. Again, the requirements are different for systems employed for faithful music reproduction, or for the sound supply to a large audience in open air or in halls.

The old saying that the loudspeaker is the weakest link in the electroacoustic transmission chain is still valid today despite the high technical standard this component has reached. This has a simple physical reason: the acoustical output is generated by the vibration of a limited diaphragm

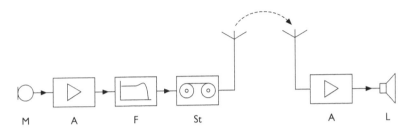

Figure 20.1 General scheme of an electroacoustic transmission system (M: microphone, A: amplifier, F: frequency filter, St: sound store, L: loudspeaker).

and is distributed from here into a very extended region in which the sound signal is still expected to be sufficiently loud. So the concentration of acoustical energy within the loudspeaker is relatively large, and therefore its diaphragm must not be too small. Of equal importance is its ability to perform vibrations with high amplitudes. Hence it is evident that a loudspeaker reaches much sooner the limit of linear operation or even may be destroyed by overload. This is in contrast to the construction of electrical power amplifiers which offers no fundamental difficulties.

Of course, it is impossible to describe all kinds of transmission systems in this account. Instead, some special issues will be selected which have not yet been dealt with or received only a brief mention so far. One of them is stereophony which is used to convey not only the signal itself but also information on its spatial or directional structure. Furthermore, the fundamentals of sound recording techniques will be presented. And a final section is devoted to the peculiarities of large systems such as are employed for sound reinforcement in the open air or in large auditoria.

20.1   Stereophony

As mentioned in Chapter 12 the human hearing is able to localise the direction of sound incidence and, in connection with this ability, to recognise the spatial structure of sound fields and sound sources up to a certain degree. This is a very important property of our hearing organ which, for example, facilitates our understanding of speech in the presence of disturbing noise or in a highly reverberant environment, or to distinguish between different speakers at a party. Since any performance in a closed space produces a complicated sound field which is responsible for the listener's subjective impression of spaciousness, a perfect electroacoustic transmission must also include the spatial structure of the sound field.

To 'transplant' a spatial sound field from one room to another by electroacoustic means is a very old idea indeed. According to an early proposal this can be done by picking up the sound signals with numerous microphones which are distributed over an imaginary surface within the original room. After suitable amplification their output voltages are supplied to a corresponding loudspeaker arrangement in the room where the sound field is to be reproduced. It is only in more recent time that this idea which can be conceived as a realisation of Huyghens' principle has been taken up again in somewhat different form by *A. J. Berkhout and D. de Vries*.

20.1.1 *Conventional stereophony*

The main problem of a 'stereophonic sound transmission' according to this scheme is the high technical expenditure needed for creating and operating a large number of independent transmission channels. Nowadays, in the most common realisation of stereophony the number of channels is reduced to the absolute minimum permissible, namely, to just two of them. This suggests itself by the fact that in our hearing localisation also relies on two independent sensors, namely, on both ears.

As explained in Section 12.7 our ability to localise sound sources is brought about by the fact that a sound signal emitted by a single sound source arrives at both our ears with different intensities and delays and that these interaural differences depend on the lateral angle of sound incidence. It is interesting to note that differences either of intensity or of delay alone are sufficient to create a directional impression. Depending on the kind of difference to which more emphasis is given in sound recording we speak of intensity or delay stereophony. With the former, the sound is picked up with two directional microphones, for instance, with gradient microphones which are arranged close to each other but oriented in different directions thus covering different parts of an extended sound source (see Fig. 20.2a). Of course, there will always be some overlap of both directional characteristics. To produce stereophonic signals with delays depending on the direction of sound incidence two microphones with equal directivity are used which are placed at different locations as shown in Figure 20.2b. In contrast to intensity stereophony this procedure is not 'mono-compatible' which means that a monophonic signal cannot be obtained just by adding the signals in both channels.

With both methods a noticeable stereophonic effect can only be achieved if the intensity or delay differences are larger than those occurring in natural hearing. For example, the interaural delay differences caused by our head never exceed about 0.63 milliseconds, while two microphones at a distance of 1 metre produce delays of up to about 2.9 milliseconds. Such exaggerations are necessary because of the imperfections of the reproducing system. This consists usually of two equal loudspeakers which are placed in such a way

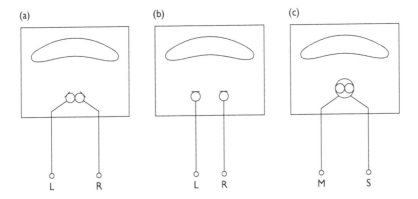

(a)   (b)   (c)

L   R   L   R   M   S

Figure 20.2 Stereophonic sound recording: (a) intensity stereophony, (b) delay stereophony, (c) MS stereophony.

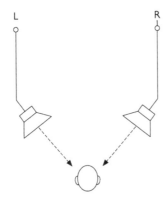

L   R

Figure 20.3 Stereophonic reproduction with loudspeakers.

that they subtend an angle at the listener of ±30°. Each of them is connected
to one of both transmission channels. However, it is inevitable that the left
ear receives not only the signal from the left loudspeaker but also the signal
produced by the right loudspeaker and vice versa, however modified and
attenuated by the head (see Fig. 20.3). Shortly, we shall return to this kind
of 'cross-talk'. A further cause of signal confusion are wall reflections within
the reproduction room.

Another method of stereophonic recording (middle-side or MS-method)
uses – as the simple intensity stereophony – two microphones placed at virtually the same position. One of them is omnidirectional while the other one
has a figure-of-eight characteristic oriented sideways (see Fig. 20.2c). We
denote the signals they pick up with M and S. The signals for the right and

the left channel are obtained by adding and subtracting them:

$$L = M + S$$
$$R = M - S$$

Other methods employ much more than just two recording microphones; the signals intended for the right and the left loudspeaker are obtained by electronic mixing.

None of these recording procedures is free of some arbitrariness, and none of them is capable of conveying a completely true impression of the spatial structure of the sound field. Often, this is not intended by the sound engineers; on the contrary, stereophonic recordings are sometimes used to create effects which never occur in the original sound field. Likewise, it is inevitable that sometimes things are overdone, for instance, when a symphony orchestra or a choir is recorded. Thus it is in doubt as to whether it was the intention of a composer when a listener in his living room hears one group of instruments from the right loudspeaker whereas the sound from another group are reproduced by the left one.

20.1.2   Binaural sound recording and reproduction

If one intends to transmit auditory impressions as faithfully as possible quite a different approach can be taken. Instead of transferring a sound field from one room into another we may attempt to transmit the sound signals which would occur at the ears of a listener in the original room to the ears of some other person. This approach is based on the idea that every person's impression of the acoustical environment is brought about merely by the sound pressure acting on both his eardrums. Hence the sounds are picked up by two small microphones which are mounted in the ear canals of a dummy simulating the human head in shape and dimensions. When placed in a sound field this dummy will generate about the same diffraction field as the head of a real listener. The output signals of the microphones are supplied to the ears of a listener. The easiest way to achieve this is by using a pair of earphones. In this way indeed a very naturally sounding reproduction is obtained.

However, this method also has its limitations. At first the shapes of human heads (including pinnae) vary widely. These differences go hand in hand with a considerable spread of individual head-related transfer functions which are decisive for our spatial hearing. In other words, every person has become used to hearing with his own ears. At best one can try to find a dummy shape with head-related transfer functions which represent as many persons as possible. An example of such a dummy head is shown in Figure 20.4. One

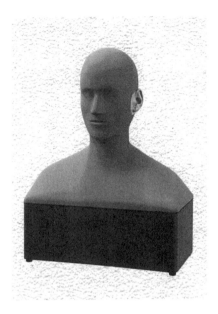

Figure 20.4 Dummy head.

has to accept then, of course, that not all listeners will be perfectly satisfied with the result.

Second, earphones are afflicted with what is called 'in-head-localisation' which means that the sound source, although appearing somehow spatial, seems to be located within the head or very close to it. In particular, it is difficult to convey or to create the impression of a frontally incident sound wave. It seems that a satisfactory explanation of this phenomenon is still to be found.

On the other hand, reproduction of the stereo signals by loudspeakers is afflicted with the cross-talk effects mentioned earlier. In the next section a method is described to get rid of these.

20.1.3 Cross-talk compensation

The cancellation of crosstalk between loudspeaker signals is achieved by a method which was first proposed and demonstrated by *B. S. Atal* and *M. R. Schroeder*. Its principle is explained by Figure 20.5 for a symmetrical loudspeaker arrangement. We suppose that each channel transmits a short impulse. The upper diagrams show the electrical signals supplied to the left and right loudspeaker, the original impulses S_L and S_R are represented by

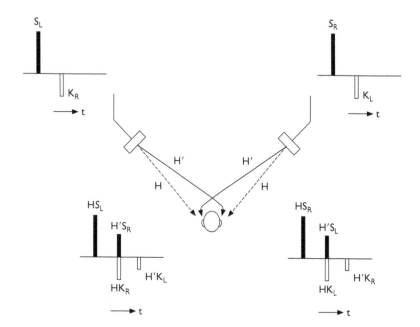

Figure 20.5 Loudspeaker reproduction of stereophonic signals: principle of cross-talk cancellation.

full bars. In the lower diagrams the signals arriving at the listener's ears are indicated. At first each ear receives the signal which is destined for it, attenuated by a factor H. A little bit later both ears receive the cross-talk signals from each opposite loudspeaker, now diminished by another factor H′ (full bars). They can be removed as a first step by supplying properly adjusted correction signals K_R and K_L to the loudspeakers (empty bars in upper diagrams), with K_R being derived from S_R and K_L from S_L. However, both correction signals in turn produce second order cross-talk signals at the listener's ears (empty bars in the lower diagrams) which can be eliminated in a second step by additional correction signals and so on. Obviously, the cancellation is an iterative process which converges the faster the more pronounced the shadow cast by the head. In practice, however, a reasonable result is obtained with just one compensation step; it is pretty good when two steps are carried out. Of course, each additional iteration step improves the quality of the cancellation. In practice, the compensation is achieved by passing the electrical loudspeaker signals through a filter the general structure of which is shown in Figure 20.6 and which can be realised as a digital FIR filter the response of which is based on head-related transfer functions.

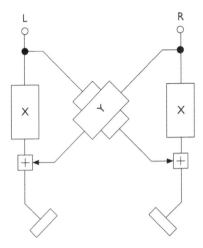

Figure 20.6 General structure of a cross-talk cancellation filter.

In combination with a good dummy head this kind of reproduction yields a sound transmission of unsurpassed fidelity. But even this chalice contains a drop of bitterness: selecting particular head-related transfer functions is tantamount to selecting a particular listener position relative to the loud-speakers; furthermore, the listener is not supposed to move his head when listening. In principle, these limitations can be overcome by controlling the compensation filter according to the actual head position which is automatically measured. Less critical is another condition, namely, that the sound should be reproduced in an anechoic environment. In fact, quite good results can be obtained in living rooms provided they are not too reverberant. Nevertheless, this method of two-channel sound reproduction will probably find its main application not in the normal consumer scene but in psychoacoustic research and in auralisation (see Section 13.6).

20.2 Sound recording

Sound recording is an important and popular way to retain sound signals which are transitory according to their very nature, with the aim of reproducing them at any place and at a later time. One important aspect is that stored records can be duplicated ad lib. Although the modern techniques of sound recording have not much to do with acoustics – what is actually stored are electrical signals – even so a brief account will be given of the different methods of sound recording.

20.2.1 Disc recording

Media for storage of sound signal are either magnetic or optical; furthermore, digital stores find increasing use. However, the oldest medium is a solid carrier, that is, a rotating drum or disc into which an oscillogram of the sound signal, in effect, is engraved. For reproducing the sound signal the groove on the carrier is traced with a fine needle. In old systems of this kind (phonograph, gramophone) both the recording and the reproducing process was performed by purely mechanical means; accordingly, the cutting tool or the tracing needle was attached to a diaphragm which was at the narrow end of a horn and was set into motion by the arriving sound waves, or from which the vibration of the needle was radiated into the environment.

With the advent of electrical amplifier techniques new possibilities were opened up. For 'cutting' of a disc record a specially designed electroacoustic transducer is used now which converts the output voltage of a microphone into motions of a cutting stylus. By slowly moving the chisel-ended cutter over the rotating disc a spiral groove is produced which is modulated by the vibrations of the cutting stylus. Figure 20.7 shows various possibilities of undulating the groove and hence of storing the acoustical information on it. Vertical motion of the stylus (Fig. 20.7a) generates a modulation in depth. This method, however, is obsolete and has been superseded long ago by lateral undulation (Fig. 20.7b). Figure 20.7c shows how the two-channel information needed for recording stereophonic signals can be stored in the flanks of a groove. A major step forward in disc recording was the invention of a technique by which the distance of neighbouring grooves is continuously adapted to the actual signal.

The original disc is mechanically very delicate and is converted in several steps into stampers by which commercial records are pressed. For reproducing the sound signals from the rotating disc an electroacoustic pickup carrying a fine spherical-ended stylus of sapphire or diamond is set into the groove to convert its undulations into an electrical signal.

This kind of disc record had attained a high standard of performance. One drawback of them, however, was the limited dynamic range. In fact, the

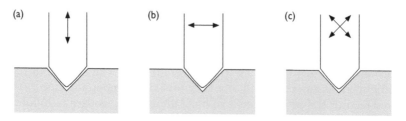

Figure 20.7 Various forms of disc recording: (a) vertical modulation of the groove, (b) lateral modulation and (c) 45° modulation.

level difference between the maximum amplitudes which could be recorded without distortions and the noise caused by the grainy structure of the disc material was 60 dB at best while the maximum separation of both stereo channels was 30 dB. Moreover, a disc record is subject to mechanical wear each time it is replayed.

Today, the 'black disc' has been superseded almost completely by the digital disc, the Compact Disc (CD). Here the signal is not stored in the form of an undulation which is more or less a replica of the signal but as a sequence of pits of varying lengths which have uniform depth and width and which are impressed into the plastic substrate (see Fig. 20.8). For this purpose, the electrical signal must be 'digitised' at first, that is, it must be converted into a sequence of ones and zeros. At first the signals of both stereo channels are sampled at a rate of 44.1 kHz; then the samples are quantisised into 2^{16} amplitude steps of equal size. In this process a certain quantisation error has to accepted which is characterised by the ratio

step size / maximum amplitude

in the present case 2^{-16}. It corresponds to a signal-to-noise ratio of more than 90 dB which certainly meets the highest standards. To the data words derived from the original signal so-called parity bits are added which allow an automatic error correction during reproduction. To increase the density of storage these data are embedded in a certain code in such a way that each pit edge represents a binary 1 and the flat areas in between represent binary 0s.

In the recording process the pits are impressed by optical etching into a thin layer of photoresist on a glass plate. They have a width of 0.6 μm and are 0.12 μm deep; the mutual distance of adjacent traces which are arranged along a spiral as with the 'black' disc record is 1.6 μm. (If this figure is compared with the average trace distance of about 55 μm of an analogue

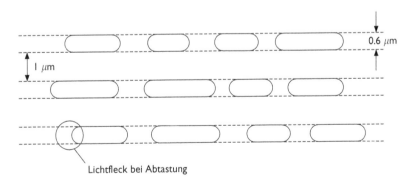

Figure 20.8 Surface of a compact disc. The depth of pits is 0.12 μm.

disc the enormous increase in storage density becomes evident.) The trace velocity of the CD in the recording process (and also in reproduction) is kept constant at about 1.3 m/s. In a complicated process comprising several copying steps the final CD is manufactured. With a diameter of 120 mm and a thickness of 1.2 mm, it consists mainly of a transparent substrate (poly carbonate) containing on one side the pits which carry the information. This side is covered with a light reflecting layer of aluminium, silver or gold which in turn is provided with a thin protective layer with a thickness of 10 to 30 μm.

For reproduction the stored data are retrieved with an optical pickup which scans the trace at constant speed. Its main component is a laser diode which illuminates the reflecting metal layer with a convergent light bundle through the substrate. Its focus on the reflecting layer has a diameter of 1 μm. When it strikes a flat part ('land') the incident light will be almost perfectly reflected. However, when the focus lies on a pit about one half of it will be reflected from its bottom while the other half is reflected from the surrounding land. Since the optical wavelength within the substrate is about 0.5 μm both parts cancel each other by destructive interference. The light reflected from the metal layer is focussed on four photo diodes which not only yield an electrical signal corresponding to the sequence of pits but also correction signals which indicate tracking errors or imperfect focussing and which are used to readjust the position of the pickup. The whole device is no larger than about 45 mm × 12 mm.

The audio signal is restored by decoding and digital-to-analogue conversion. During decoding minor errors are immediately eliminated by using the parity bits. If larger errors are detected the signal is reconstructed as far as possible by linear interpolation. If this fails, the amplifier gain is gradually reduced to zero where it is kept for the duration of the error.

The compact disc permits sound recordings of highest quality. The linear and non-linear distortions are vanishingly small. Since the reproduction pickup is free of any mechanical contact a compact disc is not subject to any wear. Thanks to the high storage density the maximum playing time of a CD is more than one hour.

From about 1992 a smaller digital disc has become commercially available, the so-called MiniDisc (MD), both as pre-recorded carriers and for sound recording and erasing by the consumer. In the former version the techniques of recording and reproducing is very similar to that of the common CD, and in the second one the information is stored in the form of remanent magnetisation which is converted into modulated light in the reproduction process. In both cases the playing time is similar to that of a conventional CD. This is made possible by exploiting masking effects as described in Section 12.5: those spectral components which are expected to be masked by other components and therefore are inaudible need not be stored and hence are eliminated prior to recording. It is remarkable that modern storage techniques which are not very closely related with acoustics recently include properties of our hearing to such a high degree.

20.2.2 Sound motion picture

The preceding section describes the development of the disc record as marked by the transition from mechanical to optical recording techniques. However, there is a much earlier application of optical recording techniques, namely, in motion pictures. In optical sound recording a flat light ray is directed to a moving photographic film tape. The signal to be recorded modulates the light flux and hence the optical transparency of the exposed film material. The sound signal is recorded on a sound track which is about 2 mm wide and which is at one side of the film. For stereophonic recordings the film carries two sound tracks, sometimes there are even more sound tracks for producing special sound effects.

The bundle of light rays is produced by focusing an illuminated slit on the sound track. The modulation of the light flux can be achieved in two ways: either by controlling the light intensity or by varying the width of the exposed region on the sound track. Accordingly, one distinguishes variable-density recording and variable-area recording. The latter is obtained with a triangular mask operated by an electroacoustical transducer, and it varies the actual length of the slit when moved perpendicular to it as sketched in Figure 20.9a. Modulation of intensity can also be achieved with an electro-optical converter such as a Kerr cell, or again with a variable aperture which is now placed in a different position of the optical path. For reproduction the moving sound track is illuminated with a slit-shaped bundle of light rays; the light passing the film is detected by a photo cell or photo diode which yields an output voltage proportional to the received light flux. Of course, there must be an offset on the film between a particular picture and the associated spot of the sound track since for reproducing the sound the film must move at constant velocity while projection of a picture requires that the film stands still for a little while.

The variable-density method is no longer in use today since the sensitivity of film materials used in producing copies must meet strict requirements in order to avoid non-linear distortions. The variable-area method can be modified in many ways; by using more complicated masks multiple traces can be created (see Fig. 20.9b).

Scanning the sound track with a light bundle of finite width gives rise to a characteristic linear signal distortion which is known as 'slit effect'. Each illuminated length element of the film contributes to the light flux passing the film according to its transparency, hence the flux is proportional to the integral over the width b of the slit. Suppose that the transparency carries a sinusoidal modulation according to $A + B \sin(Kx)$ with $A > B$. When the film passes the slit with the speed v (usually 45.6 cm/s), each element dx of

(a)

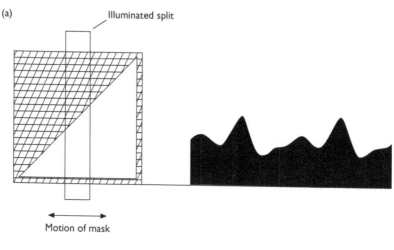

(b)

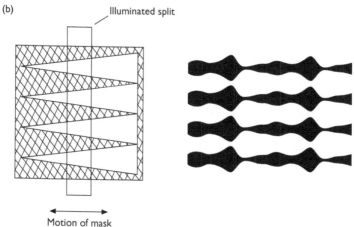

Figure 20.9 Sound motion picture, variable-area recording: (a) simple track, (b) multiple track.

the slit contributes an amount proportional to $A + B \sin K(x - vt)$ to the light flux ϕ. Integrating this within the limits $\pm b/2$ with respect to x yeilds (with $K = \omega/v$):

$$\phi \propto A + B \frac{\sin(b\omega/2v)}{b\omega/2v} \cdot \sin \omega t \qquad (20.1)$$

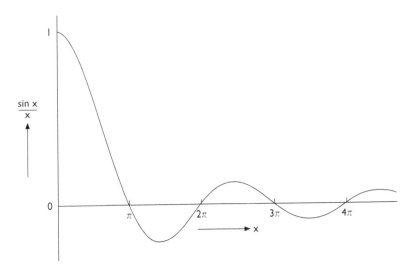

Figure 20.10 The function sin x/x.

The function sin x/x is shown in Figure 20.10. Evidently, the finite width of the sensing ray bundle causes a suppression of high frequency sound components. At certain frequencies it may even happen that the time-dependent part of the light flux completely vanishes. This is the case when the argument of the sin x/x-function becomes an integral multiple of π. Then the width b agrees with an integral number of wavelengths $\Lambda = 2\pi/K$ of the recorded transparency variation. We define the upper limit f_{max} of useful frequencies by the argument x at which the function in Figure 20.10 has the value $1/\sqrt{2}$. This yields

$$f_{max} = 0.442 \frac{v}{b} \tag{20.2}$$

It should be noted that the width b cannot be reduced ad lib but is limited by the onset of diffraction.

An alternative to optical sound recording is the techniques to be described in the subsequent section which employs a magnetic sound track embedded in the film material.

20.2.3   Magnetic sound recording

The storage medium used in magnetic sound recording is a thin ferromagnetic layer deposited on one side of a plastic tape. During recording the electrical signal produces in this layer a varying magnetisation which in the current techniques is parallel to the direction of the tape motion.

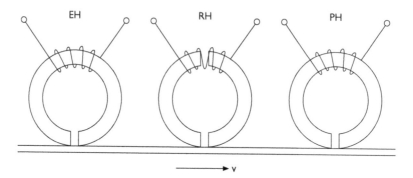

Figure 20.11 Erasing head EH, recording head RH and playback head PH of a tape recorder.

Figure 20.11 shows schematically the essential components of a tape recorder, namely, the magnetic 'heads' used for recording, reproducing and erasing signals. Each of them consists of a nearly closed core of highly permeable material which carries a winding and has a small gap. The latter is in contact with the tape which is pulled past the head with constant speed. For recording the electrical signal is supplied to the recording head RH where it creates a magnetic flux. In the gap region the ferromagnetic layer on the tape attracts this flux because of its high permeability. After leaving the recording gap the layer carries longitudinally magnetised sections of varying length and magnetisation.

When a pre-recorded tape is reproduced the field lines originating from the magnetised section are drawn into the playback head PH since the core of this head has a lower magnetic resistance than the surrounding air. They penetrate the winding and induce a voltage which is proportional to the temporal variation of the magnetic flux Φ:

$$U = w\frac{d\Phi}{dt} = j\omega w\Phi \qquad (20.3)$$

(w = number of turns). This relation indicates a linear rise of the output voltage with increasing frequency. Apart from this rise due to the induction law, there are other effects which alter the frequency response particularly at high frequencies. In the first place we should remember the slit effect already described in the preceding section. Another drop in sensitivity at high frequencies is caused by eddy currents in the core of the playback head and by the fact that the ferromagnetic layer on the tape is not magnetised in its whole thickness at high frequencies, and that the distance of this layer from the head is very small but finite. All these influences must be eliminated by a suitable equalising filter.

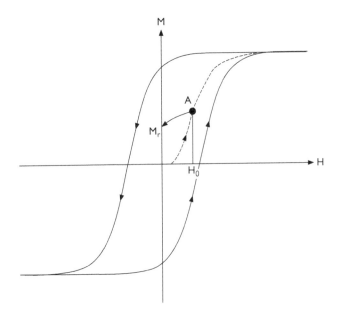

Figure 20.12 Magnetic sound recording: initial magnetisation curve (broken line) and hysteresis loop (H: magnetic field strength, M: magnetisation, M_r: residual magnetisation).

The recording procedure is somewhat more involved because of the complicated processes occurring in the magnetisation of a ferromagnetic material. As is well-known these materials show the phenomenon of hysteresis (see Fig. 20.12): if an originally non-magnetic material is exposed to a magnetic field of increasing strength, its magnetisation grows monotonically until saturation as indicated by the broken line. If we then reduce the field strength, the magnetisation will not diminish according to that curve but according to the left branch of the so-called hysteresis loop (solid curve). If the sense of field variation is again changed after reaching the negative saturation the magnetisation increases following the right branch of the loop. The initial magnetisation curve (dashed curve in Fig. 20.12) will never be reached again unless the magnetisation of the material is completely erased.

Now suppose the field strength H_0 of the magnetising field to which the material is exposed is too small for reaching saturation but magnetises it only up to some point A in Figure 20.12. When the field is switched off some residual magnetisation M_r remains in the material. Unfortunately, M_r is by no means proportional to the field strength H_0 at point A. Hence, if the recording head were fed just by the signal current the residual magnetisation stored on the tape would not be a replica of the signal but would suffer from strong distortions.

Some linearisation of the recording process could be achieved by superposing a DC current to the signal prior to recording. In fact, this method which we encountered in the discussion of certain electroacoustic transducers has been applied in the early days of magnetic recording. However, in this way only signals with small amplitudes can be recorded without noticeable distortion.

A much wider dynamic range is attained when the DC bias is replaced with a high amplitude AC current with a frequency of 30–150 kHz. Then a section of the tape within the range of the recording gap is alternately magnetised until positive and negative saturation, that is, the magnetisation runs repeatedly along the whole hysteresis loop. When this section is pulled away from the gap the field strength gradually diminishes; consequently, the loop contracts itself towards a point on the M-axis leaving a residual magnetisation which is virtually proportional to the actual value of the signal. We refrain from presenting a rigorous proof of this fact because it is too long-winded. For removing any magnetisation the erasing head is fed merely with the AC current.

We conclude this discussion by presenting a few technical specifications. The width of a magnetic tape for consumer's use is 6.3 mm; however, smaller tapes are also very common. Thus, for instance, the tape of a cassette recorder is only 3.81 mm wide. The total thickness of a tape is between 12 and 50 μm, and the active layer is 4 to 15 μm thick. It consists of small particles of iron oxide or chromium dioxide; pure iron or mixtures of iron and chromium are applied. The standardised tape speed is 15 inches per second; other speeds are obtained by repeatedly halving this figure (7.5, 3.75, etc. inches per second). In the popular cassette recorder the tape runs with a speed of 1.875 inches per second. For stereophonic recordings tapes with two parallel tracks are employed but tapes with more tracks are also in use.

The cores of the recording and the playback head are made of permalloy laminations or of ferrite. A recording head has – apart from recording head next to the tape – a second air gap at its rear side which is to increase its magnetic resistance and hence to make its magnetic properties less sensitive against fluctuations of the tape distance. The gap of the playback head is only 3–8 μm wide in order to keep the loss of high-frequency spectral components due to the slit effect within limits (see Fig. 20.10). Often, one and the same head is used for recording and playback.

The signal-to-noise ratio may reach 70 dB in master recorders. In consumer recorders, for instance, in cassette recorders, it is smaller, but can be considerably improved by dynamic compression (Dolby).

As in disc recording, digital techniques have entered magnetic recording. In both fields the main advantage is the absence of non-linear distortions since it is just two signal values which are stored on the carrier. Consequently, the digital tape is always magnetised until saturation, and the digital information is contained in the sign of the magnetisation. Digitising and coding the analogue signals are very similar to the procedures employed in the compact disc. The high bandwidth required is achieved – as in a video recorder – by arranging the recording and the reproduction head on a drum which rotates at 2000 revolutions per minute. The tape is partially wound around the drum the rotation axis of which is not exactly perpendicular to the motion of the tape. In this way a sequence of oblique recording tracks is produced on the tape.

20.3   Sound reinforcement systems

Systems for sound reinforcement are used whenever large and extended audiences must be supplied with sound. The need for such systems follows from the fact that acoustical power produced by the human voice or by some other natural source is often too small to reach distant listeners. Another factor which may impair natural sound transmission is the ubiquitous noise originating either from technical sources (road traffic, air conditioning) or produced by the restless audience itself.

In large sporting events, congresses and conventions, mass demonstrations and so forth the necessity for sound reinforcement is immediately evident. Moreover, electroacoustic sound reinforcement is not only used for securing the intelligibility of speech but also for increasing the loudness in certain musical performances particularly of the popular entertainment. Thus, the presentation of a musical to a large audience, for instance, is inconceivable without electroacoustic amplification. The same holds for open-air theatre performance. And in a large cathedral with its long reverberation sufficient speech intelligibility cannot be achieved without electroacoustic support. On the other hand, it is often believed that even in small meeting rooms or lecture rooms electroacoustic reinforcment is indispensable although it only serves the comfort of the speaker and the listeners: many speakers do not bother to speak loud enough and to articulate clearly, and also the listener is used to his TV or radio where he has just to reach for the volume control to avoid any listening effort.

Achieving good speech intelligibility or clear and transparent sound in a musical performance requires in the first place sufficient loudness of the sound signal. Of equal importance, however, is that the sound signal reaching the listener is free of artefacts and distortions as may be caused either by imperfect technical equipment or by acoustical peculiarities of the room in which the system is operated. Moreover, the sound reinforcement should sound as natural as possible which implies that the acoustical localisation coincides with the visual one. In the ideal case the listener should not even notice that the sound he hears originates from a loudspeaker and not from a speaker's mouth.

20.3.1   Design of sound reinforcement systems

For speech reproduction the level at the listener's place should be about 70 to 75 dB provided there is no significant ambient noise. Otherwise, it must exceed the noise level by at least level 10 dB. For calculating the necessary acoustical power eq. (13.17) can be applied completed on the right by a factor γ, the gain of the loudspeaker as defined in eq. (5.16). If we express the energy density w_d in the former equation by $\tilde{p}^2/\rho_0 c^2$ and the effective sound pressure $\tilde{p}$ by the sound pressure level after eq. (3.34) we obtain for the power:

$$P = 4\pi r^2 \cdot \frac{p_b^2}{\gamma Z_0} \cdot 10^{0.1L} \tag{20.4}$$

with $p_b = 2 \cdot 10^{-5}$ Pa. According to this equation a point source or a non-directional loudspeaker would have to supply an acoustical power of about 50 mW if a level of 70 dB is to be produced at a distance of 20 m. If, however, the level to be achieved amounts to 100 dB the required power output would be 50 W – quite a high value which requires correspondingly high amplifier output, taking into account the low efficiency of loudspeakers. It can be significantly reduced by employing a directional loudspeaker provided the listener is within the main lobe of its directivity.

This consideration applies to free field propagation. Matters are more complicated if the system is operated in a hall. Here reflections from the walls and the ceiling may help to improve the sound supply to listeners. On the other hand, the sound emitted by a loudspeaker is not only projected towards the audience but excites also the reverberation of the room which is detrimental for the intelligibility of speech. Any increase in the loudspeaker output will also increase the reverberant energy and hence does not improve the situation. The only remedy is to employ directional loudspeakers (or to reduce the reverberation time of the auditorium). In the ideal case a loud-speaker would direct the whole energy onto the highly absorbent audience thus preventing it from impinging on the reflecting boundary. This, however, is impossible, mainly because of the frequency dependence of the loudspeaker directivity. Particularly at low frequencies a good deal of the emitted energy will feed the reverberant field.

The following consideration is to quantify the situation. We assume that the reverberant energy decays according to an exponential law and consider that part of the decay as detrimental which arrives at the listener with a delay in excess of 100 ms with respect to the direct sound which is the fraction

$$\int_{0.1s}^{\infty} e^{-2\delta t}dt \bigg/ \int_{0}^{\infty} e^{-2\delta t}dt = e^{-0.2\delta} \approx 2^{-2/T} \tag{20.5}$$

of the reverberant energy. Now we require that the energy density of the direct component surpasses the detrimental part of the reverberant energy. According to eq. (13.19a) the total steady-state energy density in a room is

$$w_{tot} = \frac{P}{4\pi c}\left(\frac{\gamma}{r^2} + \frac{1}{r_c^2}\right)$$

with P denoting the power output of the sound source. The first term contains the energy density of the direct sound while the second one corresponds to the reverberant one. The quantity r_c is the diffuse-field distance at which both components would be equal if the sound source were non-directional (see Section 13.4). Hence the requirement mentioned earlier reads

$$r < 2^{1/T}\sqrt{\gamma}\cdot r_c = r_{max} \qquad (20.6)$$

r_{max} is the maximum distance at which satisfactory transparency of sound, especially speech intelligibility, can be expected. After expressing the diffuse-field distance by the reverberation time we obtain

$$r_{max} \approx 0.06\sqrt{\gamma V}\cdot\frac{2^{1/T}}{\sqrt{T}} \qquad (20.7)$$

The latter fraction which depends merely on the reverberation time is 5.66 for T = 0.5 s; at a reverberation time of 1 s it has dropped to 2. This underlines the influence of the reverberation time of an auditorium.

Equation (20.7) is somewhat too pessimistic inasmuch it neglects the sound absorption of the audience to which the loudspeaker is usually directed. In fact, the sound falling onto the audience does not excite reverberation. However, this beneficial effect is not present at low frequencies for which the absorption of the audience is small. Both influences together, the low audience absorption and insufficient loudspeaker directivity in the low-frequency range, are responsible for the muffled loudspeaker sound which is observed so often and is so unfavourable for good speech intelligibility. The only remedy is to suppress the low-frequency components of the signal as far as possible which are anyway irrelevant for intelligibility.

Now an estimate of the necessary minimum loudspeaker power can be carried out using eq. (20.4) in which r is to be replaced with r_{max}.

20.3.2 Loudspeaker arrangement

The amplified sound signal can be projected onto the audience by one single loudspeaker or with several loudspeakers combined closely together. Then we speak of a central loudspeaker system. The counterpart of it is a decentralised system consisting of several or even many loudspeakers located at different positions. Which of both possibilities is more useful depends on the local situation as well as on the kind and purpose of the sound system. Sound reinforcement in meeting halls which are mainly or exclusively used to amplify the human voice as in speeches, lectures etc. are preferably designed as a central system. The same holds true, of course, for mobile sound systems used in meetings or pop concerts in open air. In principle, they produce a more natural listening impression since the direction of the loudspeaker deviates little from that of the natural source. On the contrary, in large sport stadia installations with locally distributed loudspeakers may be more favourable. They have the general advantage of operating with less total power since the required output power of a loudspeaker increases with the square of the distance to be covered. Furthermore, they ensure a more uniform sound supply to the audience. This contrasts with the disadvantage that there may be regions where the listener perceives sounds arriving from two (or more) loudspeakers. This happens when the path difference between the contributions of two loudspeakers exceeds 17 m (corresponding to a time difference of 50 ms) and, at the same time, the level difference between both sounds is less than 10 dB. But even when there is no 'double hearing' the listener identifies the next loudspeaker as the relevant sound source, according to the law of the first wavefront (see Section 12.7).

In Figure 20.13 a central loudspeaker system is schematically depicted. By suitable arrangement and orientation of the loudspeaker a uniform distribution of direct sound over the audience should be strived at. To check this

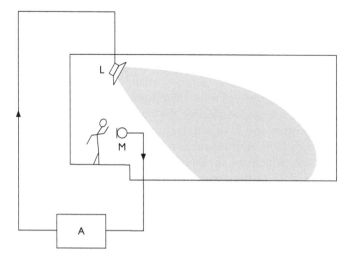

Figure 20.13 Central loudspeaker system (M: microphone, L: loudspeaker, A: amplifier).

condition experimentally it is advantageous to use short sound impulses as test signals because this makes it easy to separate the direct contribution of the loudspeaker signal from reverberation or echoes.

In Section 12.7 we mentioned the Haas effect according to which the level of the amplified loudspeaker signal may surpass the level of the original source by up to 10 dB without destroying the illusion that all the received sound is produced by the original source, for instance, by a speaker. The condition for this to happen is that at the listener's position the latter component precedes the loudspeaker's contribution by 10 to 15 milliseconds. This can often be achieved by choosing a slightly more remote loudspeaker position. If this solution is not possible, the same effect can be achieved by properly delaying the electrical loudspeaker signal. For this purpose electronic delay devices are available nowadays. If the natural direct sound is too weak it can be moderately enhanced by so-called 'simulation projectors', that is, by loudspeakers which are arranged close to the source, for instance, in the front facing panel of a speaker's desk.

In very large or long halls it may be difficult to fulfil the condition in eq. (20.7) with just one loudspeaker or loudspeaker cluster. Then it will be more practical to subdivide the whole distance by use of several, suitably distributed loudspeakers into several sections each with much smaller r_{max} as shown for two loudspeakers in Figure 20.14. It is obvious that the electrical signal feeding the loudspeaker which is closer to the listener must be delayed according to the different distance. The same holds true for any auxiliary loudspeakers used for supplying more remote parts of the audience, for instance, of the region beneath a deep balcony.

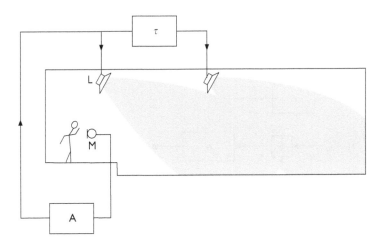

Figure 20.14 System with distributed loudspeakers (τ = delay).

20.3.3 Acoustical feedback

Very often the microphone which is to pick up the original signal as, for instance, the voice of a speaker is in the same room in which the sound reinforcement system is operated. It is inevitable that the loudspeaker projects a portion of the amplified signal towards this microphone which enters again the amplifier–loudspeaker chain. This process is known as 'acoustical feedback'. It is not only responsible for a change in the signal spectrum but may also lead to self-sustained oscillations of the whole system which are heard as ringing, howling or whistling.

The situation is depicted in Figure 20.15a. As we know the transmission of a sound signal in a room is characterised by the impulse response $g(t)$ of the room or, alternatively, by its Fourier transform, the frequency transfer function $G(\omega)$. In Figure 20.15 there are two such transfer functions, namely, the function $\overline{G}(\omega)$ describing the transmission from the loudspeaker to a particular listener while the other one denoted by $G(\omega)$ describes the transmission path connecting the loudspeaker with the microphone. Both have the complicated structure described in Section 9.7 (see also Fig. 9.12).

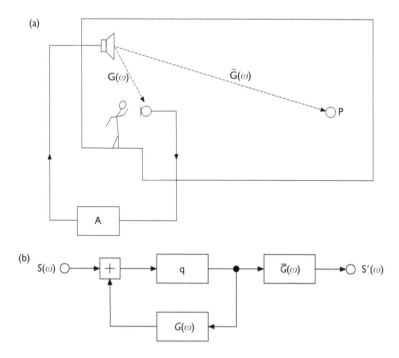

Figure 20.15 Acoustic feedback in a room: (a) transmission paths (A = amplifier), (b) block diagram.

Figure 20.15b is a more abstract version of Figure 20.15a. $S(\omega)$ and $S'(\omega)$ are the Fourier transforms of the original sound signal and of the signal received by the listener. Obviously, the feedback loop consists of the elements q and $G(\omega)$. From this block diagram, in which q stands for the gain of the amplifier, we see immediately that

$$S' = \overline{G}\left(qS + q^2GS + q^3G^2S + \cdots\right) = \frac{q\overline{G}}{1 - qG}S \qquad (20.8)$$

The second version is obtained by applying the summation rule for geometric series. Hence the transfer function modified by feedback reads:

$$G'(\omega) = \frac{q\overline{G}(\omega)}{1 - qG(\omega)} \qquad (20.9)$$

The product qG is called the open-loop gain of the system. If it is small compared to unity the modified transfer function differs only by the factor q from the natural transfer function G. With growing open-loop gain the differences between both transfer functions become more significant.

This is illustrated in Figure 20.16. It shows a section of a particular 'frequency response curve', that is, the absolute value $|G'|$ in logarithmic presentation, and the variations it undergoes when the amplifier gain is increased

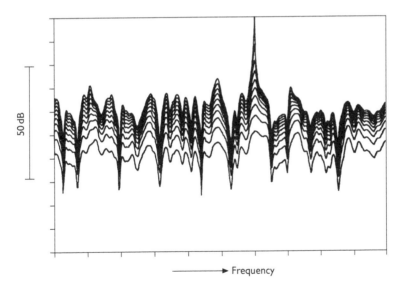

Figure 20.16 Variations of a frequency response curve by acoustical feedback (simulation). The open-loop gain varies from $-20\,dB$ to $0\,dB$ in steps of $2\,dB$ with respect to the stability limit.

in steps of 2 dB. Most obvious is a peak of increasing height and sharpness which grows out of one particular maximum. This causes a distortion of a transmitted signal spectrum which is subjectively perceived as change in timbre. In the time domain this peak corresponds to increased reverberance of this particular spectral component perceived as 'ringing' since it dies out very slowly during its repeated round trips within the loop. In the uppermost curve the magnitude of the open-loop gain in eq. (20.9) has reached the value 1. With any further increase of gain the system becomes unstable, that is, it starts to oscillate at the frequency of the peak. A more rigorous discussion of the system's stability should include the phase of the transfer function. For practical purposes it is sufficient to characterise the limit of stability by

$$q\,|G|_{max} = 1 \tag{20.10}$$

Here $|G|_{max}$ is the maximum of the transfer function; it corresponds to the maximum level as expressed by eq. (9.41). The colouration caused by feedback is unnoticeable as long as the open-loop gain remains at least 5 dB for speech, 12 dB for music, below the critical value given by eq. (20.10).

To improve the stability of a sound reinforcement system, that is, to avoid the negative effects of acoustical feedback the magnitude of $G(\omega)$ should be kept as low as possible. Sometimes this can be achieved by suitable arrangement of the loudspeaker(s), the directional characteristics of which should exclude the microphone as far as possible. A similar effect is brought about by using a microphone with suitable directivity. The feedback stability of the system can be further improved by shifting the whole spectrum of the electrical signal by a few Hertz towards higher or lower frequencies, prior to feeding it to the power amplifier and the loudspeaker (*M. R. Schroeder*). In this way every spectral component falls with each round trip on another point of the frequency response curve. This is tantamount to smoothing the curve, in particular, to removing its maxima. After eq. (9.41) it should be expected that this measure yields an increase in stability of about 10 dB. The practical improvement of useful gain is 4–6 dB. In the case of speech the frequency shift is inaudible. However, this method is inapplicable to music because even a small frequency shift alters the musical intervals in an unacceptable way.

Literature

Ahnert, W. and F. Steffen: *Sound Reinforcement Engineering – Fundamentals and Practice*. E & FN Spon, London 1999.

Benade, A. H.: *Fundamentals of Musical Acoustics*, 2nd Edition. Dover Publications, New York, 1990.

Beranek, L. L.: *Concert Halls and Opera Houses – Music, Acoustics and Architecture* 2nd edition. Springer, New York, 2004.

Beranek, L. L. and I. L. Vér (Herausg.): *Noise and Vibration Control Engineering*. John Wiley, New York 1992.

Blauert, J.: *Spatial Hearing – Revised Edition: The Psychoacoustics of Human Sound Localisation Spatial Hearing*. MIT Press, Cambridge, MA 1997.

Brian C. J. Moore: *An Introduction to the Psychology of Hearing*, 5th Edition. Academic Press, Amsterdam 2003.

Cremer, L. and H. A. Müller: *Principles and Applications of Room Acoustics*, two volumes. Chapman & Hall, London 1982.

Cremer, L., M. Heckl and B. A. T. Peterson: *Structural Vibrations and Sound Radiation*. Springer-Verlag, Berlin 2005.

Crocker, M. J. (Herausg.): *Encyclopedia of Acoustics*, four volumes. John Wiley, New York 1997.

Fastl, H. and E. Zwicker: *Psychoacoustics, Facts and Models*, 3rd edition. Springer, Berlin 2006.

Fletcher, N. H. and T. D. Rossing: *The Physics of Musical Instruments*, 2nd Edition. Springer-Verlag, New York 1998.

Kuttruff, H.: *Room Acoustics*, 4th Edition. Spon Press, London 2000.

Lurton, X.: *An Introduction to Underwater Acoustics: Principles and Applications*. Springer-Verlag, Heidelberg 2003.

Mechel, F.: *Schallabsorber*, three volumes (in German). S. Hirzel Verlag, Stuttgart 1989/95/98.

Morse, P. M. and U. Ingard: *Theoretical Acoustics*. McGraw-Hill, New York 1968.

Pohlmann, K. C.: *Principles of Digital Audio*, 3rd Edition. McGraw-Hill, New York 1995.

Skudrzyk, E.: *The Foundations of Acoustics*. Springer-Verlag Wien, New York 1971.

Trevor J. Cox: *Acoustic Absorbers and Diffusors. Theory, Design and Application*. Spon Press, London 2004.

Wille, P. C.: *Sound Images of the Ocean in Research and Monitoring*. Springer-Verlag, Heidelberg 2005.

Zwicker, E. and H. Fastl: *Psychoacoustics*. Springer-Verlag, Berlin 1999.

Index

Absorption 55, 271; *see also* Attenuation
Absorption area, equivalent 268
Absorption coefficient 103, 266
Accelerometer 398
Acoustical feedback 448
Acoustical short-circuit 80, 205, 318, 411
Admittance 14
Ambient noise 338
Amplitude 8
Anechoic room 280
Angular: frequency 9; wave number 53
A-Scan 348
Attenuation 55; classical 58; in gases 57; in liquids 62; molecular 58; in pipes 140; in sea water 335; in solids 63, 206, 306
Audiometer 243
Auralisation 279

Babinet's principle 133
Backscattering cross section 334, 337
Basilar membrane 236
Bass-reflex cabinet 413
Beam steering 341, 418
Beats 12
Bending 197; stiffness 198; wave 201
Bernoulli's law 224
Bimorph element 388
Binaural transmission 430
Bone conduction 239, 331
Boundary layer 139, 272
Brass instrument 226
Breathing sphere 84
B-scan 348
Building acoustics 283

Calibration 399
Cavitation 314, 349
Cent 214
Characteristic frequency *see* Eigenfrequency
Characteristic function *see* Eigenfunction
Characteristic impedance 51
Clarity index 263
Cochlea 236
Coherence region 314
Coincidence 292
Compact disc (CD) 435
Compliance 15
Compound partition 287
Compound transducer 351
Compressional wave 190
Compression modulus 55
Condensor: loudspeaker 408; microphone 382
Consonance 212
Consonant 210, 232
Coupling factor 375
Critical: distance 269; frequency 205, 292; frequency band 247
Cross section: backscattering 334, 337; scattering 121, 337
Cross-talk compensation 434

Decay: constant 19, 182; time *see* Reverberation, time
Deep channel 337
Definition 267
Delta function 27
Density wave 190

Difference limen: of pitch 241; of sound level 244
Difference tone 238
Diffraction 64, 118; by apertures 124, 126, 127; by half plane 129; of light by ultrasonic wave 349; by rigid sphere 121
Diffraction grating 349
Diffraction wave 120
Diffuse: reflection 137; sound field 265, 281
Diffuse-field distance 269
Diffusion 135
Dilatation 43
Dipole 79, 125, 314
Dirac function 27
Directional: diagram 77; factor 77
Directivity of: a dipole 80; a horn 417; a linear array 82; loudspeakers 417; microphones 392; a rigid piston 89
Direct sound 259
Disk recording 434
Dislocation 63
Dispersion 60, 162, 203, 337
Displacement 7, 34
Distortion: linear 360, 379, 403; non-linear 32, 379, 403, 409
Doppler effect 74, 334, 348, 408
Dummy head 430

Ear 234; muff 332; plug 331
Earphone 403, 419
Echo 94, 263, 277, 330, 337
Eddy 223, 313; see also Vortex
Edge tone 223, 313
Eigenfrequency 166
Eigenfunction 179
Eigenvalue 179
Elastic: losses 206; stress 36
Electroacoustic: system 426; transducer see Transducer
Electromechanical analogies 19
Elongation 7
Enclosure (of noise sources) 321
End correction 127, 274
Energy density 44, 51, 268
Energy flux density see Sound, intensity
Equally tempered scale 214
Equivalent circuit 21
Equivalent noise level 311
Error correction 435
E-transducer 361

Extensional wave 200
Eyring formula 271

Far field 88, 126
Feedback: acoustical 448; stability 450
Flanking transmission 285
Floating floor 305
Floor covering 305
Flow noise 312
Flow resistance 112; specific 109
Formant 230
Fourier: analysis 23; coefficient 24; integral 26; spectrum 24; transform 26
Frequency: analyser 29; response curve 185, 449; shifter 450
Fresnel zone 92
Fundamental: frequency 24; tone 210

Gain 77, 84, 91
Gap effect see Slit effect
Group velocity 164, 203
Gyrator 369

Haas effect 254, 447
Hair cells 238
Half-width: of main lobe 78, 392; of resonance curves 17, 180
Harmonic: oscillation 8; wave 51
Harmonics 24
Head-related transfer function 245
Hearing 233; loss 309; protector 311
Hearing threshold 243; masked 249
Heat conduction 58, 139
Helmholtz: equation 178; resonator 149
Hooke's law 7
Horn 150; Bessel 227; conical 151, 168; equation 151; exponential 153; loudspeaker 414
Huygens' principle 119, 124, 428
Hydrophone 341, 379, 396

Image source 260
Impact noise 302; insulation 305
Impact sound level 303; normalised 303; weighted 304
Impedance 14; characteristic 51; specific 14; transformation 142, 145, 236; tube 106; wall 100
Impulse echo method 345
Impulse response 29, 182, 261
Infrasound 5

In-head localisation 431
Intensity 44, 51; differential 265
Interference 76, 87, 105, 348
Interval, musical 212
Irradiation density 137, 265

Jet 223, 313

Kirchhoff integral 124
Kundt's tube 106

Lambert's law 136
Lamé constants 42
Large-room condition 185, 258
Law of the first wavefront 254, 264
Level difference 47
Linear: array 81; superposition 30;
  system 30
Linearisation 365, 372
Localisation 252, 264
Local reaction 103, 266
Loss factor 206
Losses: elastic 206; frictional 139;
  viscous 275
Loudness 244; level 245;
  measurement 252
Loudspeaker 403; array 417; cabinet
  412; cone 405; dynamic 405;
  electrostatic 408; horn 414;
  magnetic 410
Low-pass filter, acoustical 149

Mach cone 315
Masking 248
Mean free path length 267
Mel 240
Microphone 379; array 395;
  calibration 399; carbon 391;
  cardioid 393; condensor 382;
  directional 392; electret 385;
  gradient 393; piezoelectric 387;
  sensitivity 379, 399
MiniDisc (MD) 436
Motional feedback 407
Motion picture 437
Muffler 326
Multipole 80
Musical instruments 214

Near field 88, 126
Needle hydrophone 396
Nodal surface 170, 179

Node 105
Noise 209, 309; criteria 310;
  generation 311; white 28
Noise control 309; by absorption 326;
  by barriers 323; by vegetation 325
Non-destructive testing 346
Non-linearity parameter 66
Normal mode 167, 179
N-wave 316

Octave 213; filter 29
Ohm's law 247
Organ 215
Organ of Corti 238
Oscillation see Vibration
Overtone 210

Panel 115; absorber 273; perforated
  146, 318
Paris formula 266
Partial 24; tone 210; vibration 24
Particle velocity 34
Partition: compound 287; double-leaf
  296; single-leaf 289
Perforated panel 146, 318
Period 8
Phase angle 9
Phased array 341
Phase hearing 247
Phase velocity 158, 164, 203
Phasor 11, 14
Phonem 210, 228
Phonon 356
Piano 222
Piezoelectric: constant 361; effect 361;
  modulus 362; transducer 361
Piezomagnetic constant 374
Pinna 234
Pipe 138; discontinuity 143
Piston 86, 175
Pitch 209, 211; psychoacoustic 239;
  virtual 242
Point source 71
Poisson's ratio 197
Polarisation 191
Porosity 109, 274, 318
Porous absorber 108, 272
Porous layer 108, 113
Power 22, 281; spectrum 28
Precedence effect 254
Pressure chamber 400, 404

Psychoacoustics 239
Pulsating sphere *see* Breathing sphere

Quadrupole 80, 314
Quality factor, Q-factor 16, 19, 208
Quasi-longitudinal wave *see*
    Extensional wave

Radiation: balance 344; impedance 78;
    mass 85, 92; pressure 66, 344
Radiation resistance 78; of a breathing
    sphere 84; of a rigid piston 91
Rayleigh: distribution 185; model 109;
    wave 195
Ray tracing 279
Reciprocity 72, 322, 378, 392;
    calibration 400; parameter 401
Reference for: air-borne sound
    insulation 286; impact sound level
    303; pitch 211
Reflection 94, 194; diffuse 136; factor
    99, 144; specular 136, 258; useful
    262, 277
Refraction 94, 194
Relaxation 59; radiation 357;
    structural 63; thermal 60, 62
Residue 242
Resonance 14; absorber 274; box
    214; curve 17; frequency 17;
    system 15
Resonator 15, 116, 148, 215
Reverberant sound field 269
Reverberation 184, 277, 347; room
    (or - chamber) 281; time 185, 277;
    in underwater sound 339
Room acoustics 257
Root-mean-square pressure 47

Sabine formula 270
Sandwich plate 318
Scattering 64; cross section 121, 337;
    from rough walls 135; multiple 134
Schroeder frequency 185
Sector scanner 349
Semitone 214
Semivowel 232
Shear: modulus 199; stress 36;
    wave 191
Shock wave, shock front 315, 320
Silencer 326; dissipative 328;
    reactive 326

Slit effect 437, 440
Snell's law 95, 194
Solid, isotropic 36, 40, 189
Sonar 333
Sone 246
Sonic boom 316, 320
Sonochemistry 350
Sonography 345, 348
Sonoluminescence 350
Sonotrode 351
Sound: absorption 55, 271; intensity
    44; level meter 251; reduction index
    284; temperature 35, 57;
    transmission loss 284; velocity 49,
    54, 190
Sound power 28, 84, 91, 281; level 78
Sound pressure 35
Sound pressure level 47; weighted 251
Sound particle 135, 259
Sound radiation: from a breathing
    sphere 84; from a linear array 81;
    from a rigid piston 86; from a
    vibrating plate 204
Sound ray 96, 259
Sound recording 433; magnetic 440;
    mechanical (on discs) 434; optical
    (motion picture) 437
Sound wave 3; longitudinal 43;
    transverse 43
Spaciousness 427
Spatial hearing 252
Spatial impression 278
Specific: flow resistance 109; impedance
    see Impedance; mass 115, 146, 202,
    274, 285
Speckles 348
Spectral density 26
Spectral power density 28
Spectrum 24
Speech 210; intelligibility 263, 277;
    melody 231
Standing wave 105, 167, 176
Steepening 66, 315
Stereophony 427
Stiffness 15
Strain gauge 398
Stress 36
String instruments 215
Structure-borne sound 283, 301, 306
Superconductivity 356
Surface wave 195

Tapping machine 302
Target 333, 344
Thermal sensor 344
Third-octave band 29, 285, 303
Threshold of pain 244
Timbre 210
Time gain control 348
Tone 209; fundamental 210; interval
  212; scale 213
Torsional wave 193
Total reflection 96, 195
Trace fitting 95, 204, 291
Trace velocity 95
Transducer 359; array 340; compound
  424; constant 362, 368; dynamic
  368; electroacoustic 359;
  electromechanical 359; electrostatic
  365; magnetic 371; magnetostrictive
  340, 373, 424; piezoelectric 340,
  361, 421; reversible 340, 359
Transfer function 29, 182
Transmission: line 141; system 30
Transparency 263
Tunnel effect 357
Turbulence 313, 320
Two-port equations 377

Ultrasonic: cleaning 350; drilling 352;
  joining 350
Ultrasound 333, 342, 421
Underwater sound 333, 421

Velocity: particle 34; potential 35;
  sound 49; transformer 351,
  353, 424
Vibration 7; harmonic 8; periodic 9;
  pickup 396; random 10
Violin 215
Virtual: pitch 242; sound source 260
Viscosity 58, 275
Vocal cords 229
Vocal tract 230
Voice 228; chords 229; coil 405
Volume velocity 71
Vortex 313, 320
Vowel 210, 228, 230

Wall impedance see Impedance
Wave: bending 202, 284; equation 42,
  69, 190; extensional 200, 284;
  longitudinal 43, 190; normal 49
  plane 48; shear 191; spherical 69;
  standing 105, 167, 176; surface 49;
  torsional 193; transverse 190
Waveguide 138
Wavelength 53
Wavenumber, angular 53
Webster's equation 151
Weighting filter 251
Weighted: impact sound level 304;
  sound pressure level 252; sound
  reduction index 286
Wind instruments 223

Young' modulus 197